SYSTEM MODELING AND OPTIMIZATION XX

IFIP – The International Federation for Information Processing

IFIP was founded in 1960 under the auspices of UNESCO, following the First World Computer Congress held in Paris the previous year. An umbrella organization for societies working in information processing, IFIP's aim is two-fold: to support information processing within its member countries and to encourage technology transfer to developing nations. As its mission statement clearly states,

> *IFIP's mission is to be the leading, truly international, apolitical organization which encourages and assists in the development, exploitation and application of information technology for the benefit of all people.*

IFIP is a non-profitmaking organization, run almost solely by 2500 volunteers. It operates through a number of technical committees, which organize events and publications. IFIP's events range from an international congress to local seminars, but the most important are:

- The IFIP World Computer Congress, held every second year;
- Open conferences;
- Working conferences.

The flagship event is the IFIP World Computer Congress, at which both invited and contributed papers are presented. Contributed papers are rigorously refereed and the rejection rate is high.

As with the Congress, participation in the open conferences is open to all and papers may be invited or submitted. Again, submitted papers are stringently refereed.

The working conferences are structured differently. They are usually run by a working group and attendance is small and by invitation only. Their purpose is to create an atmosphere conducive to innovation and development. Refereeing is less rigorous and papers are subjected to extensive group discussion.

Publications arising from IFIP events vary. The papers presented at the IFIP World Computer Congress and at open conferences are published as conference proceedings, while the results of the working conferences are often published as collections of selected and edited papers.

Any national society whose primary activity is in information may apply to become a full member of IFIP, although full membership is restricted to one society per country. Full members are entitled to vote at the annual General Assembly, National societies preferring a less committed involvement may apply for associate or corresponding membership. Associate members enjoy the same benefits as full members, but without voting rights. Corresponding members are not represented in IFIP bodies. Affiliated membership is open to non-national societies, and individual and honorary membership schemes are also offered.

SYSTEM MODELING AND OPTIMIZATION XX

IFIP TC7 20th Conference on System Modeling and Optimization July 23–27, 2001, Trier, Germany

Edited by

E. W. Sachs
Department of Mathematics
University of Trier / Virginia Polytechnic Institute and State University
Germany / USA

R. Tichatschke
Department of Mathematics
University of Trier
Germany

SPRINGER SCIENCE+BUSINESS MEDIA, LLC

Library of Congress Cataloging-in-Publication Data

A C.I.P. Catalogue record for this book is available from the Library of Congress.

System Modeling and Optimization XX
Edited by E. W. Sachs and R. Tichatschke

DOI 10.1007/978-0-387-35699-0

Printed on acid-free paper.
www.springer.com/mycopy

Contents

Foreword

The 20th IFIP TC7 Conference on System Modeling and Optimization took place at the University of Trier, Germany, from July 23 to 27, 2001. This volume contains selected papers written by participants of the conference, where some of the authors made invited presentations.

The conference was attended by 128 participants from 28 countries and four continents. The organizers are grateful to all participants for their contribution to the success of this conference. During the five days of the meeting 10 invited and 94 contributed presentations were given. The talks were of high scientific quality and displayed the wide range of the area of system modeling and optimization. During the course of the 20 TC7 meetings held at a biannual rate, the conferences document the progress and development of this important research area. Also during this conference, one could follow important research of well established areas, but also become acquainted with new research fields in optimization.

The conference was supported by the International Federation for Information Processing (IFIP), in particular through the Technical Committee 7, which selected the conference site, and the program committee, which put together an interesting program. Their support and help is greatly appreciated. Financial and technical support for this meeting came from the host institution, the University of Trier, and the government of the home state, Rheinland-Pfalz. Furthermore, the Deutsche Forschungsgemeinschaft (DFG), Siemens AG, and GeneralColgone Re Capital generously supported this conference.

Many committees, institutions and individuals contributed to the success of this conference. We thank the program committee of TC7, in particular P. Kall (chairman of the TC7), and the administration of the University of Trier. The organization of the conference would not have been possible without the help and support of many individuals: Among those were H. Beewen, F. Leibfritz, J. Maruhn, U. Morbach, M. Pick, M. Ries, M. Schulze, C. Schwarz, T. Voetmann. We also appreciate the valuable assistance of the publisher, in particular Y. Lambert, in the preparation of the proceedings.

Trier, May 2003
E. W. Sachs and R. Tichatschke

ON APPROXIMATE ROBUST COUNTERPARTS OF UNCERTAIN SEMIDEFINITE AND CONIC QUADRATIC PROGRAMS

Aharon Ben-Tal
Faculty of Industrial Engineering and Management, Technion
Israel Institute of Technology *
abental@ie.technion.ac.il

Arkadi Nemirovski
Faculty of Industrial Engineering and Management, Technion
Israel Institute of Technology
nemirovs@ie.technion.ac.il

Abstract We present efficiently verifiable sufficient conditions for the validity of specific NP-hard semi-infinite systems of semidefinite and conic quadratic constraints arising in the framework of Robust Convex Programming and demonstrate that these conditions are "tight" up to an absolute constant factor. We discuss applications in Control on the construction of a quadratic Lyapunov function for linear dynamic system under interval uncertainty.

1. Introduction

The subject of this paper are "tractable approximations" of intractable semi-infinite convex optimization programs arising as *robust counterparts* of *uncertain* conic quadratic and semidefinite problems. We start with specifying the relevant notions. Let $\mathbf{K}$ be a cone in $\mathbf{R}^N$ (closed, pointed, convex and with a nonempty interior). A *conic program* asso-

*This research was partially supported by the Israeli Ministry of Science grant # 0200-1-98 and the Israel Science Foundation grant # 683/99-10.0. Part of this research was conducted while the authors were visiting the Faculty of Information Technology and Systems (ITS), Department ISA.

ciated with $\mathbf{K}$ is an optimization program of the form

$$\min_x\{f^T x \mid Ax - b \in \mathbf{K}\}; \tag{CP}$$

here $x \in \mathbf{R}^n$. An *uncertain* conic problem is a family

$$\left\{\min_x \left\{f^T x \mid Ax - b \in \mathbf{K}\right\} |(f, A, b) \in \mathcal{U}\right\} \tag{UCP}$$

of conic problems with common $\mathbf{K}$ and data (f, A, B) running through a given *uncertainty set* $\mathcal{U}$. In fact, we always can get rid of uncertainty in f and assume that f is "certain", i.e., common for all data from $\mathcal{U}$); indeed, we always can rewrite the problems of the family as

$$\min_{t,x}\left\{t \mid \begin{bmatrix} Ax - b \\ t - f^T x \end{bmatrix} \in \bar{\mathbf{K}} = \mathbf{K} \times \mathbf{R}_+\right\}.$$

Thus, we lose nothing (and gain a lot, as far as notation is concerned) when assuming from now on that f is certain, so that $n, \mathbf{K}, f$ form the common "structure" of problems from the family, while A, b are the data of particular problems ("instances") from the family.

The Robust Optimization methodology developed in [1, 2, 3, 5, 8, 9] associates with (UCP) its *Robust Counterpart* (RC)

$$\min_x \left\{f^T x \mid Ax - b \in \mathbf{K} \quad \forall (c, A, b) \in \mathcal{U}\right\}. \tag{R}$$

Feasible/optimal solutions of (R) are called *robust feasible*, resp., *robust optimal* solutions of the uncertain problem (UCP); the importance of these solutions is motivated and illustrated in [1, 2, 3, 5, 8, 9].

Accepting the concept of robust feasibile/optimal solutions, the crucial question is how to build these solutions. Note that (R) is a *semi-infinite* conic program and as such can be computationally intractable. In this respect, there are "good cases", where the RC is equivalent to an explicit computationally tractable convex optimization program, as well as "bad cases", where the RC is NP-hard (see [3, 5] for "generic examples" of both types). In "bad cases", the Robust Optimization methodology recommends replacing the computationally intractable robust counterpart by its tractable approximation. An *approximate robust counterpart* of (UCP) is a conic problem

$$\min_{x,u}\left\{f^T x \mid Px + Qu + r \in \widehat{\mathbf{K}}\right\} \tag{AR}$$

such that the projection $\mathcal{X}(\text{AR})$ of the feasible set of (AR) onto the plane of x-variables is contained in the feasible set of (R); thus, (AR) is "more conservative" than (R). An immediate question is how to measure the "conservativeness" of (AR), with the ultimate goal to use a "moderately conservative" computationally tractable approximate RCs instead of the "true" (intractable) RCs. A natural way to measure the quality of an approximate RC is as follows. Assume that the uncertainty set $\mathcal{U}$ is of the form

$$\mathcal{U} = \{(A, b) = (A^{\mathrm{n}}, b^{\mathrm{n}}) + \mathcal{V}\},$$

where $(A^{\mathrm{n}}, b^{\mathrm{n}})$ is the "nominal data" and $\mathcal{V}$ is the *perturbation set* which we assume from now on to be a convex compact set symmetric w.r.t. the origin. Under our assumptions, (UCP) can be treated as a member of the *parametric family*

$$\left\{\min_x \left\{f^T x \mid Ax - b \in \mathbf{K}\right\} : (A, b) \in \mathcal{U}_\rho = \{(A, b) = (A^{\mathrm{n}}, b^{\mathrm{n}}) + \rho\mathcal{V}\}\right\} \tag{UCP$_\rho$}$$

of uncertain conic problems, where $\rho \geq 0$ can be viewed as the "level of uncertainty". Observing that the robust feasible set $\mathcal{X}_\rho$ of (UCP$_\rho$) shrinks as ρ increases and that (AR) is an approximation of (R) if and only if $\mathcal{X}(\mathrm{AR}) \subset \mathcal{X}_1$, a natural way to measure the quality of (AR) is to look at the quantity

$$\rho(\mathrm{AR{:}R}) = \inf\{\rho \geq 1 : \mathcal{X}(\mathrm{AR}) \supset \mathcal{X}_\rho\},$$

which we call the *conservativeness* of the approximation (AR) of (R). Thus, the fact that (AR) is an approximation of (R) with the conservativeness $< \alpha$ means that

(i) If x can be extended to a feasible solution of (AR), then x is a robust feasible solution of (UCP);

(ii) If x cannot be extended to a feasible solution of (AR), then x is *not* robust feasible for the uncertain problem (UCP$_\alpha$) obtained from (UCP)≡(UCP$_1$) by increasing the level of uncertainty by the factor α.

Note that in real-world applications the level of uncertainty normally is known "up to a factor of order of 1"; thus, we have basically the same reasons to use the "true" robust counterpart as to use its approximation with $\rho(\mathrm{AR{:}R})$ of order of 1.

The goal of this paper is to overview recent results on tractable approximate robust counterparts with "$O(1)$-conservativeness", specifically, the results on semidefinite problem affected by box uncertainty and on conic quadratic problem affected by ellipsoidal uncertainty. We present the approximation schemes, discuss their quality, illustrate the results by some applications (specifically, in Lyapunov Stability Analysis/Synthesis for uncertain linear dynamic systems with interval uncertainty) and establish links of some of the results with recent developments in the area of semidefinite relaxations of difficult optimization problems.

2. Uncertain SDP with box uncertainty

Let $\mathbf{S}^m$ be space of real symmetric $m \times m$ matrices and $\mathbf{S}^m_+$ be the cone of positive semidefinite $m \times m$ matrices, and let $A^\ell[x] = A^{\ell 0} + \sum_{j=1}^{n} x_j A^{\ell j} : \mathbf{R}^n \to \mathbf{S}^m$ be affine mappings, $\ell = 0, ..., L$. Let also $C[x]$ be a symmetric matrix affinely depending on x. Consider the uncertain semidefinite program

$$\left\{ \min_x \left\{ f^T x : \begin{array}{c} C[x] \succeq 0 \\ A[x] \succeq 0 \end{array} \right\} \middle| \exists (u : \| u \|_\infty \leq \rho) : A[x] = A^0[x] + \sum_{\ell=1}^{L} u_\ell A^\ell[x] \right\} \tag{USD[ρ]}$$

here an in what follows, for $A, B \in \mathbf{S}^m$ the relation $A \succeq B$ means that $A - B \in \mathbf{S}^m_+$. Note that (USD[ρ]) is the general form of an uncertain semidefinite program affected by "box" uncertainty (one where the uncertainty set is an affine image of a multi-dimensional cube). Note also that the Linear Matrix Inequality (LMI) $C[x] \succeq 0$ represents the part of the constraints which are "certain" – not affected by the uncertainty.

The robust counterpart of (USD[ρ]) is the semi-infinite semidefinite program

$$\min_x \left\{ f^T x : \begin{array}{l} C[X] \succeq 0 \\ A^0[x] + \sum_{\ell=1}^{L} u_\ell A^\ell[x] \succeq 0 \quad \forall (u : \| u \|_\infty \leq \rho) \end{array} \right\}. \tag{R[ρ]}$$

It is known (see, e.g., [11]) that in general (R[ρ]) is NP-hard; this is so already for the associated analysis problem "Given x, check whether it is feasible for (R[ρ])", and even in the case when all the "edge matrices" $A^\ell[x]$, $\ell = 1, ..., L$, are of rank 2. At the same time, we can easily point out a tractable approximation of (R[ρ]), namely, the semidefinite program

$$\min_{x, \{X_\ell\}} \left\{ f^T x : \begin{array}{l} C[x] \succeq 0 \\ X^\ell \succeq \pm A^\ell[x], \ \ell = 1, ..., L, \\ A^0[x] - \rho \sum_{\ell=1}^{L} X^\ell \succeq 0, \end{array} \right\}. \tag{AR[ρ]}$$

This indeed is an approximation – the x-component of a feasible solution to (AR[ρ]) clearly is feasible for (R[ρ]). Surprisingly enough, this fairly simplistic approximation turns out to be pretty tight, provided that the edge matrices $\mathcal{A}^\ell[x]$, $\ell \geq 1$, are of small rank:

Theorem 1 [6] *Let $A_0, ..., A_L$ be $m \times m$ symmetric matrices. Consider the following two predicates:*

$$\begin{array}{l} (I[\rho]) : A_0 + \sum\limits_{\ell=1}^{L} u_\ell A_\ell \succeq 0 \quad \forall (u : \| u \|_\infty \leq \rho); \\ (II[\rho]) : \exists X_1, ..., X_L : X_\ell \succeq \pm A_\ell,\ \ell = 1, ..., L,\ \rho \sum\limits_{\ell=1}^{L} X_\ell \preceq A_0; \end{array} \tag{1}$$

here $\rho \geq 0$ is a parameter.

Then

(i) *If $(II[\rho])$ is valid, so is $(I[\rho])$;*

(ii) *If $(II[\rho])$ is not valid, so is $(I[\vartheta(\mu)\rho])$, where*

$$\mu = \max_{1 \leq \ell \leq L} \mathrm{Rank}(A_\ell)$$

(note $1 \leq \ell$ in the max*) and $\vartheta(\mu)$ is a universal function of μ given by*

$$\frac{1}{\vartheta(k)} = \min_{\alpha} \left\{ \int\limits_{\mathbf{R}^k} \left| \sum_{i=1}^{k} \alpha_1 u_i^2 \right| (2\pi)^{-k/2} \exp\left\{ -\frac{u^T u}{2} \right\} du \middle| \sum_{i=1}^{k} |\alpha_i| = 1 \right\}. \tag{2}$$

Note that

$$\vartheta(1) = 1, \vartheta(2) = \frac{\pi}{2} \approx 1.57..., \vartheta(3) = 1.73..., \vartheta(4) = 2;\ \vartheta(\mu) \leq \frac{\pi\sqrt{\mu}}{2} \quad \forall \mu. \tag{3}$$

Corollary 2.1 *Consider the robust counterpart* $(\mathrm{R}[\rho])$ *of an uncertain SDP affected by interval uncertainty along with the approximated robust counterpart* $(\mathrm{AR}[\rho])$ *of the problem, and let*

$$\mu = \max_{1 \leq \ell \leq L} \max_{x} \mathrm{Rank}(\mathcal{A}^\ell[x])$$

(note $1 \leq \ell$ in the max*). Then* $(\mathrm{AR}[\rho])$ *is at most $\vartheta(\mu)$-conservative approximation of* $(\mathrm{R}[\rho])$*, where ϑ is given by* (2)*. In particular,*

- *The suprema $\rho^\star$ and $\widehat{\rho}$ of those $\rho \geq 0$ for which* $(\mathrm{R}[\rho])$*, respectively,* $(\mathrm{AR}[\rho])$ *is feasible, are linked by the relation*

$$\widehat{\rho} \leq \rho^\star \leq \vartheta(\mu)\widehat{\rho};$$

- *The optimal values* $f^\star(\rho)$, $\widehat{f}(\rho)$ *of* (R[ρ]), *respectively,* (AR[ρ]) *are linked by the relation*

$$f^\star(\rho) \le \widehat{f}(\rho) \le f^\star(\vartheta(\mu)\rho), \quad \rho \ge 0.$$

The essence of the matter is that the quality of the approximate robust counterpart as stated by Corollary 2.1 depends solely on the ranks of the "basic perturbation matrices" $A^\ell[x]$, $\ell \ge 1$, and is independent of any other sizes of the problem. Fortunately, there are many applications where the ranks of the basic perturbations are small, so that the quality of the approximation is not too bad. As an important example, consider the Lyapunov Stability Analysis problem.

Lyapunov Stability Analysis. Consider an uncertain time-varying linear dynamic system

$$\dot{z}(t) = A(t)z(t) \tag{4}$$

where all we know about the matrix $A(t)$ of the system is that it is a measurable function of t taking values in a given compact set $\mathcal{U}$ which, for the sake of simplicity, is assumed to be an interval matrix:

$$A(t) \in \mathcal{U} = \mathcal{U}_\rho = \{A \in \mathbf{R}^{n\times n} : |A_{ij} - \mathbf{A}_{ij}| \le \rho \mathbf{D}_{ij},\ i,j = 1,...,n\}; \tag{5}$$

here $\mathbf{A}$ corresponds to the "nominal" time-invariant system, and $\mathbf{D}$ is a given "scale matrix" with nonnegative entries.

In applications, the very first question about (4) is whether the system is *stable*, i.e., whether it is true that whatever is a measurable function $A(\cdot)$ taking values in $\mathcal{U}_\rho$, every trajectory $z(t)$ of (4) converges to 0 as $t \to \infty$. The standard *sufficient* condition for the stability of (4) – (5) is the existence of a common *quadratic Lyapunov stability certificate* for all matrices $A \in \mathcal{U}_\rho$, i.e., the existence of an $n \times n$ matrix $X \succ 0$ such that

$$A^T X + XA \prec 0 \quad \forall A \in \mathcal{U}_\rho.$$

Indeed, if such a certificate X exists, then

$$A^T X + XA \preceq -\alpha X$$

for certain $\alpha > 0$ and all $A \in \mathcal{U}_\rho$. As an immediate computation demonstrates, the latter inequality implies that

$$\frac{d}{dt}(z^T(t)Xz(t)) \le -\alpha z^T(t)Xz(t)$$

for all t and all trajectories of (4), provided that $A(t) \in \mathcal{U}_\rho$ for all t. The resulting differential inequality, in turn, implies that $z^T(t)Xz(t) \le$

$\exp\{-\alpha t\}(z^T(0)Xz(0)) \to 0$ as $t \to +\infty$; since $X \succ 0$, it follows that $z(t) \to 0$ as $t \to \infty$.

Note that by homogeneity reasons a stability certificate, if it exists, always can be normalized by the requirements

$$\begin{array}{lrcl} (a) & X & \succeq & I, \\ (b) & A^T X + XA & \preceq & -I \quad \forall A \in \mathcal{U}_\rho. \end{array} \tag{L[ρ]}$$

Thus, whenever (L[ρ]) is solvable, we can be sure that system (4) – (5) is stable. Although this condition is in general only sufficient and not necessary for stability (it is necessary only when $\rho = 0$, i.e., in the case of a certain time-invariant system), it is commonly used to certify stability. This, however, is not always easy to check the condition itself, since (L[ρ]) is a semi-infinite system of LMIs. Of course, since the LMI (L[ρ].b) is linear in A, this semi-infinite system of LMIs is equivalent to the usual – finite – system of LMIs

$$\begin{array}{lrcl} (a) & X & \succeq & I, \\ (b) & A_v^T X + XA_v & \preceq & -I \quad \forall v = 1, ..., 2^N, \end{array} \tag{6}$$

where N is the number of uncertain entries in the matrix A (i.e., the number of pairs ij such that $\mathbf{D}_{ij} \neq 0$) and $A_1, ..., A_{2^N}$ are the vertices of $\mathcal{U}_\rho$. However, the size of (6) is not polynomial in n, except for the (unrealistic) case when N is once for ever fixed or logarithmically slow grows with n. In general, it is NP-hard to check whether (6), or, which is the same, (L[ρ]) is feasible [11].

Now note that with the interval uncertainty (5), the troublemaking semi-infinite LMI (L[ρ].b) is nothing but the robust counterpart

$$\left[I - \mathbf{A}^T X - X\mathbf{A}\right] + \sum_{i,j=1}^{n} u_{ij} \mathbf{D}_{ij} \left[e_j (Xe_i)^T + (Xe_i) e_j^T\right] \succeq 0 \tag{7}$$
$$\forall (u = \{u_{ij}\} : -\rho \leq u_{ij} \leq \rho)$$

of the uncertain LMI

$$\left\{\left\{A^T X + XA \preceq -I\right\} : A \in \mathcal{U}_\rho\right\};$$

here e_i are the standard basic orths in $\mathbf{R}^n$. Consequently, we can approximate (L[ρ]) with a tractable system of LMIs

$$\begin{array}{ll} (a) & X \succeq I \\ (b_1) & X^{ij} \succeq \pm \underbrace{\mathbf{D}_{ij}\left[e_j (Xe_i)^T + (Xe_i) e_j^T\right]}_{A^{ij}[X]}, \ \forall (i,j : \mathbf{D}_{ij} > 0) \\ (b_2) & \underbrace{\left[I - \mathbf{A}^T X - X\mathbf{A}\right]}_{A^0[X]} - \rho \sum\limits_{i,j:\mathbf{D}_{ij}>0} X^{ij} \succeq 0 \end{array} \tag{AL[ρ]}$$

in matrix variables $X, \{X^{ij}\}_{i,j:\mathbf{D}_{ij}>0}$.

Invoking Corollary 2.1, we see that the relations between (AL[ρ]) and (L[ρ]) are as follows:

1 Whenever X can be extended to a feasible solution of the system (AL[ρ]), X is feasible for (L[ρ]) and thus certifies the stability of the uncertain dynamic system (4), the uncertainty set for the system being $\mathcal{U}_\rho$;

2 If X cannot be extended to a feasible solution of the system (AL[ρ]), then X does not solve the system (L[$\frac{\pi}{2}\rho$]) (note that the ranks of the basic perturbation matrices $A^{ij}[X]$ are at most 2, and $\vartheta(2) = \frac{\pi}{2}$).

It follows that the supremum $\widehat{\rho}$ of those $\rho \geq 0$ for which (AL[ρ]) is solvable is a lower bound for the *Lyapunov stability radius* of uncertain system (4), i.e., for the supremum $\rho^\star$ of those $\rho \geq 0$ for which all matrices from $\mathcal{U}_\rho$ share a common Lyapunov stability certificate, and that this lower bound is tight up to factor $\frac{\pi}{2}$:

$$1 \leq \frac{\rho^\star}{\widehat{\rho}} \leq \frac{\pi}{2},$$

provided, of course, that $\mathbf{A}$ is stable (or, which is the same, that $\rho^\star > 0$).

Note that the bound $\widehat{\rho}$ on the Lyapunov stability radius is efficiently computable; this is the optimal value in the Generalized Eigenvalue Problem of maximizing ρ in variables $\rho, X, \{X^{ij}\}$ under the constraints (LA[ρ]).

We have considered a specific application of Theorem 1 in Control. There are many other applications of this theorem to systems of LMIs arising in Control and affected by an interval data uncertainty. Usually the structure of such a system ensures that when perturbing a single data entry, the right hand side of every LMI is perturbed by a matrix of a small rank, which is the favourable case for our approximation scheme.

Simplifying the approximation. A severe computational shortcoming of the approximation (AR[ρ]) is that its sizes, although polynomial in the sizes of the approximated system (R[ρ]) and the uncertainty set, are pretty large, since the approximation has an additional $m \times m$ matrix variable X_ℓ and two $m \times m$ LMIs $X_\ell \succeq \pm A^\ell[x]$ per every one of the basic perturbations. It turns out that under favourable circumstances the sizes of the approximation can be reduced dramatically. This size reduction is based upon the following two facts:

Lemma 2.1 [6] (i) *Let $a \neq 0, b$ be two vectors. A matrix X satisfies the relation*

$$X \succeq \pm[ab^T + ba^T]$$

if and only if there exists $\lambda \geq 0$ such that

$$X \succeq \lambda aa^T + \frac{1}{\lambda} bb^T$$

(when $\lambda = 0$, the left hand side, by definition, is the zero matrix when $b = 0$ and is undefined otherwise).

(ii) *Let S be a symmetric $m \times m$ matrix of rank $k > 0$, so that $S = P^T RP$ with a nonsingular $k \times k$ matrix R and $k \times m$ matrix P of rank k.*

A matrix X satisfies the relation

$$X \succeq \pm S$$

if and only if there exists a $k \times k$ matrix Y such that $Y \succeq \pm R$ and $X \succeq P^T Y P$.

The simplification of (AR[ρ]), based on Lemma 2.1, is as follows. Let us partition the basic perturbation matrices $A^\ell[x]$ into two groups: those with $A^\ell[x]$ actually depending on x, and those with $A^\ell[x]$ independent of x. Assume that

(A) *The basic perturbation matrices depending on x*, let them be $A^1[x], ..., A^M[x]$, *are of the form*

$$A^\ell[x] = a_\ell b_\ell^T[x] + b_\ell[x] a_\ell^T, \quad \ell = 1, ..., M, \tag{8}$$

where a_ℓ, $b_\ell[x]$ are vectors and $b_\ell[x]$ are affine in x.

Note that the assumption holds true, e.g., in the case of the Lyapunov Stability Analysis problem under interval uncertainty, see (AL[ρ]).

The basic perturbation matrices A^ℓ with $\ell > M$ are independent of x, and we can represent these matrices as

$$A^\ell = P_\ell^T B_\ell P_\ell, \tag{9}$$

where B_ℓ are nonsingular symmetric $k_\ell \times k_\ell$ matrices and $k_\ell = \mathrm{Rank}(A^\ell)$.

Observe that when speaking about the approximate robust counterpart (AR[ρ]), we are not interested at all in the additional matrix variables X_ℓ; all which matters is the projection of the feasible set of (AR[ρ]) on the plane of the original design variables x. In other words, as far as the approximating properties are concerned, we lose nothing when replacing the constraints in (AR[ρ]) with any other system $\mathcal{S}$ of constraints in variables x and, perhaps, additional variables, provided that the projection of the feasible set of the new system on the plane of x-variables is the same as the one for (AR[ρ]). Invoking Lemma 2.1, we see that the latter property is possessed by the system of constraints

$$\begin{array}{ll} (a) & C[x] \succeq 0 \\ (b) & Y_\ell \succeq \pm B_\ell,\ \ell = M+1,...,L, \\ (c) & \lambda_\ell \geq 0,\ \ell = 1,...,M, \\ (d) & A^0[x] - \rho\left[\sum_{\ell=1}^{M}\left[\lambda_\ell a_\ell a_\ell^T + \lambda_\ell^{-1} b_\ell[x] b_\ell^T[x]\right] + \sum_{\ell=M+1}^{L} P_\ell^T Y_\ell P_\ell\right] \succeq 0, \end{array} \tag{10}$$

in variables $x, \{\lambda_\ell\}, \{Y_\ell\}$. By the Schur Complement Lemma, (10) is equivalent to the system of LMIs

$$\begin{array}{ll} (a) & C[x] \succeq 0 \\ (b) & Y_\ell \succeq \pm B_\ell,\ \ell = M+1,...,L, \\ (c) & \left[\begin{array}{c|cccc} A^0[x] - \rho\sum_{\ell=1}^{M}\lambda_\ell a_\ell a_\ell^T - \rho\sum_{\ell=M+1}^{L} P_\ell^T Y_\ell P_\ell & b_1[x] & b_2[x] & \cdots & b_M[x] \\ \hline b_1^T[x] & \lambda_1 & & & \\ b_2^T[x] & & \lambda_2 & & \\ \vdots & & & \ddots & \\ b_M^T[x] & & & & \lambda_M \end{array}\right] \succeq 0 \end{array} \tag{11}$$

in variables $x, \{\lambda_\ell \in \mathbf{R}\}_{\ell=1}^M, \{Y_\ell \in \mathbf{S}^{k_\ell}\}_{\ell=M+1}^L$. Consequently, (AR[$\rho$]) is equivalent to the semidefinite program of minimizing the objective $f^T x$ under the constraints (11). Note that the resulting problem (A[ρ]) is typically much better suited for numerical processing than (AR[ρ]). Indeed, the first M of the $m \times m$ matrix variables X_ℓ arising in the original problem are now replaced with M *scalar* variables λ_ℓ, while the remaining $L - M$ of X_ℓ's are replaced with $k_\ell \times k_\ell$ matrix variables Y_ℓ; normally, the ranks k_ℓ of the basic perturbation matrices A^ℓ are much smaller than the sizes m of these matrices, so that this transformation reduces quite significantly the design dimension of the problem. As about LMIs, the $2L$ "large" (of the size $m \times m$) LMIs $X_\ell \succeq \pm A^\ell[x]$ of the original problem are now replaced with $2(L - M)$ "small" (of the sizes $k_\ell \times k_\ell$) LMIs (11.b) and a single "very large" – of the size $(m + M) \times (m + M)$ – LMI (11.c). Note that the latter LMI, although large, is of a very simple "arrow" structure and is extremely sparse.

Links with quadratic maximization over the unit cube. It turns out that Theorem 1 has direct links with the problem of maximizing a positive definite quadratic form over the unit cube. The link is given by the following simple observation:

Proposition 2.1 *Let Q be a positive definite $m \times m$ matrix. Then the reciprocal $\rho(Q)$ of the quantity*

$$\omega(Q) = \max_x \{x^T Q x : \| x \|_\infty \leq 1\}$$

equals to the maximum of those $\rho > 0$ for which all matrices from the matrix box

$$\mathcal{C}_\rho = Q^{-1} + \{A = A^T : |A_{ij}| \leq \rho\}$$

are positive semidefinite.

Proof. $\omega(Q)$ is the minimum of those ω for which the ellipsoid $\{x : x^T Q x \leq \omega\}$ contains the unit cube, or, which is the same, the minimum of those ω for which the polar of the ellipsoid (which is the ellipsoid $\{\xi : \xi^T Q^{-1} \xi \leq \omega^{-1}\}$) is contained in the polar of the cube (which is the set $\{\xi : \| \xi \|_1 \leq 1\}$). In other words,

$$\rho(Q) \equiv \omega^{-1}(Q) = \max\left\{\rho : \xi^T Q^{-1} \xi \geq \rho \| \xi \|_1^2 \ \forall \xi\right\}.$$

Observing that by evident reasons

$$\| \xi \|_1^2 = \max_A \left\{\xi^T A \xi : A = A^T, |A_{ij}| \leq 1,\, i,j = 1,...,m\right\},$$

we conclude that

$$\rho(Q) = \max\{\rho : Q^{-1} \succeq \rho A \quad \forall (A = A^T : |A_{ij}| \leq 1,\, i,j = 1,...,m)\},$$

as claimed. ∎

Since the "edge matrices" of the matrix box $\mathcal{C}_\rho$ are of ranks 1 or 2 (these are the basic symmetric matrices $E^{ij} = \begin{cases} e_i e_i^T, & i = j \\ e_i e_j^T + e_j e_i^T, & i < j \end{cases}$, $1 \leq i \leq j \leq m$, where e_i are the standard basic orths), Theorem 1 says that the efficiently computable quantity

$$\widehat{\rho} = \sup\left\{\rho : \exists \{X^{ij}\} : \begin{array}{c} X^{ij} \succeq \pm E^{ij},\, 1 \leq i \leq j \leq m \\ Q^{-1} - \rho \sum\limits_{1 \leq i \leq j \leq m} X^{ij} \succeq 0 \end{array}\right\}$$

is a lower bound, tight within the factor $\vartheta(2) = \frac{\pi}{2}$, for the quantity $\rho(Q)$, and consequently the quantity $\widehat{\omega}(Q) = \frac{1}{\widehat{\rho}(Q)}$ is an upper bound, tight within the same factor $\frac{\pi}{2}$, for the maximum $\omega(Q)$ of the quadratic form $x^T Q x$ over the unit cube:

$$\omega(Q) \leq \widehat{\omega}(Q) \leq \frac{\pi}{2}\omega(Q). \tag{12}$$

On a closest inspection (see [6]), it turns out that $\widehat{\omega}(Q)$ is nothing but the standard semidefinite bound

$$\begin{aligned} \widehat{\omega}(Q) &= \max\left\{\mathrm{Tr}(QX) : X \succeq 0, X_{ii} \leq 1,\ i = 1, ..., m\right\} \\ &= \min\left\{\sum_i \lambda_i : \mathrm{Diag}\{\lambda\} \succeq Q\right\} \end{aligned} \tag{13}$$

on $\omega(Q)$. The fact that bound (13) satisfies (12) was originally established by Yu. Nesterov [13] via a completely different construction which heavily exploits the famous MAXCUT-related "random hyperplane" technique of Goemans and Williamson [10]. Surprisingly, the re-derivation of (12) we have presented, although uses randomization arguments, seems to have nothing in common with the random hyperplane technique.

Theorem 1: Sketch of the proof.. We believe that not only the statement, but also the proof of Theorem 1 is rather instructive, this is why we sketch the proof here. We intend to focus on the most nontrivial part of the Theorem, which is the claim that when $(II[\rho])$ is not valid, so is $(I[\vartheta(\mu)\rho])$ (as about the remaining statements of Theorem 1, note that (i) is evident, while the claim that the function (2) satisfies (3) is more or less straightforward). Thus, assume that $(II[\rho])$ is not valid; we should prove that then $(I[\vartheta(\mu)\rho])$ also is not valid, where $\vartheta(\cdot)$ is given by (2).

The fact that $(II[\rho])$ is not valid means exactly that the optimal value in the semidefinite program

$$\min_{t,\{X_\ell\}} \left\{ t : X_\ell \succeq \pm A_\ell,\ \ell = 1, ..., L;\ \rho \sum_{\ell=1}^{L} X_\ell \preceq A_0 + tI \right\}$$

is positive. Applying semidefinite duality (which is a completely mechanical process) we, after simple manipulations, conclude that in this case there exists a matrix $W \succeq 0$, $\mathrm{Tr}(W) = 1$, such that

$$\sum_{\ell=1}^{L} \| \lambda(W^{1/2} A_\ell W^{1/2}) \|_1 > \mathrm{Tr}(W^{1/2} A_0 W^{1/2}), \tag{14}$$

where $\lambda(B)$ is the vector of eigenvalues of a symmetric matrix B. Now observe that if B is a symmetric $m \times m$ matrix and ξ is an m-dimensional Gaussian random vector with zero mean and unit covariance matrix, then

$$\mathbf{E}\left\{|\xi^T B \xi|\right\} \geq \frac{1}{\vartheta(\mathrm{Rank}(B))} \| \lambda(B) \|_1 . \tag{15}$$

Indeed, in the case when B is diagonal, this relation is a direct consequence of the definition of $\vartheta(\cdot)$; the general case can be reduced immediately to the case of diagonal B due to the rotational invariance of the distribution of ξ.

Since the matrices $W^{1/2}A_\ell W^{1/2}$ are of ranks not exceeding $\mu = \max\limits_{1\le\ell\le L} \mathrm{Rank}(A_\ell)$, (14) combines with (15) to imply that

$$[\rho\vartheta(\mu)]\sum_{\ell=1}^{L}\mathbf{E}\{|\xi^T W^{1/2}A_\ell W^{1/2}\xi|\} > \mathrm{Tr}(W^{1/2}A_0W^{1/2}) = \\ = \mathbf{E}\{\xi^T W^{1/2}A_0W^{1/2}\xi\}.$$

It follows that there exists $\zeta = W^{1/2}\xi$ such that

$$[\rho\vartheta(\mu)]\sum_{\ell=1}^{L}|\zeta^T A_\ell\zeta| > |\zeta^T A_0\zeta|;$$

setting $\epsilon_\ell = \mathrm{sign}(\zeta^T A_\ell \zeta)$, we can rewrite the latter relation as

$$\zeta^T\left[\sum_{\ell=1}^{L}\underbrace{(\rho\vartheta(\mu)\epsilon_\ell)}_{u_\ell}A_\ell\right]\zeta > \zeta^T A_0\zeta,$$

and we see that the matrix $A_0 - \sum\limits_\ell u_\ell A_\ell$ is not positive definite, while by construction $|u_\ell| \le \vartheta(\mu)\rho$. Thus, $(I[\vartheta(\mu)\rho])$ indeed is not valid.

3. Approximate robust counterparts of uncertain convex quadratic problems

Recall that a generic convex quadratically constrained problem is

$$\min_x\left\{f^Tx : x^TA_i^TA_ix \le 2b_i^Tx + c_i,\ i=1,...,m\right\} \tag{QP}$$

here $x \in \mathbf{R}^n$, and A_i are $m_i \times n$ matrices. The data of an instance is $(c, \{A_i, b_i, c_i\}_{i=1}^m)$. When speaking about uncertain problems of this type, we may focus on the robust counterpart of the system of constraints (since we have agreed to treat the objective as certain). In fact we can restrict ourselves with building an (approximate) robust counterpart of a *single* convex quadratic constraint, since the robust counterpart is a "constraint-wise" construction. Thus, we intend to focus on building an approximate robust counterpart of a single constraint

$$x^TA^TAx \le 2b^Tx + c \tag{16}$$

with the data (A, b, c). We assume that the uncertainty set is "parameterized" by a vector of perturbations:

$$\mathcal{U} \equiv \mathcal{U}_\rho = \left\{ (A, b, c) = (A^{\rm n}, b^{\rm n}, c^{\rm n}) + \sum_{\ell=1}^{L} \zeta_\ell (A^\ell, b^\ell, c^\ell) : \zeta \in \rho\mathcal{V} \right\}, \quad (17)$$

here $\mathcal{V}$ is a convex compact symmetric w.r.t. the origin set in $\mathbf{R}^L$ ("the set of standard perturbations") and $\rho \geq 0$ is the "perturbation level".

In what follows, we shall focus on the case when $\mathcal{V}$ is given as an intersection of ellipsoids centered at the origin:

$$\mathcal{V} = \left\{ \zeta \in \mathbf{R}^L \mid \zeta^T Q_i \zeta \leq 1,\ i = 1, ..., k \right\}, \quad (18)$$

where $Q_i \succeq 0$ and $\sum_{i=1}^{k} Q_i \succ 0$. We will be interested also in two particular cases of general ellipsoidal uncertainty (18), namely, in the cases of

- *simple ellipsoidal uncertainty* $k = 1$;
- *box uncertainty:* $k = L$ and $\zeta^T Q_i \zeta \equiv \zeta_i^2$, $i = 1, ..., L$.

Note that the ellipsoidal robust counterpart of an uncertain quadratic constraint affected by uncertainty (18) is, in general, NP-hard. Indeed, already in the case of box uncertainty to verify robust feasibility of a given candidate solution is not easier than to maximize a convex quadratic form over the unit cube, which is known to be NP-hard. Thus, all we can hope for in the case of uncertainty (18) is a "computationally tractable" approximate robust counterpart with a moderate level of conservativeness, and we are about to build such a counterpart.

3.1 Building the robust counterpart of (16) – (18)

We build an approximate robust counterpart via the standard *semidefinite relaxation scheme*. For $x \in \mathbf{R}^n$, let

$$a[x] = A^{\rm n} x,\ A[x] = \rho \left[A^1 x, A^2 x, ..., A^L x \right],\ b[x] = \rho \begin{pmatrix} x^T b^1 \\ \vdots \\ x^T b^L \end{pmatrix},$$
$$d = \frac{\rho}{2} \begin{pmatrix} c^1 \\ \vdots \\ c^L \end{pmatrix},\ e[x] = 2x^T b^{\rm n} + c^{\rm n}, \quad (19)$$

so that for all ζ one has

$$x^T \left[A^{\rm n} + \rho \sum_\ell \zeta_\ell A^\ell \right]^T \left[A^{\rm n} + \rho \sum_\ell \zeta_\ell A^\ell \right] x =$$
$$= (a[x] + A[x]\zeta)^T (a[x] + A[x]\zeta),$$

$$2x^T \left[b^{\rm n} + \rho \sum_{\ell=1}^{L} \zeta_\ell b^\ell \right] + \left[c^{\rm n} + \rho \sum_{\ell=1}^{L} \zeta_\ell c^\ell \right] = 2\,(b[x] + d)^T \zeta + e[x].$$

From these relations it follows that

$$\begin{array}{ll}
(a) & x \text{ is robust feasible for (16) – (18)} \\
& \Updownarrow \\
& (a[x] + A[x]\zeta)^T (a[x] + A[x]\zeta) \le 2\,(b[x] + d)^T \zeta + e[x] \\
& \qquad \forall(\zeta : \zeta^T Q_i \zeta \le 1, i = 1, ..., k) \\
& \Updownarrow \\
& \zeta^T A^T[x] A[x]\zeta + 2\zeta^T [A[x]a[x] - b[x] - d] \le e[x] - a^T[x]a[x] \\
& \qquad \forall(\zeta : \zeta^T Q_i \zeta \le 1, i = 1, ..., k) \\
& \Updownarrow \\
& \zeta^T A^T[x] A[x]\zeta + 2t\zeta^T [A[x]a[x] - b[x] - d] \le e[x] - a^T[x]a[x] \\
& \qquad \forall(\zeta, t : \zeta^T Q_i \zeta \le 1, i = 1, ..., k, t^2 = 1) \\
& \Updownarrow \\
(b) & \zeta^T A^T[x] A[x]\zeta + 2t\zeta^T [A[x]a[x] - b[x] - d] \le e[x] - a^T[x]a[x] \\
& \qquad \forall(\zeta, t : \zeta^T Q_i \zeta \le 1, i = 1, ..., k, t^2 \le 1).
\end{array} \tag{20}$$

Thus, x *is robust feasible if and only if the quadratic inequality* (20.b) *in variables* ζ, t *is a consequence of the system of quadratic inequalities* $\zeta^T Q_i \zeta \le 1, i = 1, ..., k, t^2 \le 1$. An evident *sufficient condition* for (20.b) to hold true is the possibility to majorate the left hand side of (20.b) *for all* ζ, t by a sum $\sum\limits_i \lambda_i \zeta^T Q_i \zeta + \mu t^2$ with nonnegative weights λ_i, μ satisfying the relation $\sum\limits_i \lambda_i + \mu \le e[x] - a^T[x]a[x]$. Thus, we come to the implication

$$\begin{array}{ll}
(a) & \exists(\mu \ge 0, \{\lambda_i \ge 0\}) : \quad \begin{array}{l} \sum\limits_{i=1}^{k} \lambda_i + \mu \le e[x] - a^T[x]a[x], \\ \sum\limits_{i=1}^{k} \lambda_i \zeta^T Q_i \zeta + \mu t^2 \ge \zeta^T A^T[x]A[x]\zeta \\ +2t\zeta^T [A[x]a[x] - b[x] - d] \;\; \forall(\zeta, t) \end{array} \\
& \Downarrow \\
& \zeta^T A^T[x] A[x]\zeta + 2t\zeta^T [A[x]a[x] - b[x] - d] \le e[x] - a^T[x]a[x] \\
& \qquad \forall(\zeta, t : \zeta^T Q_i \zeta \le 1, i = 1, ..., k, t^2 \le 1) \\
& \Downarrow \\
(b) & x \text{ is robust feasible}
\end{array} \tag{21}$$

A routine processing of condition (21.a) which we skip here demonstrates that the condition is equivalent to the solvability of the system of LMIs

$$\begin{pmatrix} e[x] - \sum\limits_{i=1}^{k} \lambda_i & [-b[x]-d]^T & a^T[x] \\ [-b[x]-d] & \sum\limits_{i=1}^{k} \lambda_i Q_i & -A^T[x] \\ a[x] & -A[x] & I \end{pmatrix} \succeq 0, \qquad (22)$$
$$\lambda_i \geq 0,\ i = 1, ..., k,$$

in variables x, λ. We arrive at the following

Proposition 3.1 *The system of LMIs* (22) *is an approximate robust counterpart of the uncertain convex quadratic constraint* (16) – (18).

The level of conservativeness Ω of (22) can be bounded as follows:

Theorem 2 [7]
(i) *In the case of a general-type ellipsoidal uncertainty* (18), *one has*

$$\Omega \leq \widetilde{\Omega} \equiv \sqrt{3.6 + 2\ln\left(\sum_{i=1}^{k} \mathrm{Rank}(Q_i)\right)}. \qquad (23)$$

Note that the right hand side in (23) *is* < 6, *provided that*

$$\sum_{i=1}^{k} \mathrm{Rank}(Q_i) \leq 10,853,519.$$

(ii) *In the case of box uncertainty:*

$$\zeta^T Q_i \zeta = \zeta_i^2, 1 \leq i \leq k = L \equiv \dim \zeta,\ \ \Omega \leq \tfrac{\pi}{2}$$

(iii) *In the case of simple* ($k = 1$) *ellipsoidal uncertainty* (18), $\Omega = 1$ (22) *is equivalent to the robust counterpart of* (16) – (18).

An instrumental role in the proof is played by the following fact which seems to be interesting by its own right:

Theorem 3 [7] *Let* $R, R_0, R_1, ..., R_k$ *be symmetric* $n \times n$ *matrices such that* $R_1, ..., R_k \succeq 0$ *and there exist nonnegative weights* λ_i *such that* $\sum\limits_{i=0}^{k} \lambda_i R_i \succ 0$. *Consider the optimization program*

$$\mathrm{OPT} = \max_y \left\{ y^T R y : y^T R_i y \leq 1,\ i = 0, ..., k \right\} \qquad (24)$$

along with the semidefinite program

$$\mathrm{SDP} = \min_{\mu} \left\{ \sum_{i=0}^{k} \mu_i : \sum_{i=0}^{k} \mu_i R_i \succeq R, \ \mu \geq 0 \right\}. \qquad (25)$$

Then (25) *is solvable, its optimal value majorates the one of* (24), *and there exists a vector* y_* *such that*

$$y_*^T R y_* = \mathrm{SDP}; \quad y_*^T R_0 y_* \leq 1; \quad y_*^T R_i y_* \leq \widetilde{\Omega}^2, \ i = 1, ..., k,$$

$$\widetilde{\Omega} = \begin{cases} \sqrt{3.6 + 2\ln\left(\sum\limits_{i=1}^{k} \mathrm{Rank}(Q_i)\right)}, & R_0 = q^T q \text{ is dyadic} \\ \sqrt{8\ln 2 + 4\ln n + 2\ln\left(\sum\limits_{i=1}^{k} \mathrm{Rank}(Q_i)\right)}, & \text{otherwise} \end{cases} \qquad (26)$$

In particular,

$$\mathrm{OPT} \leq \mathrm{SDP} \leq \widetilde{\Omega}^2 \cdot \mathrm{OPT}.$$

4. Approximate robust counterparts of uncertain conic quadratic problems

The constructions and results of the previous section can be extended to the essentially more general case of *conic quadratic* problems. Recall that a generic conic quadratic problem (another name: SOCP – Second Order Cone Problem) is

$$\min_{x} \left\{ f^T x : \| A_i x + b_i \|_2 \leq \alpha_i^T x + \beta_i, \ i = 1, ..., m \right\} \qquad \text{(CQP)}$$

here $x \in \mathbf{R}^n$, and A_i are $m_i \times n$ matrices; the data of (CQP) is the collection $(f, \{A_i, b_i, \alpha_i, \beta_i\}_{i=1}^m)$. As always, we assume the objective to be "certain" and thus may restrict ourselves with building an approximate robust counterpart of a *single* conic quadratic constraint

$$\| Ax + b \|_2 \leq \alpha^T x + \beta \qquad (27)$$

with data (A, b, α, β).

We assume that the uncertainty set is parameterized by a vector of perturbations and *that the uncertainty is "side-wise"*: the perturbations affecting the left- and the right hand side data of (27) run independently of each other through the respective uncertainty sets:

$$\begin{aligned}
\mathcal{U} &\equiv \mathcal{U}_\rho^{\rm left} \times \mathcal{U}_\rho^{\rm right}, \\
\mathcal{U}_\rho^{\rm left} &= \left\{ (A,b) = (A^{\rm n}, b^{\rm n}) + \sum_{\ell=1}^{L} \zeta_\ell (A^\ell, b^\ell) \,\middle|\, \zeta \in \rho \mathcal{V}^{\rm left} \right\}, \\
\mathcal{U}_\rho^{\rm right} &= \left\{ (\alpha,\beta) = (\alpha^{\rm n}, \beta^{\rm n}) + \sum_{r=1}^{R} \eta_r (\alpha^\ell, \beta^\ell) \,\middle|\, \eta \in \rho \mathcal{V}^{\rm right} \right\}.
\end{aligned} \tag{28}$$

In what follows, we focus on the case when $\mathcal{V}^{\rm left}$ is given as an intersection of ellipsoids centered at the origin:

$$\mathcal{V}^{\rm left} = \left\{ \zeta \in \mathbf{R}^L \mid \zeta^T Q_i \zeta \leq 1,\ i = 1, ..., k \right\}, \tag{29}$$

where $Q_i \succeq 0$ and $\sum\limits_{i=1}^{k} Q_i \succ 0$. We will be interested also in two particular cases of general ellipsoidal uncertainty (29), namely, in the cases of

- *simple ellipsoidal uncertainty* $k = 1$;
- *box uncertainty:* $k = L$ and $\zeta^T Q_i \zeta \equiv \zeta_i^2$, $i = 1, ..., L$.

As about the "right hand side" perturbation set $\mathcal{V}^{\rm right}$, we allow for a much more general geometry, namely, we assume only that $\mathcal{V}^{\rm right}$ is bounded, contains 0 and is *semidefinite-representable*:

$$\eta \in \mathcal{V}^{\rm right} \Leftrightarrow \exists u: \ P(\eta) + Q(u) - S \succeq 0, \tag{30}$$

where $P(\eta)$, $Q(u)$ are symmetric matrices linearly depending on η, u, respectively. We assume also that the LMI in (30) is strictly feasible, i.e., that

$$P(\bar{\eta}) + Q(\bar{u}) - S \succ 0$$

for appropriately chosen $\bar{\eta}$, $\bar{u}$.

4.1 Building approximate robust counterpart of (27) – (30)

For $x \in \mathbf{R}^n$, let

$$a[x] = A^{\rm n} x + b^{\rm n},\ A[x] = \rho \left[A^1 x + b^1, A^2 x + b^2, ..., A^L x + b^L \right], \tag{31}$$

so that for all ζ one has

$$\left[A^{\rm n} + \rho \sum_\ell \zeta_\ell A^\ell \right] x + \left[b^{\rm n} + \rho \sum_\ell \zeta_\ell b^\ell \right] = a[x] + A[x]\zeta.$$

Since the uncertainty is side-wise, x is robust feasible for (27) – (30) if and only if there exists τ such that the left hand side in (27), evaluated at

x, is at most τ for all left-hand-side data from $\mathcal{U}_\rho^{\text{left}}$, while the right hand side, evaluated at x, is at least τ for all right-hand-side data from $\mathcal{U}_\rho^{\text{right}}$. The latter condition can be processed straightforwardly via *semidefinite duality*, with the following result:

Proposition 4.1 *A pair (x,τ) is such that $\tau \leq \alpha^T x + \beta$ for all $(\alpha,\beta) \in \mathcal{U}_\rho^{\text{right}}$ if and only if it can be extended to a solution of the system of LMIs*

$$\tau \leq x^T\alpha^{\text{n}} + \beta^{\text{n}} + \text{Tr}(SV), \quad P^*(V) = \gamma[x] \equiv \begin{pmatrix} \rho[x^T\alpha^1 + \beta^1] \\ \vdots \\ \rho[x^T\alpha^R + \beta^R] \end{pmatrix} \qquad (32)$$
$$Q^*(V) = 0, \quad V \succeq 0$$

in variables x,τ,V. Here for a linear mapping $A(z) = \sum_{i=1}^{k} z_k A_k : \mathbf{R}^k \to \mathbf{S}^m$ taking values in the space $\mathbf{S}^m$ of $m\times m$ symmetric matrices, $A^*(Z) = \begin{pmatrix} \text{Tr}(ZA_1) \\ \vdots \\ \text{Tr}(ZA_1) \end{pmatrix} : \mathbf{S}^m \to \mathbf{R}^k$ is the mapping conjugate to $A(\cdot)$.

In view of the above observations, we have

$$\begin{array}{ll}
(a) & x \text{ is robust feasible for } (27) - (30) \\
& \Updownarrow \\
& \exists(\tau,V): \begin{cases} (x,\tau,V) \text{ solves } (32) \\ \| a[x] + A[x]\zeta \|_2 \leq \tau \quad \forall(\zeta : \zeta^T Q_i \zeta \leq 1, i=1,...,k) \end{cases} \\
& \Updownarrow \\
& \exists(\tau,V): \begin{cases} (x,\tau,V) \text{ solves } (32) \\ \| \pm a[x] + A[x]\zeta \|_2 \leq \tau \quad \forall(\zeta : \zeta^T Q_i \zeta \leq 1, i=1,...,k) \end{cases} \\
& \Updownarrow \\
& \exists(\tau,V): \begin{cases} (x,\tau,V) \text{ solves } (32) \\ \| ta[x] + A[x]\zeta \|_2 \leq \tau \\ \quad \forall(\zeta,t : \zeta^T Q_i \zeta \leq 1, i=1,...,k, t^2 \leq 1) \end{cases} \\
& \Updownarrow \\
(b) & \exists(\tau,V): \begin{cases} (1) & (x,\tau,V) \text{ solves } (32) \\ (2) & \tau \geq 0 \\ (3) & \| ta[x] + A[x]\zeta \|_2^2 \leq \tau^2 \\ & \quad \forall(\zeta,t : \zeta^T Q_i \zeta \leq 1, i=1,...,k, t^2 \leq 1) \end{cases}
\end{array} \qquad (33)$$

Thus, x is robust feasible *if and only if* (33.b) holds true. Now observe that

$$\boxed{\begin{array}{ll} \exists(\mu,\lambda_1,...,\lambda_k): & \\ (a) & \mu\geq 0, \lambda_i\geq 0,\ i=1,...,k, \\ (b) & \mu+\sum\limits_{i=1}^k\lambda_i\leq\tau, \\ (c)\quad \tau\left(\mu t^2+\sum\limits_{i=1}^k\lambda_i\zeta^TQ_i\zeta\right)\geq\parallel ta[x]+A[x]\zeta\parallel_2^2 & \forall(t,\zeta) \end{array}} \qquad (34)$$

$$\Downarrow$$

$$\parallel ta[x]+A[x]\zeta\parallel_2^2\leq\tau^2\quad\forall(\zeta,t:\zeta^TQ_i\zeta\leq 1, i=1,...,k, t^2\leq 1).$$

Via Schur complement, for nonnegative $\tau,\mu,\{\lambda_i\}$ the inequality

$$\tau\left(\mu t^2+\sum_{i=1}^k\lambda_i\zeta^TQ_i\zeta\right)\geq\parallel ta[x]+A[x]\zeta\parallel_2^2$$

holds true for a given pair (t,ζ) if and only if

$$\begin{array}{rcl} 0 & \preceq & \left(\begin{array}{c|c} \left[\begin{array}{cc} t & \zeta^T\end{array}\right]\left[\begin{array}{c|c}\mu & \\ \hline & \sum\limits_i\lambda_iQ_i\end{array}\right]\left[\begin{array}{c} t\\ \zeta\end{array}\right] & \left[\begin{array}{cc} t & \zeta^T\end{array}\right]\left[\begin{array}{c} a^T[x]\\ A^T[x]\end{array}\right] \\ \hline \left[\begin{array}{cc} a[x] & A[x]\end{array}\right]\left[\begin{array}{c} t\\ \zeta\end{array}\right] & \tau I\end{array}\right) \\ & = & \left(\begin{array}{c|c|c}\tau & \zeta^T & \\ \hline & & I\end{array}\right)\left(\begin{array}{c|c|c}\mu & & a^T[x] \\ \hline & \sum\limits_{i=1}^k\lambda_iQ_i & A^T[x] \\ \hline a[x] & A[x] & \tau I\end{array}\right)\left(\begin{array}{c|c|c}\tau & \zeta^T & \\ \hline & & I\end{array}\right)^T. \end{array}$$

In view of this observation combined with the fact that the union, over all τ,ζ, of the image spaces of the matrices $\left(\begin{array}{c|c} & \tau \\ \hline & \zeta \\ \hline I & \end{array}\right)\in\mathbf{R}^{(1+L+m)\times(m+1)}$ is the entire $\mathbf{R}^{1+L+m}$, we conclude that in the case of nonnegative $\tau,\mu,\{\lambda_i\}$ the relation (34.c) is equivalent to

$$\left(\begin{array}{c|c|c}\mu & & a^T[x] \\ \hline & \sum\limits_{i=1}^k\lambda_iQ_i & A^T[x] \\ \hline a[x] & A[x] & \tau I\end{array}\right)\succeq 0.$$

Looking at this relation, (34) and (33), we see that the following implication is valid:

$$\boxed{\begin{array}{lc} \exists(\tau,V,\mu,\lambda_1,...,\lambda_k): & \\ (a) & \mu\geq 0, \lambda_i\geq 0,\ i=1,...,k, \\ (b) & \sum\limits_{i=1}^{k}\lambda_i+\mu\leq\tau, \\ (c) & \left(\begin{array}{c|c|c} \mu & & a^T[x] \\ \hline & \sum\limits_{i=1}^{k}\lambda_i Q_i & A^T[x] \\ \hline a[x] & A[x] & \tau I \end{array}\right)\succeq 0, \\ (d) & (x,\tau,V)\text{ solves (32)} \end{array}} \tag{35}$$

$$\Downarrow$$

$$x \text{ is robust feasible for (27) – (30).}$$

We can immediately eliminate μ from the premise of (35), thus arriving at the following result:

Proposition 4.2 *The system of LMIs in variables* $x, \tau, V, \{\lambda_i\}_{i=1}^{k}$

$$\left(\begin{array}{c|c|c} \tau-\sum\limits_i\lambda_i & & a^T[x] \\ \hline & \sum\limits_{i=1}^{k}\lambda_i Q_i & A^T[x] \\ \hline a[x] & A[x] & \tau I \end{array}\right)\succeq 0,\quad P^*(V)=\left(\begin{array}{c}\rho[x^T\alpha^1+\beta^1] \\ \vdots \\ \rho[x^T\alpha^R+\beta^R]\end{array}\right),$$
$$\lambda_i\geq 0,\ i=1,...,k,\quad \tau\leq x^T\alpha^{\rm n}+\beta^{\rm n}+{\rm Tr}(SV),\quad Q^*(V)=0,\quad V\succeq 0, \tag{36}$$

where $a[x], A[x]$ *are given by* (31), *is an approximate robust counterpart of the uncertain conic quadratic constraint* (27) – (30).

The level of conservativeness Ω of (36) can be bounded as follows:

Theorem 4 [7] (i) *In the case of a general-type ellipsoidal uncertainty* (29), *one has*

$$\Omega\leq\tilde{\Omega}\equiv\sqrt{3.6+2\ln\left(\sum_{i=1}^{k}{\rm Rank}(Q_i)\right)}. \tag{37}$$

(ii) *In the case of box uncertainty:* $\zeta^T Q_i\zeta=\zeta_i^2,\ 1\leq i\leq k=L\equiv$ dim ζ, $\Omega\leq\frac{\pi}{2}$.

(iii) *In the case of simple* ($k=1$) *ellipsoidal uncertainty* (29), $\Omega=1$ – *The system* (36) *is equivalent to the robust counterpart of* (27) – (30).

References

[1] A. Ben-Tal, A. Nemirovski, "Stable Truss Topology Design via Semidefinite Programming" – *SIAM Journal on Optimization* **7** (1997), 991-1016.

[2] A. Ben-Tal, A. Nemirovski, "Robust solutions to uncertain linear programs" – *OR Letters* **25** (1999), 1-13.

[3] A. Ben-Tal, A. Nemirovski, "Robust Convex Optimization" – *Mathematics of Operations Research* **23** (1998).

[4] S. Boyd, L. El Ghaoui, F. Feron, V. Balakrishnan, *Linear Matrix Inequalities in System and Control Theory* – volume 15 of *Studies in Applied Mathematics*, SIAM, Philadelphia, 1994.

[5] A. Ben-Tal, L. El Ghaoui, A. Nemirovski, "Robust Semidefinite Programming" – in: R. Saigal, H. Wolkowicz, L. Vandenberghe, Eds. *Handbook on Semidefinite Programming*, Kluwer Academis Publishers, 2000, 139-162.

[6] A. Ben-Tal, A. Nemirovski, "On tractable approximations of uncertain linear matrix inequalities affected by interval uncertainty" – *SIAM J. on Optimization*, 2001, to appear.

[7] A. Ben-Tal, A. Nemirovski, C. Roos, "Robust solutions of uncertain quadratic and conic quadratic problems" – Research Report #2/01, Minerva Optimization Center, Technion – Israel Institute of Technology, Technion City, Haifa 32000, Israel. http://iew3.technion.ac.il:8080/subhome.phtml?/Home/research

[8] L. El-Ghaoui, H. Lebret, "Robust solutions to least-square problems with uncertain data matrices" – *SIAM J. of Matrix Anal. and Appl.* **18** (1997), 1035-1064.

[9] L. El-Ghaoui, F. Oustry, H. Lebret, "Robust solutions to uncertain semidefinite programs" – *SIAM J. on Optimization* **9** (1998), 33-52.

[10] M.X. Goemans, D.P. Williamson, "Improved approximation algorithms for Maximum Cut and Satisfiability problems using semidefinite programming" – *Journal of ACM* **42** (1995), 1115-1145.

[11] A. Nemirovski, "Several NP-hard problems arising in Robust Stability Analysis" – *Math. Control Signals Systems* **6** (1993), 99-105.

[12] A. Nemirovski, C. Roos, T. Terlaky, "On maximization of quadratic form over intersection of ellipsoids with common center" – *Mathematical Programming* **86** (2000), 463-473.

[13] Yu. Nesterov, "Semidefinite relaxation and non-convex quadratic optimization" – *Optimization Methods and Software* **12** (1997), 1-20.

[14] Yu. Nesterov, "Nonconvex quadratic optimization via conic relaxation", in: R. Saigal, H. Wolkowicz, L. Vandenberghe, Eds. *Handbook on Semidefinite Programming*, Kluwer Academis Publishers, 2000, 363-387.

[15] Y. Ye, "Approximating quadratic programming with bounds and quadratic constraints" – *Math. Programming* **47** (1999), 219-226.

GLOBAL CONVERGENCE OF A HYBRID TRUST-REGION SQP-FILTER ALGORITHM FOR GENERAL NONLINEAR PROGRAMMING

Nick Gould
Rutherford Appleton Laboratory
Computational Science and Engineering Department
Chilton, Oxfordshire, England
gould@rl.ac.uk

Philippe L. Toint
Department of Mathematics
University of Namur
61, rue de Bruxelles, B-5000 Namur, Belgium
philippe.toint@fundp.ac.be

Abstract Global convergence to first-order critical points is proved for a variant of the trust-region SQP-filter algorithm analyzed in (Fletcher, Gould, Leyffer and Toint). This variant allows the use of two types of step strategies: the first decomposes the step into its normal and tangential components, while the second replaces this decomposition by a stronger condition on the associated model decrease.

1. Introduction

We analyze an algorithm for solving optimization problems where a smooth objective function is to be minimized subject to smooth nonlinear constraints. No convexity assumption is made. More formally, we consider the problem

$$\begin{array}{ll} \text{minimize} & f(x) \\ \text{subject to} & c_{\mathcal{E}}(x) = 0 \\ & c_{\mathcal{I}}(x) \geq 0, \end{array} \tag{1.1}$$

where f is a twice continuously differentiable real valued function of the variables $x \in \mathbb{R}^n$ and $c_{\mathcal{E}}(x)$ and $c_{\mathcal{I}}(x)$ are twice continuously differentiable functions from $\mathbb{R}^n$ into $\mathbb{R}^m$ and from $\mathbb{R}^n$ into $\mathbb{R}^p$, respectively. Let $c(x)^T = (c_{\mathcal{E}}(x)^T \;\; c_{\mathcal{I}}(x)^T)$.

The class of algorithms that we discuss belongs to the class of trust-region methods and, more specifically, to that of *filter methods* introduced by Fletcher and Leyffer (1997), in which the use of a penalty function, a common feature of the large majority of the algorithms for constrained optimization, is replaced by the introduction of a so-called "filter".

A global convergence theory for an algorithm of this class is proposed in Fletcher, Leyffer and Toint (1998), in which the objective function is locally approximated by a linear function, leading, at each iteration, to the (exact) solution of a linear program. This algorithm therefore mixes the use of the filter with sequential linear programming (SLP). Similar results are shown in Fletcher, Leyffer and Toint (2000), where the approximation of the objective function is quadratic, leading to sequential quadratic programming (SQP) methods, but at the price of finding a global minimizer of the possibly nonconvex quadratic programming subproblem, which is known to be a very difficult task. Convergence of SQP filter methods was also considered in Fletcher, Gould, Leyffer and Toint (1999), where the SQP step was decomposed in "normal" and "tangential" components. Although this latter procedure is computationally well-defined and considerably less complex than finding the global minimum of a general quadratic program, it may sometimes be costly, and a simpler strategy, where the step is computed "as a whole" can also be of practical interest whenever possible. The purpose of this paper, a companion of Fletcher et al. (1999), is to analyze a hybrid algorithm that uses the decomposition of the step into normal and tangential components as infrequently as possible.

2. A Hybrid Trust-Region SQP-Filter Algorithms

For the sake of completeness and clarity, we review briefly the main constituent parts of the SQP algorithm discussed in Fletcher et al. (1999). Sequential quadratic programming methods are iterative. At a given iterate x_k, they implicitly apply Newton's method to solve (a local version of) the first-order necessary optimality conditions by solving the quadratic programming subproblem $\mathrm{QP}(x_k)$ given by

$$\begin{array}{ll} \text{minimize} & f_k + \langle g_k, s\rangle + \frac{1}{2}\langle s, H_k s\rangle \\ \text{subject to} & c_{\mathcal{E}}(x_k) + A_{\mathcal{E}}(x_k)s = 0 \\ & c_{\mathcal{I}}(x_k) + A_{\mathcal{I}}(x_k)s \geq 0, \end{array} \tag{2.1}$$

where $f_k = f(x_k)$, $g_k = g(x_k) \stackrel{\text{def}}{=} \nabla_x f(x_k)$, where $A_{\mathcal{E}}(x_k)$ and $A_{\mathcal{I}}(x_k)$ are the Jacobians of the constraint functions $c_{\mathcal{E}}$ and $c_{\mathcal{I}}$ at x_k and where H_k is a symmetric matrix. We will not immediately be concerned about how H_k is obtained, but we will return to this point in Section 3. Assuming a suitable value of H_k can be found, the solution of $\mathrm{QP}(x_k)$ then yields a step s_k. If $s_k = 0$, then x_k is first-order critical for problem (1.1).

2.1 The filter

Unfortunately, due to the locally convergent nature of Newton's iteration, the step s_k may not always be very good. Thus, having computed (in a so far unspecified manner) a step s_k from our current iterate x_k, we need to decide whether the trial point $x_k + s_k$ is any better than x_k as an approximate solution to our original problem (1.1). This is achieved by using the notion of a filter, itself based on that of dominance.

If we define the feasibility measure

$$\theta(x) = \max\left[0, \max_{i\in\mathcal{E}} |c_i(x)|, \max_{i\in\mathcal{I}} -c_i(x)\right], \tag{2.2}$$

we say that a point x_1 *dominates* a point x_2 whenever

$$\theta(x_1) \leq \theta(x_2) \;\text{ and }\; f(x_1) \leq f(x_2).$$

Thus, if iterate x_k dominates iterate x_j, the latter is of no real interest to us since x_k is at least as good as x_j on account of both feasibility and optimality. All we need to do now is to remember iterates that are not dominated by any other iterates using a structure called a filter. A *filter* is a list $\mathcal{F}$ of pairs of the form (θ_i, f_i) such that either

$$\theta_i < \theta_j \;\text{ or }\; f_i < f_j$$

for $i \neq j$. Fletcher et al. (1999) propose to accept a new trial iterate $x_k + s_k$ only if it is not dominated by any other iterate in the filter and x_k. In the vocabulary of multi-criteria optimization, this amounts to building elements of the efficient frontier associated with the bi-criteria problem of reducing infeasibility and the objective function value. We may describe this concept by associating with each iterate x_k its (θ, f)-pair (θ_k, f_k) and accept $x_k + s_k$ only if its (θ, f)-pair does not lie, in the two-dimensional space spanned by constraint violation and objective function value, above and on the right of a previously accepted pair (including that associated with x_k).

While the idea of not accepting dominated trial points is simple and elegant, it needs to be refined a little in order to provide an efficient algorithmic tool. In particular, we do not wish to accept $x_k + s_k$ if its (θ, f)-pair is arbitrarily close to that of x_k or that of a point already in the filter. Thus Fletcher et al. (1999) set a small "margin" around the border of the dominated part of the (θ, f)-space in which we shall also reject trial points. Formally, we say that a point x is *acceptable for the filter* if and only if

$$\theta(x) < (1 - \gamma_\theta)\theta_j \quad \text{or} \quad f(x) < f_j - \gamma_\theta \theta_j \quad \text{for all } (\theta_j, f_j) \in \mathcal{F}, \tag{2.3}$$

for some $\gamma_\theta \in (0, 1)$. We also say that x is "acceptable for the filter and x_k" if (2.3) holds with $\mathcal{F}$ replaced by $\mathcal{F} \cup (\theta_k, f_k)$. We thus consider moving from x_k to $x_k + s_k$ only if $x_k + s_k$ is acceptable for the filter and x_k.

As the algorithm progresses, Fletcher et al. (1999) add (θ, f)-pairs to the filter. If an iterate x_k is acceptable for $\mathcal{F}$, this is done by adding the pair (θ_k, f_k) to the filter and by removing from it every other pair (θ_j, f_j) such that $\theta_j \geq \theta_k$ and $f_j - \gamma_\theta \theta_j \geq f_k - \gamma_\theta \theta_k$. We also refer to this operation as "adding x_k to the filter" although, strictly speaking, it is the (θ, f)-pair which is added.

We conclude this introduction to the notion of a filter by noting that, if a point x_k is in the filter or is acceptable for the filter, then any other point x such that

$$\theta(x) \leq (1 - \gamma_\theta)\theta_k \quad \textit{and} \quad f(x) \leq f_k - \gamma_\theta \theta_k$$

is also be acceptable for the filter and x_k.

2.2 The composite SQP step

Of course, the step s_k must be computed, typically by solving, possibly approximately, a variant of (2.1). In the trust-region approach, one takes into account the fact that (2.1) only approximates our original problem locally: the step s_k is thus restricted in norm to ensure that $x_k + s_k$ remains in a *trust-region* centred at x_k, where we believe this approximation to be adequate. In other words, we replace QP(x_k) by the subproblem TRQP(x_k, Δ_k) given by

$$\begin{array}{ll} \text{minimize} & m_k(x_k + s) \\ \text{subject to} & c_{\mathcal{E}}(x_k) + A_{\mathcal{E}}(x_k)s = 0, \\ & c_{\mathcal{I}}(x_k) + A_{\mathcal{I}}(x_k)s \geq 0, \\ \text{and} & \|s\| \leq \Delta_k, \end{array} \tag{2.4}$$

for some (positive) value of the *trust-region radius* Δ_k, where we have defined

$$m_k(x_k + s) = f_k + \langle g_k, s\rangle + \tfrac{1}{2}\langle s, H_k s\rangle, \tag{2.5}$$

and where $\|\cdot\|$ denotes the Euclidean norm. This latter choice is purely for ease of exposition. We could equally use a family of iteration dependent norms $\|\cdot\|_k$, so long as we require that all members of the family are uniformly equivalent to the Euclidean norm. The interested reader may verify that all subsequent developments can be adapted to this more general case by introducing the constants implied by this uniform equivalence wherever needed.

Remarkably, most of the existing SQP algorithms assume that an exact local solution of QP(x_k) or TRQP(x_k, Δ_k) is found, although attempts have been made by Dembo and Tulowitzki (1983) and Murray and Prieto (1995) to design conditions under which an approximate solution of the subproblem is acceptable. In contrast, the algorithm of Fletcher et al. (1999) is in spirit to the composite-step SQP methods pionneered by Vardi (1985), Byrd, Schnabel and Shultz (1987), and Omojokun (1989) and later developed by several authors, including Biegler, Nocedal and Schmid (1995), El-Alem (1995, 1999), Byrd, Gilbert and Nocedal (2000*a*), Byrd, Hribar and Nocedal (2000*b*), Bielschowsky and Gomes (1998), Liu and Yuan (1998) and Lalee, Nocedal and Plantenga (1998). It decomposes the step s_k into the sum of two distinct components, a *normal step* n_k, such that $x_k + n_k$ satisfies the constraints of TRQP(x_k, Δ_k), and a *tangential step* t_k, whose purpose is to obtain reduction of the objective function's model while continuing to satisfy those constraints. The step s_k is then called *composite*. More formally, we write

$$s_k = n_k + t_k \tag{2.6}$$

and assume that

$$c_{\mathcal{E}}(x_k) + A_{\mathcal{E}}(x_k)n_k = 0, \quad c_{\mathcal{I}}(x_k) + A_{\mathcal{I}}(x_k)n_k \geq 0, \tag{2.7}$$

$$\|s_k\| \leq \Delta_k, \tag{2.8}$$

and

$$c_{\mathcal{E}}(x_k) + A_{\mathcal{E}}(x_k)s_k = 0, \quad c_{\mathcal{I}}(x_k) + A_{\mathcal{I}}(x_k)s_k \geq 0. \tag{2.9}$$

Of course, this is a strong assumption, since in particular (2.7) or (2.8)/(2.9) may not have a solution. We shall return to this possibility shortly. Given our assumption, there are many ways to compute n_k and t_k. For instance, we could compute n_k from

$$n_k = P_k[x_k] - x_k, \tag{2.10}$$

where P_k is the orthogonal projector onto the feasible set of QP(x_k). No specific choice for n_k is made, but one instead assumes that n_k exists when the maximum violation of the nonlinear constraints at the k-th iterate $\theta_k \stackrel{\text{def}}{=} \theta(x_k)$ is sufficiently small, and that n_k is then reasonably scaled with respect to the values of the constraints. In other words, Fletcher et al. (1999) assume that

$$n_k \text{ exists and } \|n_k\| \leq \kappa_{\text{usc}}\theta_k, \quad \text{whenever } \theta_k \leq \delta_n, \tag{2.11}$$

for some constants $\kappa_{\text{usc}} > 0$ and $\delta_n > 0$. This assumption is also used by Dennis, El-Alem and Maciel (1997) and Dennis and Vicente (1997) in the context of problems with equality constraints only. It can be shown not to impose conditions on the constraints or the normal step itself that are unduly restrictive (see Fletcher et al. (1999) for a discussion).

Having defined the normal step, we are in position to use it if it falls within the trust-region, that is if $\|n_k\| \leq \Delta_k$. In this case, we write

$$x_k^{\text{N}} = x_k + n_k, \tag{2.12}$$

and observe that n_k satisfies the constraints of TRQP(x_k, Δ_k) and thus also of QP(x_k). It is crucial to note, at this stage, that such an n_k may fail to exist because the constraints of QP(x_k) may be incompatible, in which case P_k is undefined, or because all feasible points for $QP(x_k)$ may lie outside the trust region.

Let us continue to consider the case where this problem does not arise, and a normal step n_k has been found with $\|n_k\| \leq \Delta_k$. We then have to find a tangential step t_k, starting from x_k^{N} and satisfying (2.8) and (2.9), with the aim of decreasing the value of the objective function. As always in trust-region methods, this is achieved by computing a step that produces a sufficient decrease in m_k, which is to say that we wish $m_k(x_k^{\text{N}}) - m_k(x_k + s_k)$ to be "sufficiently large". Of course, this is only possible if the maximum size of t_k is not too small, which is to say that x_k^{N} is not too close to the trust-region boundary. We formalize this condition by replacing our condition that $\|n_k\| \leq \Delta_k$ by the stronger requirement that

$$\|n_k\| \leq \kappa_\Delta \Delta_k \min[1, \kappa_\mu \Delta_k^{\mu_k}], \tag{2.13}$$

for some $\kappa_\Delta \in (0, 1]$, some $\kappa_\mu > 0$ and some $\mu_k \in [0, 1)$. If condition (2.13) does not hold, Fletcher et al. (1999) presume that the computation of t_k is unlikely to produce a satisfactory decrease in m_k, and proceed just as if the feasible set of TRQP(x_k, Δ_k) were empty. If n_k can be computed and (2.13) holds, TRQP(x_k, Δ_k) is said to be *compatible for* μ. In this case at least a sufficient model decrease seems possible, in the form of a familiar Cauchy-point condition. In order to formalize this notion, we recall that the feasible set of $\mathrm{QP}(x_k)$ is convex, and we can therefore introduce the first-order criticality measure

$$\chi_k = \Big| \min_{\substack{A_{\mathcal{E}}(x_k)t=0 \\ c_{\mathcal{I}}(x_k)+A_{\mathcal{I}}(x_k)(n_k+t)\geq 0 \\ \|t\|\leq 1}} \langle g_k + H_k n_k, t\rangle \Big| \tag{2.14}$$

(see Conn, Gould, Sartenaer and Toint, 1993). Note that this function is zero if and only if x_k^{N} is a first-order critical point of the linearized "tangential" problem

$$\begin{array}{ll} \text{minimize} & \langle g_k + H_k n_k, t\rangle + \frac{1}{2}\langle H_k t, t\rangle \\ \text{subject to} & A_{\mathcal{E}}(x_k)t = 0 \\ & c_{\mathcal{I}}(x_k) + A_{\mathcal{I}}(x_k)(n_k + t) \geq 0, \end{array} \tag{2.15}$$

which is equivalent to $\mathrm{QP}(x_k)$ with $s = n_k + t$. The sufficient decrease condition then consists in assuming that there exists a constant $\kappa_{\mathrm{tmd}} > 0$ such that

$$m_k(x_k^{\mathrm{N}}) - m_k(x_k^{\mathrm{N}} + t_k) \geq \kappa_{\mathrm{tmd}} \chi_k \min\left[\frac{\chi_k}{\beta_k}, \Delta_k\right], \tag{2.16}$$

whenever $\mathrm{TRQP}(x_k, \Delta_k)$ is compatible, where $\beta_k = 1 + \|H_k\|$. We know from Toint (1988) and Conn et al. (1993) that this condition holds if the model reduction exceeds that which would be obtained at the generalized Cauchy point, that is the point resulting from a backtracking curvilinear search along the projected gradient path from x_k^{N}, that is

$$x_k(\alpha) = P_k[x_k^{\mathrm{N}} - \alpha \nabla_x m_k(x_k^{\mathrm{N}})].$$

This technique therefore provides an implementable algorithm for computing a step that satisfies (2.16) (see Gould, Hribar and Nocedal, 1998 for an example in the case where $c(x) = c_{\mathcal{E}}(x)$, or Toint, 1988 and Moré and Toraldo, 1991 for the case of bound constraints), but, of course, reduction of m_k beyond that imposed by (2.16) is often possible and desirable if fast convergence is sought. Also note that the minimization problem of the right-hand side of (2.14) reduces to a linear programming problem if we choose to use a polyhedral norm in its definition at iteration k.

If $\mathrm{TRQP}(x_k, \Delta_k)$ is not compatible for μ, that is when the feasible set determined by the constraints of $\mathrm{QP}(x_k)$ is empty, or the freedom left to reduce m_k within the trust region is too small in the sense that (2.13) fails, solving $\mathrm{TRQP}(x_k, \Delta_k)$ is most likely pointless, and we must consider an alternative. Observe that, if $\theta(x_k)$ is sufficiently small and the true nonlinear constraints are locally compatible, the linearized constraints should also be compatible, since they approximate the nonlinear constraints (locally) correctly. Furthermore, the feasible region for the linearized constraints should then be close enough to x_k for there to be some room to reduce m_k, at least if Δ_k is large enough. If the nonlinear constraints are locally incompatible, we have to find a neighbourhood where this is not the case, since the problem (1.1) does not make sense in the current one. Fletcher et al. (1999) thus rely on a *restoration procedure*, whose aim is to produce a new point $x_k + r_k$ for which TRQP$(x_k + r_k, \Delta_{k+1})$ is compatible for some $\Delta_{k+1} > 0$—another condition will actually be needed, which we will discuss shortly.

The idea of the restoration procedure is to (approximately) solve

$$\min_{x \in \mathbb{R}^n} \theta(x) \tag{2.17}$$

starting from x_k, the current iterate. This is a non-smooth problem, but there exist methods, possibly of trust-region type (such as that suggested by Yuan, 1994), which can be successfully applied to solve it. Thus we will not describe the restoration procedure in detail. Note that we have chosen here to reduce the infinity norm of the constraint violation, but we could equally well consider other norms, such as ℓ_1 or ℓ_2, in which case the methods of Fletcher and Leyffer (1998) or of El-Hallabi and Tapia (1995) and Dennis, El-Alem and Williamson (1999) can respectively be considered. Of course, this technique only guarantees convergence to a first-order critical point of the chosen measure of constraint violation, which means that, in fact, the restoration procedure may fail as this critical point may not be feasible for the constraints of (1.1). However, even in this case, the result of the procedure is of interest because it typically produces a local minimizer of $\theta(x)$, or of whatever other measure of constraint violation we choose for the restoration, yielding a point of locally-least infeasibility.

There seems to be no easy way to circumvent this drawback, as it is known that finding a feasible point or proving that no such point exists is a global optimization problem and can be as difficult as the optimization problem (1.1) itself. One therefore has to accept two possible outcomes of the restoration procedure: either the procedure fails in that it does not produce a sequence of iterates converging to feasibility, or a point $x_k + r_k$ is produced such that $\theta(x_k + r_k)$ is as small as desired.

2.3 An alternative step

Is it possible to find a cheaper alternative to computing a normal step, finding a generalized Cauchy point and explicitly checking (2.16)?

Suppose, for now, that it is possible to compute a point $x_k + s_k$ directly to satisfy the constraints of $\text{TRQP}(x_k, \Delta_k)$ and for which

$$m_k(x_k) - m_k(x_k + s_k) \geq \kappa_{\text{tmd}} \min[\pi_k, \Delta_k] \tag{2.18}$$

and $\pi_k = \pi(x_k)$, where π is a continuous function of its argument that is a criticality measure for $\text{TRQP}(x_k, \Delta_k)$. Such a s_k could for instance be computed by applying any efficient method to this latter problem (we might think of interior point methods of the type described in Conn, Gould, Orban and Toint, 2000) for instance, and its performance could be assessed by computing

$$\pi(x) = \min_{y | y_{\mathcal{I}} \geq 0} \|g(x) - A(x)^T y\|.$$

Of course, nothing guarantees that such an s_k exists (depending on our choice of $\pi(x)$) or is cheaply computable for each x_k, which means that we may have to resort to the normal-tangential strategy of Fletcher et al. (1999) if such problems arise. However, if we can find s_k at a fraction of the cost of computing n_k and t_k, can we use it inside an SQP-filter algorithm and maintain the desirable convergence to first-order critical points?

Obviously, the answer to that question depends on the manner in which the use of s_k is integrated into a complete algorithm.

2.4 A hybrid SQP-filter Algorithm

We have now discussed the main ingredients of the class of algorithms we wish to consider, and we are now ready to define it formally as Algorithm 2.1:

Algorithm 2.1: Hybrid SQP-filter Algorithm

Step 0: Initialization. Let an initial point x_0, an initial trust-region radius $\Delta_0 > 0$ and an initial symmetric matrix H_0 be given, as well as constants $\gamma_0 < \gamma_1 \leq 1 \leq \gamma_2$, $0 < \eta_1 \leq \eta_2 < 1$, $\gamma_\theta \in (0,1)$, $\kappa_\theta \in (0,1)$, $\kappa_\Delta \in (0,1]$, $\kappa_\mu > 0$, $\mu \in (0,1)$, $\psi > 1/(1+\mu)$, $\kappa_{\text{u}} > 0$ and $\kappa_{\text{tmd}} \in (0,1]$. Compute $f(x_0)$ and $c(x_0)$. Set $\mathcal{F} = \emptyset$ and $k = 0$.

Step 1: Test for optimality. If $\theta_k = 0$ and either $\chi_k = 0$ or $\pi_k = 0$, stop.

Step 2: Alternative step. If

$$\theta_k > \kappa_u \Delta_k \min[1, \Delta_k^{\mu}], \tag{2.19}$$

set $\mu_k = \mu$ and go to Step 3. Otherwise, attempt to compute a step s_k that satisfies the constraints of TRQP(x_k, Δ_k) and (2.18). If this succeeds, go to Step 4. Otherwise, set $\mu_k = 0$.

Step 3: Composite step.

Step 3a: Normal component. Attempt to compute a normal step n_k. If TRQP (x_k, Δ_k) is compatible for μ_k, go to Step 3b. Otherwise, include x_k in the filter and compute a restoration step r_k for which TRQP$(x_k + r_k, \Delta_{k+1})$ is compatible for some $\Delta_{k+1} > 0$, and $x_k + r_k$ is acceptable for the filter. If this proves impossible, stop. Otherwise, define $x_{k+1} = x_k + r_k$ and go to Step 7.

Step 3b: Tangential component. Compute a tangential step t_k and set $s_k = n_k + t_k$.

Step 4: Tests to accept the trial step.

- Evaluate $c(x_k + s_k)$ and $f(x_k + s_k)$.
- If $x_k + s_k$ is not acceptable for the filter and x_k, set $x_{k+1} = x_k$, choose $\Delta_{k+1} \in [\gamma_0 \Delta_k, \gamma_1 \Delta_k]$, set $n_{k+1} = n_k$ if Step 3 was executed, and go to Step 7.
- If

$$m_k(x_k) - m_k(x_k + s_k) \geq \kappa_\theta \theta_k^{\psi}, \tag{2.20}$$

and

$$\rho_k \stackrel{\text{def}}{=} \frac{f(x_k) - f(x_k + s_k)}{m_k(x_k) - m_k(x_k + s_k)} < \eta_1, \tag{2.21}$$

again set $x_{k+1} = x_k$, choose $\Delta_{k+1} \in [\gamma_0 \Delta_k, \gamma_1 \Delta_k]$, set $n_{k+1} = n_k$ if Step 3 was executed, and go to Step 7.

Step 5: Test to include the current iterate in the filter. If (2.20) fails, include x_k in the filter $\mathcal{F}$.

Step 6: Move to the new iterate. Set $x_{k+1} = x_k + s_k$ and choose Δ_{k+1} such that

$$\Delta_{k+1} \in [\Delta_k, \gamma_2 \Delta_k] \text{ if } \rho_k \geq \eta_2 \text{ and (2.20) holds.}$$

Step 7: Update the Hessian approximation. Determine H_{k+1}. Increment k by one and go to Step 1.

This algorithm differs from that of Fletcher et al. (1999) in that it contains the alternative step strategy, but also because it allows the normal step to satisfy (2.13) with $\mu = 0$ whenever (2.19) holds, that is whenever the current iterate is sufficiently feasible. (As we will see later, (2.13) with $\mu > 0$ can be viewed as an implicit technique to impose (2.19).)

As in Fletcher and Leyffer (1997) and Fletcher and Leyffer (1998), one may choose $\psi = 2$ (Note that the choice $\psi = 1$ is always possible because $\mu > 0$). Reasonable values for the constants might then be

$$\gamma_0 = 0.1, \quad \gamma_1 = 0.5, \quad \gamma_2 = 2, \quad \eta_1 = 0.01, \quad \eta_2 = 0.9, \quad \gamma_\theta = 10^{-4},$$
$$\kappa_\Delta = 0.7, \quad \kappa_\mu = 100, \quad \mu = 0.01, \quad \kappa_\theta = 10^{-4}, \text{ and } \kappa_{\text{tmd}} = 0.01.$$

but it is too early to know if these are even close to the best possible choices.

As in Fletcher et al. (1999), some comments on this algorithm are now in order. Observe first that, by construction, every iterate x_k must be acceptable for the filter at the beginning of iteration k, irrespective of the possibility that it is added to the filter later. Also note that the restoration step r_k cannot be zero, that is restoration cannot simply entail enlarging the trust-region radius to ensure (2.13), even if n_k exists. This is because x_k is added to the filter before r_k is computed, and $x_k + r_k$ must be acceptable for the filter which now contains x_k. Also note that the restoration procedure cannot be applied on two successive iterations, since the iterate $x_k + r_k$ produced by the first of these iterations is both compatible and acceptable for the filter.

For the restoration procedure in Step 3a to succeed, we have to evaluate whether TRQP$(x_k + r_k, \Delta_{k+1})$ is compatible for a suitable value of Δ_{k+1}. This requires that a suitable normal step be computed which successfully passes the test (2.13). Of course, once this is achieved, this normal step may be reused at iteration $k + 1$, if the composite step strategy is used.

As it stands, the algorithm is not specific about how to choose Δ_{k+1} during a restoration iteration. On one hand, there is an advantage to choosing a large Δ_{k+1}, since this allows a large step and one hopes good progress. It also makes it easier to satisfy (2.13). On the other, it may be unwise to choose it to be too large, as this may possibly result in a large number of unsuccessful iterations, during which the radius is reduced, before the algorithm can make any progress. A possible choice might be to restart from the radius obtained during the restoration iteration itself, if it uses a trust-region method. Reasonable alternatives would be to use the average radius observed during past successful iterations, or to apply the internal doubling strategy of Byrd et al. (1987) to increase the new radius, or even to consider the technique described by Sartenaer (1997). However, we recognize that extensive numerical experience will remain the ultimate measure of any suggestion at this level.

The role of condition (2.20) may be interpreted as follows. If this condition fails, then one may think that the constraint violation is significant and that one should aim to improve on this situation in the future, by inserting the current point in the filter. Fletcher and Leyffer (1997) use the term of "θ-step" in such circumstances, to indicate that the main preoccupation is to improve feasibility . On the other hand, if condition (2.20) holds, then the reduction in the objective function predicted by the model is more significant than the current constraint violation and it is thus appealing to let the algorithm behave as if it were unconstrained. Fletcher and Leyffer (1997) use the term of "f-step" to denote the step generated, in order to reflect the dominant role of the objective function f in this case. In this case, it is important that the predicted decrease in the model is realized by the actual decrease in the function, which is why we also require that (2.21) does not hold. In particular, if the iterate x_k is feasible, then (2.19) and (2.11) imply that $x_k = x_k^{\mathrm{N}}$ and we obtain that

$$\kappa_\theta \theta_k^\psi = 0 \leq m_k(x_k^{\mathrm{N}}) - m_k(x_k + s_k) = m_k(x_k) - m_k(x_k + s_k). \quad (2.22)$$

As a consequence, the filter mechanism is irrelevant if all iterates are feasible, and the algorithm reduces to a classical unconstrained trust-region method. Another consequence of (2.22) is that *no feasible iterate is ever included in the filter*, which is crucial in allowing finite termination of the restoration procedure. Indeed, if the restoration procedure is required at iteration k of the filter algorithm and produces a sequence of points $\{x_{k,j}\}$ converging to feasibility, there must be an iterate $x_{k,j}$ for which

$$\theta_{k,j} \stackrel{\text{def}}{=} \theta(x_{k,j}) \leq \min\left[(1-\gamma_\theta)\theta_k^{\min}, \frac{\kappa_\Delta}{\kappa_{\text{usc}}}\Delta_{k+1}\min[1, \kappa_\mu \Delta_{k+1}^\mu]\right],$$

for any given $\Delta_{k+1} > 0$, where

$$\theta_k^{\min} = \min_{i \in \mathcal{Z},\, i \leq k} \theta_i > 0$$

and

$$\mathcal{Z} = \{k \mid x_k \text{ is added to the filter}\}.$$

Moreover, $\theta_{k,j}$ must eventually be small enough to ensure, using our assumption on the normal step, the existence of a normal step $n_{k,j}$ from $x_{k,j}$. In other words, the restoration iteration must eventually find an iterate $x_{k,j}$ which is acceptable for the filter and for which the normal step exists and satisfies (2.13), i.e. an iterate x_j which is both acceptable and compatible. As a consequence, the restoration procedure will terminate in a finite number of steps, and the filter algorithm may then proceed. Note that the restoration step may not terminate in a finite number of iterations if we do not assume the existence of the normal step when the constraint violation is small enough, even if this violation converges to zero (see Fletcher, Leyffer and Toint, 1998, for an example).

Notice also that (2.20) ensures that the denominator of ρ_k in (2.21) will be strictly positive whenever θ_k is. If $\theta_k = 0$, then $x_k = x_k^{\text{N}}$, and the denominator of (2.21) will be strictly positive unless x_k is a first-order critical point because of (2.16).

The attentive reader will have observed that we have defined n_{k+1} in Step 4 in the cases where iteration k is unsuccessful (just before branching back to Step 2), while we may not use it if the alternative step of Step 2 is then used at iteration $k+1$. This is to keep the expression of the algorithm as general as possible: a more restrictive version would impose a branch back to Step 3b from Step 4 if iteration k is unsuccessful, but this would then prevent the use of an alternative step at iteration $k+1$. We have chosen not to impose that restriction, but we obviously require that n_{k+1} is used in Step 3a whenever it has been set at iteration k, instead of recomputing it from scratch.

Finally, note that Step 6 allows a relatively wide choice of the new trust-region radius Δ_{k+1}. While the stated conditions are sufficient for the theory developed below, one must obviously be more specific in practice. For instance, one may wish to distinguish, at this point in the algorithm, the cases where (2.20) fails or holds. If (2.20) holds, the main effect of the current iteration is not to reduce the model (which makes the value of ρ_k essentially irrelevant), but rather to reduce the constraint violation (which is taken care of by inserting the current iterate in the filter at Step 5). In this case, Step 6 imposes no further restriction on Δ_{k+1}. In practice, it may be reasonable not to reduce the trust-region radius, because this might cause too small steps towards feasibility or an unnecessary restoration phase. However, there is no compelling reason to increase the radius either, given the compatibility of TRQP(x_k,Δ_k). A reasonable strategy might then be to choose $\Delta_{k+1} = \Delta_k$. If, on the other hand, (2.20) holds, the emphasis of the iteration is then on reducing the objective function, a case akin to unconstrained minimization. Thus a more detailed rule of the type

$$\Delta_{k+1} \in \begin{cases} [\gamma_1\Delta_k, \gamma_2\Delta_k] & \text{if } \rho_k \in [\eta_1, \eta_2), \\ [\Delta_k, \gamma_2\Delta_k] & \text{if } \rho_k \geq \eta_2 \end{cases}$$

seems reasonable in these circumstances.

3. Convergence to First-Order Critical Points

We now prove that our hybrid algorithm generates a globally convergent sequence of iterates, at least if the restoration iteration always succeeds. For the purpose of our analysis, we shall consider

$$\mathcal{S} = \{k \mid x_{k+1} = x_k + s_k\},$$

the set of (indices of) successful iterations,

$$\mathcal{R} = \left\{ \begin{array}{l|l} k & \text{Step 3 is executed and} \\ & \text{either TRQP}(x_k, \Delta_k) \text{ has no feasible point} \\ & \text{or } \|n_k\| > \kappa_\Delta \Delta_k \min[1, \kappa_\mu \Delta_k^\mu] \end{array} \right\},$$

the set of *restoration* iterations,

$$\mathcal{A} = \{k \mid \text{the alternative step is used at iteration } k\},$$

and

$$\mathcal{C} = \{k \mid s_k = n_k + t_k\},$$

the set of iterations where a composite step is used (with $\mu_k \geq 0$). Note that (2.19) implies that

$$\theta_k \leq \kappa_{\text{u}} \Delta_k \min[1, \Delta_k^\mu] \leq \kappa_{\text{u}} \Delta_k^{1+\mu}, \tag{3.1}$$

for every $k \in \mathcal{A}$. Also note that $\{1, 2, \ldots\} = \mathcal{A} \cup \mathcal{C} \cup \mathcal{R}$ and that $\mathcal{R} \subseteq \mathcal{Z}$.

In order to obtain our global convergence result, we will use the assumptions

AS1: f and the constraint functions $c_\mathcal{E}$ and $c_\mathcal{I}$ are twice continuously differentiable;

AS2: there exists $\kappa_{\text{umh}} > 1$ such that

$$\|H_k\| \leq \kappa_{\text{umh}} - 1 < \kappa_{\text{umh}} \text{ for all } k,$$

AS3: the iterates $\{x_k\}$ remain in a closed, bounded domain $X \subset \mathbb{R}^n$.

If, for example, H_k is chosen as the Hessian of the Lagrangian function

$$\ell(x, y) = f(x) + \langle y_\mathcal{E}, c_\mathcal{E}(x) \rangle + \langle y_\mathcal{I}, c_\mathcal{I}(x) \rangle$$

at x_k, in that

$$H_k = \nabla_{xx} f(x_k) + \sum_{i \in \mathcal{E} \cup \mathcal{I}} [y_k]_i \nabla_{xx} c_i(x_k), \tag{3.2}$$

where $[y_k]_i$ denotes the i-th component of the vector of Lagrange multipliers $y_k^T = (y_{\mathcal{E},k}^T \;\; y_{\mathcal{I},k}^T)$, then we see from AS1 and AS3 that AS2 is satisfied when these multipliers remain bounded. The same is true if the Hessian matrices in (3.2) are replaced by bounded approximations.

A first immediate consequence of AS1–AS3 is that there exists a constant $\kappa_{\text{ubh}} > 1$ such that, for all k,

$$|f(x_k + s_k) - m_k(x_k + s_k)| \leq \kappa_{\text{ubh}} \Delta_k^2. \tag{3.3}$$

A proof of this property, based on Taylor expansion, may be found, for instance, in Toint (1988). A second important consequence of our assumptions is that AS1 and AS3 together directly ensure that, for all k,

$$f^{\min} \leq f(x_k) \leq f^{\max} \text{ and } 0 \leq \theta_k \leq \theta^{\max} \tag{3.4}$$

for some constants $f^{\min}$, $f^{\max}$ and $\theta^{\max} > 0$. Thus the part of the (θ, f)-space in which the (θ, f)-pairs associated with the filter iterates lie is restricted to the rectangle

$$\mathcal{M}_0 = [0, \theta^{\max}] \times [f^{\min}, f^{\max}],$$

whose area, surf$(\mathcal{M}_0)$, is clearly finite.

We also note the following simple consequence of (2.11) and AS3.

Lemma 1 Suppose that Algorithm 2.1 is applied to problem (1.1). Suppose also that (2.11) and AS3 hold, that $k \in \mathcal{C}$, and that

$$\theta_k \leq \delta_n.$$

Then there exists a constant $\kappa_{\text{lsc}} > 0$ independent of k such that

$$\kappa_{\text{lsc}} \theta_k \leq \|n_k\|. \tag{3.5}$$

Proof. Since $k \in \mathcal{C}$, we first obtain that n_k exists (as a consequence of (2.11)), and define

$$\mathcal{V}_k \stackrel{\text{def}}{=} \{j \in \mathcal{E} \mid \theta_k = |c_j(x_k)|\} \bigcup \{j \in \mathcal{I} \mid \theta_k = -c_j(x_k)\},$$

that is the subset of most-violated constraints. From the definitions of θ_k in (2.2) and of the normal step in (2.7) we obtain, using the Cauchy-Schwartz inequality, that

$$\theta_k \leq |\langle \nabla_x c_j(x_k), n_k \rangle| \leq \|\nabla_x c_j(x_k)\| \, \|n_k\| \tag{3.6}$$

for all $j \in \mathcal{V}_k$. But AS3 ensures that there exists a constant $\kappa_{\text{lsc}} > 0$ such that

$$\max_{j \in \mathcal{E} \cup \mathcal{I}} \max_{x \in X} \|\nabla_x c_j(x)\| \stackrel{\text{def}}{=} \frac{1}{\kappa_{\text{lsc}}}.$$

We then obtain the desired conclusion by substituting this bound in (3.6). □

Our assumptions and the definition of χ_k in (2.14) ensure that θ_k and χ_k can be used (together) to measure criticality for problem (1.1).

Lemma 2 Suppose that Algorithm 2.1 is applied to problem (1.1) and that finite termination does not occur. Suppose also that AS1 and AS3 hold, and that there exists a subsequence $\{k_i\} \not\subseteq \mathcal{R}$ such that

$$\lim_{i \to \infty} \theta_{k_i} = 0, \quad \lim_{\substack{i \to \infty \\ k_i \in \mathcal{C}}} \chi_{k_i} = 0 \text{ and } \lim_{\substack{i \to \infty \\ k_i \in \mathcal{A}}} \pi_{k_i} = 0. \tag{3.7}$$

Then every limit point of the sequence $\{x_{k_i}\}$ is a first-order critical point for problem (1.1).

Proof. Consider x_*, a limit point of the sequence $\{x_{k_i}\}$, whose existence is ensured by AS3, and assume that $\{k_\ell\} \subseteq \{k_i\}$ is the index set of a subsequence such that $\{x_{k_\ell}\}$ converges to x_*. If $\{k_\ell\}$ contains infinitely many indices of $\mathcal{A}$, the definition of π_k implies that x_* is a first-order critical point for problem (1.1). If this is not the case, the fact that $k_\ell \notin \mathcal{R}$ implies that n_{k_ℓ} satisfies (2.11) for sufficiently large ℓ and converges to zero, because $\{\theta_{k_\ell}\}$ converges to zero and the second part of this condition. As a consequence, we deduce from (2.12) that $\{x^{\text{N}}_{k_\ell}\}$ also converges to x_*. Since the minimization problem occuring in the definition of χ_{k_ℓ} (in (2.14)) is convex, we then obtain from classical perturbation theory (see, for instance, Fiacco, 1983, pp. 14-17), AS1 and the first part of (3.7) that

$$\Big| \min_{\substack{A_{\mathcal{E}}(x_*)t=0 \\ c_{\mathcal{I}}(x_*)+A_{\mathcal{I}}(x_*)t\geq 0 \\ \|t\|\leq 1}} \langle g_*, t\rangle \Big| = 0.$$

This in turn guarantees that x_* is first-order critical for problem (1.1). □

We start our analysis by examining what happens when an infinite number of iterates (that is, their (θ, f)-pairs) are added to the filter.

Lemma 3 Suppose that Algorithm 2.1 is applied to problem (1.1) and that finite termination does not occur. Suppose also that AS1 and AS3 hold and that $|\mathcal{Z}| = \infty$. Then

$$\lim_{\substack{k\to\infty \\ k\in\mathcal{Z}}} \theta_k = 0.$$

Proof. Suppose, for the purpose of obtaining a contradiction, that there exists an infinite subsequence $\{k_i\} \subseteq \mathcal{Z}$ such that

$$\theta_{k_i} \geq \epsilon \tag{3.8}$$

for all i and for some $\epsilon > 0$. At each iteration k_i, the (θ, f)-pair associated with x_{k_i}, that is (θ_{k_i}, f_{k_i}), is added to the filter. This means that no other (θ, f)-pair can be added to the filter at a later stage within the square

$$[\theta_{k_i} - \gamma_\theta \epsilon, \theta_{k_i}] \times [f_{k_i} - \gamma_\theta \epsilon, f_{k_i}],$$

or with the intersection of this square with $\mathcal{M}_0$. But the area of each of these squares is $\gamma_\theta^2 \epsilon^2$. Thus the set $\mathcal{M}_0$ is completely covered by at most a finite number of such squares. This puts a finite upper bound on the number of iterations in $\{k_j\}$, and the conclusion follows. □

We next examine the size of the constraint violation before and after a "composite iteration" where restoration did not occur.

Lemma 4 Suppose that Algorithm 2.1 is applied to problem (1.1). Suppose also that AS1 and AS3 hold and that n_k satisfies (3.5) for $k \in \mathcal{C}$. Then there exists a constant $\kappa_{\text{ubt}} > 0$ such that

$$\theta_k \leq \kappa_{\text{ubt}} \Delta_k^{1+\mu} \tag{3.9}$$

and

$$\theta(x_k + s_k) \leq \kappa_{\text{ubt}} \Delta_k^2. \tag{3.10}$$

for all $l \notin \mathcal{R}$.

Proof. Assume first that $k \in \mathcal{C}$ with $\mu_k = \mu$. Since $k \notin \mathcal{R}$, we have from (3.5) and (2.13) that

$$\kappa_{\text{lsc}} \theta_k \leq \|n_k\| \leq \kappa_\Delta \kappa_\mu \Delta_k^{1+\mu}, \tag{3.11}$$

which gives (3.9). On the other hand, (3.1) implies that an inequality of the form (3.9) holds for $k \in \mathcal{A}$ or $k \in \mathcal{C}$ with $\mu_k = 0$. Now, for any k, the i-th constraint function at $x_k + s_k$ can be expressed as

$$c_i(x_k + s_k) = c_i(x_k) + \langle e_i, A_k s_k \rangle + \tfrac{1}{2} \langle s_k, \nabla_{xx} c_i(\xi_k) s_k \rangle,$$

for $i \in \mathcal{E} \cup \mathcal{I}$, where we have used AS1, the mean-value theorem, and where ξ_k belongs to the segment $[x_k, x_k + s_k]$. Using AS3, we may bound the Hessian of the constraint functions and we obtain from (2.9), the Cauchy-Schwartz inequality, and (2.8) we have that

$$|c_i(x_k + s_k)| \leq \tfrac{1}{2} \max_{x \in X} \|\nabla_{xx} c_i(x)\| \, \|s_k\|^2 \leq \kappa_1 \Delta_k^2,$$

if $i \in \mathcal{E}$, or

$$-c_i(x_k + s_k) \leq \tfrac{1}{2} \max_{x \in X} \|\nabla_{xx} c_i(x)\| \, \|s_k\|^2 \leq \kappa_1 \Delta_k^2,$$

if $i \in \mathcal{I}$, where we have defined

$$\kappa_1 \stackrel{\text{def}}{=} \tfrac{1}{2} \max_{i \in \mathcal{E} \cup \mathcal{I}} \max_{x \in X} \|\nabla_{xx} c_i(x)\|.$$

This gives the desired bound for any

$$\kappa_{\text{ubt}} \geq \max[\kappa_1, \kappa_{\text{u}}, \kappa_\Delta \kappa_\mu / \kappa_{\text{lsc}}].$$

□

We next assess the model decrease when the trust-region radius is sufficiently small.

Lemma 5 Suppose that Algorithm 2.1 is applied to problem (1.1) and that finite termination does not occur. Suppose also that AS1–AS3 and (2.16) hold, that $k \in \mathcal{C}$, that, for some $\epsilon > 0$,

$$\chi_k \geq \epsilon. \tag{3.12}$$

Suppose furthermore that

$$\Delta_k \leq \min\left[1, \frac{\epsilon}{\kappa_{\rm umh}}, \left(2\frac{\kappa_{\rm ubg}}{\kappa_{\rm umh}\kappa_\Delta\kappa_\mu}\right)^{\frac{1}{1+\mu}}, \left(\frac{\kappa_{\rm tmd}\epsilon}{4\kappa_{\rm ubg}\kappa_\Delta\kappa_\mu}\right)^{\frac{1}{\mu}}\right] \stackrel{\rm def}{=} \delta_m, \tag{3.13}$$

where $\kappa_{\rm ubg} \stackrel{\rm def}{=} \max_{x\in X} \|\nabla_x f(x)\|$. Then

$$m_k(x_k) - m_k(x_k + s_k) \geq \tfrac{1}{2}\kappa_{\rm tmd}\epsilon\Delta_k.$$

This last inequality also holds if $k \in \mathcal{A}$, if (3.13) holds and

$$\pi_k \geq \epsilon. \tag{3.14}$$

Proof. Assume first that $k \in \mathcal{C}$. We note that, by (2.16), AS2, (3.12) and (3.13),

$$m_k(x_k^{\rm N}) - m_k(x_k + s_k) \geq \kappa_{\rm tmd}\chi_k \min\left[\frac{\chi_k}{\kappa_{\rm umh}}, \Delta_k\right] \geq \kappa_{\rm tmd}\epsilon\Delta_k. \tag{3.15}$$

Now

$$m_k(x_k^{\rm N}) = m_k(x_k) + \langle g_k, n_k\rangle + \tfrac{1}{2}\langle n_k, H_k n_k\rangle$$

and therefore, using the Cauchy-Schwartz inequality, AS2, (2.13) and (3.13) that

$$\begin{aligned} |m_k(x_k) - m_k(x_k^{\rm N})| &\leq \|n_k\|\,\|g_k\| + \tfrac{1}{2}\|H_k\|\,\|n_k\|^2 \\ &\leq \kappa_{\rm ubg}\|n_k\| + \tfrac{1}{2}\kappa_{\rm umh}\|n_k\|^2 \\ &\leq \kappa_{\rm ubg}\kappa_\Delta\kappa_\mu\Delta_k^{1+\mu} + \tfrac{1}{2}\kappa_{\rm umh}\kappa_\Delta^2\kappa_\mu^2\Delta_k^{2(1+\mu)} \\ &\leq 2\kappa_{\rm ubg}\kappa_\Delta\kappa_\mu\Delta_k^{1+\mu} \\ &\leq \tfrac{1}{2}\kappa_{\rm tmd}\epsilon\Delta_k. \end{aligned}$$

We thus conclude from this last inequality and (3.15) that the desired conclusion holds for $k \in \mathcal{C}$. If we now assume that $k \in \mathcal{A}$ (that is iteration k uses an alternative step), then (2.18), (3.13) and the inequality $\kappa_{\text{umh}} \geq 1$ directly yields that

$$m_k(x_k) - m_k(x_k + s_k) \geq \kappa_{\text{tmd}} \min[\epsilon, \Delta_k] \geq \tfrac{1}{2}\kappa_{\text{tmd}}\Delta_k$$

as desired. □

We continue our analysis by showing, as the reader has grown to expect, that iterations have to be very successful when the trust-region radius is sufficiently small.

Lemma 6 Suppose that Algorithm 2.1 is applied to problem (1.1) and that finite termination does not occur. Suppose also that AS1–AS3, (2.16) and (3.12) hold, that $k \notin \mathcal{R}$, and that

$$\Delta_k \leq \min\left[\delta_m, \frac{(1-\eta_2)\kappa_{\text{tmd}}\epsilon}{2\kappa_{\text{ubh}}}\right] \stackrel{\text{def}}{=} \delta_\rho. \tag{3.16}$$

Then

$$\rho_k \geq \eta_2.$$

Proof. Using (2.21), (3.3), Lemma 5 and (3.16), we find that

$$|\rho_k - 1| = \frac{|f(x_k + s_k) - m_k(x_k + s_k)|}{|m_k(x_k) - m_k(x_k + s_k)|} \leq \frac{\kappa_{\text{ubh}}\Delta_k^2}{\frac{1}{2}\kappa_{\text{tmd}}\epsilon\Delta_k} \leq 1 - \eta_2,$$

from which the conclusion immediately follows. □

Note that this proof could easily be extended if the definition of ρ_k in (2.21) were altered to be of the form

$$\rho_k \stackrel{\text{def}}{=} \frac{f(x_k) - f(x_k + s_k) + \Theta_k}{m_k(x_k) - m_k(x_k + s_k)} \tag{3.17}$$

provided Θ_k is bounded above by a multiple of Δ_k^2. We will comment in Section 4 why such a modification might be of interest.

Now, we also show that the test (2.20) will always be satisfied when the trust-region radius is sufficiently small.

Lemma 7 Suppose that Algorithm 2.1 is applied to problem (1.1) and that finite termination does not occur. Suppose also that AS1–AS3, (2.16) and (3.12) hold, that $k \notin \mathcal{R}$, that n_k satisfies (3.5) if $k \in \mathcal{C}$, and that

$$\Delta_k \leq \min\left[\delta_m, \left(\frac{\kappa_{\rm tmd}\epsilon}{2\kappa_\theta \kappa_{\rm ubt}^{\psi}}\right)^{\frac{1}{\psi(1+\mu)-1}}\right] \stackrel{\rm def}{=} \delta_f. \tag{3.18}$$

Then

$$m_k(x_k) - m_k(x_k + s_k) \geq \kappa_\theta \theta_k^{\psi}.$$

Proof. This directly results from the inequalities

$$\kappa_\theta \theta_k^{\psi} \leq \kappa_\theta \kappa_{\rm ubt}^{\psi} \Delta_k^{\psi(1+\mu)} \leq \tfrac{1}{2}\kappa_{\rm tmd}\epsilon\Delta_k \leq m_k(x_k) - m_k(x_k + s_k),$$

where we successively used Lemma 4, (3.18) and Lemma 5. □

We may also guarantee a decrease in the objective function, large enough to ensure that the trial point is acceptable with respect to the (θ, f)-pair associated with x_k, so long as the constraint violation is itself sufficiently small.

Lemma 8 Suppose that Algorithm 2.1 is applied to problem (1.1) and that finite termination does not occur. Suppose also that AS1–AS3, (2.16), (3.12) and (3.16) hold, that $k \notin \mathcal{R}$, that n_k satisfies (3.5) if $k \in \mathcal{C}$, and that

$$\theta_k \leq \kappa_{\rm ubt}^{-\frac{1}{\mu}} \left(\frac{\eta_2 \kappa_{\rm tmd}\epsilon}{2\gamma_\theta}\right)^{\frac{1+\mu}{\mu}} \stackrel{\rm def}{=} \delta_\theta. \tag{3.19}$$

Then

$$f(x_k + s_k) \leq f(x_k) - \gamma_\theta \theta_k.$$

Proof. Applying Lemmas 4–6—which is possible because of (3.12), (3.16), $k \notin \mathcal{R}$ and n_k satisfies (3.5) for $k \in \mathcal{C}$—and (3.19), we obtain that

$$\begin{aligned} f(x_k) - f(x_k + s_k) &\geq \eta_2[m_k(x_k) - m_k(x_k + s_k)] \\ &\geq \tfrac{1}{2}\eta_2\kappa_{\text{tmd}}\epsilon\Delta_k \\ &\geq \tfrac{1}{2}\eta_2\kappa_{\text{tmd}}\epsilon\left(\frac{\theta_k}{\kappa_{\text{ubt}}}\right)^{\frac{1}{1+\mu}} \\ &\geq \gamma_\theta\theta_k \end{aligned}$$

and the desired inequality follows. □

We now establish that if the trust-region radius and the constraint violation are both small at a non-critical iterate x_k, TRQP(x_k, Δ_k) must be compatible.

Lemma 9 Suppose that Algorithm 2.1 is applied to problem (1.1) and that finite termination does not occur. Suppose that AS1–AS3, (2.11), and (3.12) hold, that (2.16) holds for $k \notin \mathcal{R}$, and that

$$\Delta_k \leq \min\left[\gamma_0\delta_\rho, \left(\frac{1}{\kappa_\mu}\right)^{\frac{1}{\mu}}, \left(\frac{\gamma_0^2(1-\gamma_\theta)\kappa_\Delta\kappa_\mu}{\kappa_{\text{usc}}\kappa_{\text{ubt}}}\right)^{\frac{1}{1-\mu}}\right]. \tag{3.20}$$

Suppose furthermore that

$$\theta_k \leq \min[\delta_\theta, \delta_n]. \tag{3.21}$$

Then $k \notin \mathcal{R}$.

Proof. If an alternative step is used at iteration k, then $k \notin \mathcal{R}$. Assume therefore that $k \notin \mathcal{A}$. Because $\theta_k \leq \delta_n$, we know from (2.11) and Lemma 1 that (2.11) and (3.5) hold. Moreover, since $\theta_k \leq \delta_\theta$, we have that (3.19) also holds. Assume, for the purpose of deriving a contradiction, that $k \in \mathcal{R}$, which implies that

$$\|n_k\| > \kappa_\Delta\kappa_\mu\Delta_k^{1+\mu}, \tag{3.22}$$

where we have used (2.13) and the fact that $\kappa_\mu\Delta_k^{\mu_k} \leq \kappa_\mu\Delta_k^\mu \leq 1$ because of (3.20). In this case, the mechanism of the algorithm then ensures that $k-1 \notin \mathcal{R}$. Now assume that iteration $k-1$ is unsuccessful. Because of Lemmas 6 and 8, which hold at iteration $k-1 \notin \mathcal{R}$ because of (3.20), the fact that $\theta_k = \theta_{k-1}$, (2.11), and (3.19), we obtain that

$$\rho_{k-1} \geq \eta_2 \text{ and } f(x_{k-1} + s_{k-1}) \leq f(x_{k-1}) - \gamma_\theta\theta_{k-1}.$$

Hence, given that x_{k-1} is acceptable for the filter at the beginning of iteration $k-1$, if this iteration is unsuccessful, it must be because

$$\theta(x_{k-1}+s_{k-1}) > (1-\gamma_\theta)\theta_{k-1} = (1-\gamma_\theta)\theta_k.$$

But Lemma 4 and the mechanism of the algorithm then imply that

$$(1-\gamma_\theta)\theta_k \le \kappa_{\rm ubt}\Delta_{k-1}^2 \le \frac{\kappa_{\rm ubt}}{\gamma_0^2}\Delta_k^2.$$

Combining this last bound with (3.22) and (2.11), we deduce that

$$\kappa_\Delta\kappa_\mu\Delta_k^{1+\mu} < \|n_k\| \le \kappa_{\rm usc}\theta_k \le \frac{\kappa_{\rm usc}\kappa_{\rm ubt}}{\gamma_0^2(1-\gamma_\theta)}\Delta_k^2$$

and hence that

$$\Delta_k^{1-\mu} > \frac{\gamma_0^2(1-\gamma_\theta)\kappa_\Delta\kappa_\mu}{\kappa_{\rm usc}\kappa_{\rm ubt}}.$$

Since this last inequality contradicts (3.20), our assumption that iteration $k-1$ is unsuccessful must be false. Thus iteration $k-1$ is successful and $\theta_k = \theta(x_{k-1}+s_{k-1})$. We then obtain from (3.22), (2.11) and (3.10) that

$$\kappa_\Delta\kappa_\mu\Delta_k^{1+\mu} < \|n_k\| \le \kappa_{\rm usc}\theta_k \le \kappa_{\rm usc}\kappa_{\rm ubt}\Delta_{k-1}^2 \le \frac{\kappa_{\rm usc}\kappa_{\rm ubt}}{\gamma_0^2}\Delta_k^2,$$

which is again impossible because of (3.20) and because $(1-\gamma_\theta) < 1$. Hence our initial assumption (3.22) must be false, which yields the desired conclusion. □

We now distinguish two mutually exclusive cases. For the first, we consider what happens if there is an infinite subsequence of iterates belonging to the filter.

Lemma 10 Suppose that Algorithm 2.1 is applied to problem (1.1) and that finite termination does not occur. Suppose also that AS1–AS3 and (2.11) hold and that (2.16) holds for $k \notin \mathcal{R}$. Suppose furthermore that $|\mathcal{Z}| = \infty$. Then there exists an infinite subsequence $\{k_j\} \subseteq \mathcal{Z}$ such that

$$\lim_{j\to\infty} \theta_{k_j} = 0 \tag{3.23}$$

and

$$\lim_{\substack{j\to\infty\\ k_j\in\mathcal{C}}} \chi_{k_j} = 0 \quad\text{and}\quad \lim_{\substack{j\to\infty\\ k_j\in\mathcal{A}}} \pi_{k_j} = 0. \tag{3.24}$$

Proof. Let $\{k_i\}$ be any infinite subsequence of $\mathcal{Z}$. We observe that (3.23) follows from Lemma 3. Suppose now that, for some $\epsilon_2 > 0$.

$$\chi_{k_i} \geq \epsilon_2 \tag{3.25}$$

for all i such that $k_i \in \mathcal{C}$ and

$$\pi_{k_i} \geq \epsilon_2 \tag{3.26}$$

for all i such that $k_i \in \mathcal{A}$. Suppose furthermore that there exists $\epsilon_3 > 0$ such that, for all $i \geq i_0$,

$$\Delta_{k_i} \geq \epsilon_3. \tag{3.27}$$

If $k_i \notin \mathcal{A}$, (3.23) and (2.11) ensure that n_{k_i} exists for $i \geq i_0$, say, and also that

$$\lim_{i\to\infty} \|n_{k_i}\| = 0. \tag{3.28}$$

Thus (3.27) ensures that (2.13) holds for sufficiently large i and $k_i \notin \mathcal{R}$. We may then decompose the model decrease in its normal and tangential components, that is

$$m_{k_i}(x_{k_i}) - m_{k_i}(x_{k_i} + s_{k_i}) = m_{k_i}(x_{k_i}) - m_{k_i}(x^{\rm N}_{k_i}) + m_{k_i}(x^{\rm N}_{k_i}) - m_{k_i}(x_{k_i} + s_{k_i}). \tag{3.29}$$

Consider the normal component first. As we noted in the proof of Lemma 5,

$$|m_{k_i}(x_{k_i}) - m_{k_i}(x^{\rm N}_{k_i})| \leq \kappa_{\rm ubg}\|n_{k_i}\| + \tfrac{1}{2}\kappa_{\rm umh}\|n_{k_i}\|^2,$$

which in turn, with (3.28), yields that

$$\lim_{i\to\infty} [m_{k_i}(x_{k_i}) - m_{k_i}(x^{\rm N}_{k_i})] = 0. \tag{3.30}$$

If we now consider the normal component, (3.25), (3.27) (2.16) and AS2 yield that

$$m_{k_i}(x^{\rm N}_{k_i}) - m_{k_i}(x_{k_i} + s_{k_i}) \geq \kappa_{\rm tmd}\epsilon_2 \min\left[\frac{\epsilon_2}{\kappa_{\rm umh}}, \epsilon_3\right] \stackrel{\rm def}{=} \delta_1 > 0. \tag{3.31}$$

Substituting (3.30) and (3.31) into (3.29), we find that, for $k_i \in \mathcal{C}$,

$$m_{k_i}(x_{k_i}) - m_{k_i}(x_{k_i} + s_{k_i}) \geq \delta_1 > 0.$$

If, on the other hand, $k_i \in \mathcal{A}$, then (3.26), (3.27) and (2.18) give that

$$m_{k_i}(x_{k_i}) - m_{k_i}(x_{k_i} + s_{k_i}) \geq \kappa_{\rm tmd} \min[\epsilon_2, \epsilon_3] \stackrel{\rm def}{=} \delta_2 > 0.$$

Thus

$$\liminf_{i\to\infty}[m_{k_i}(x_{k_i}) - m_{k_i}(x_{k_i} + s_{k_i})] \geq \min[\delta_1, \delta_2] \stackrel{\text{def}}{=} \delta > 0. \quad (3.32)$$

We now observe that, because x_{k_i} is added to the filter at iteration k_i, the mechanism of the algorithm imposes that either iteration $k_i \in \mathcal{R}$ or (2.20) must fail. Since we already verified that $k_i \notin \mathcal{R}$ for $i \geq i_0$ sufficiently large, we obtain that (2.20) must fail for such i, that is

$$m_{k_i}(x_{k_i}) - m_{k_i}(x_{k_i} + s_{k_i}) < \kappa_\theta \theta_{k_i}^{\psi}. \quad (3.33)$$

Combining this bound with (3.32), we find that θ_{k_i} is bounded away from zero for i sufficiently large, which is impossible in view of (3.23). We therefore deduce that (3.27) cannot hold and obtain that there is a subsequence $\{k_\ell\} \subseteq \{k_i\}$ for which

$$\lim_{\ell\to\infty} \Delta_{k_\ell} = 0.$$

We now restrict our attention to the tail of this subsequence, that is to the set of indices k_ℓ that are large enough to ensure that (3.18), (3.19) and (3.20) hold, which is possible by definition of the subsequence and because of (3.23). For these indices, we may therefore apply Lemma 9, and deduce that iteration $k_\ell \notin \mathcal{R}$ for ℓ sufficiently large. Hence, as above, (3.33) must hold for ℓ sufficiently large. However, we may also apply Lemma 7, which contradicts (3.33), and therefore (3.25) and (3.26) cannot hold together, yielding the desired result. □

Thus, if an infinite subsequence of iterates is added to the filter, Lemma 2 ensures that it converges to a first-order critical point. Our remaining analysis then naturally concentrates on the possibility that there may be no such infinite subsequence. In this case, no further iterates are added to the filter for k sufficiently large. In particular, this means that the number of restoration iterations, $|\mathcal{R}|$, must be finite. In what follows, we assume that $k_0 \geq 0$ is the last iteration for which x_{k_0-1} is added to the filter.

Lemma 11 Suppose that Algorithm 2.1 is applied to problem (1.1) and that finite termination does not occur. Suppose also that AS1–AS3 and (2.11) hold and that (2.16) holds for $k \notin \mathcal{R}$. Then we have that

$$\lim_{k\to\infty} \theta_k = 0. \tag{3.34}$$

Furthermore, n_k exists and satisfies (3.5) for all $k \geq k_0$ sufficiently large.

Proof. Consider any successful iterate with $k \geq k_0$. Then we have that

$$f(x_k) - f(x_{k+1}) \geq \eta_1[m_k(x_k) - m_k(x_k + s_k)] \geq \eta_1\kappa_\theta\theta_k^\psi \geq 0. \tag{3.35}$$

Thus the objective function does not increase for all successful iterations with $k \geq k_0$. But AS1 and AS3 imply (3.4) and therefore we must have, from the first part of this statement, that

$$\lim_{\substack{k\in\mathcal{S}\\ k\to\infty}} f(x_k) - f(x_{k+1}) = 0. \tag{3.36}$$

(3.34) then immediately follows from (3.35) and the fact that $\theta_j = \theta_k$ for all unsuccessful iterations j that immediately follow the successful iteration k, if any. The last conclusion then results from (2.11) and Lemma 1. □

We now show that the trust-region radius cannot become arbitrarily small if the (asymptotically feasible) iterates stay away from first-order critical points.

Lemma 12 Suppose that Algorithm 2.1 is applied to problem (1.1) and that finite termination does not occur. Suppose also that AS1—AS3 hold and that (2.16) holds for $k \notin \mathcal{R}$. Suppose furthermore that (3.12) hold for all $k \geq k_0$. Then there exists a $\Delta_{\min} > 0$ such that

$$\Delta_k \geq \Delta_{\min}$$

for all k.

Proof. Suppose that $k_1 \geq k_0$ is chosen sufficiently large to ensure that (3.21) holds and thus that (2.11) also holds for all $k \geq k_1$, which is possible because of Lemma 11. Suppose also, for the purpose of obtaining a contradiction, that iteration j is the first iteration following iteration k_1 for which

$$\Delta_j \leq \gamma_0 \min\left[\delta_\rho, \sqrt{\frac{(1-\gamma_\theta)\theta^{\mathrm{F}}}{\kappa_{\mathrm{ubt}}}}, \Delta_{k_1}\right] \stackrel{\text{def}}{=} \gamma_0 \delta_s, \tag{3.37}$$

where

$$\theta^{\mathrm{F}} \stackrel{\text{def}}{=} \min_{i \in \mathcal{Z}} \theta_i$$

is the smallest constraint violation appearing in the filter. Note also that the inequality $\Delta_j \leq \gamma_0 \Delta_{k_1}$, which is implied by (3.37), ensures that $j \geq k_1 + 1$ and hence that $j - 1 \geq k_1$ and thus that $j - 1 \notin \mathcal{R}$. Then the mechanism of the algorithm and (3.37) imply that

$$\Delta_{j-1} \leq \frac{1}{\gamma_0} \Delta_j \leq \delta_s \tag{3.38}$$

and Lemma 6, which is applicable because (3.37) and (3.38) together imply (3.16) with k replaced by $j - 1$, then ensures that

$$\rho_{j-1} \geq \eta_2. \tag{3.39}$$

Furthermore, since $n - j - 1$ satisfies (2.11), Lemma 1 implies that we can apply Lemma 4. This together with (3.37) and (3.38), gives that

$$\theta(x_{j-1} + s_{j-1}) \leq \kappa_{\mathrm{ubt}} \Delta_{j-1}^2 \leq (1 - \gamma_\theta)\theta^{\mathrm{F}}. \tag{3.40}$$

We may also apply Lemma 8 because (3.37) and (3.38) ensure that (3.16) holds and because (3.19) also holds for $j - 1 \geq k_1$. Hence we deduce that

$$f(x_{j-1} + s_{j-1}) \leq f(x_{j-1}) - \gamma_\theta \theta_{j-1}.$$

This last relation and (3.40) ensure that $x_{j-1} + s_{j-1}$ is acceptable for the filter and x_{j-1}. Combining this conclusion with (3.39) and the mechanism of the algorithm, we obtain that $\Delta_j \geq \Delta_{j-1}$. As a consequence, and since (2.20) also holds at iteration $j - 1$, iteration j cannot be the first iteration following k_1 for which (3.37) holds. This contradiction shows that $\Delta_k \geq \gamma_0 \delta_s$ for all $k > k_1$, and the desired result follows if we define

$$\Delta_{\min} = \min[\Delta_0, \ldots, \Delta_{k_1}, \gamma_0 \delta_s].$$

□

We may now analyze the convergence of χ_k itself.

Lemma 13 Suppose that Algorithm 2.1 is applied to problem (1.1) and that finite termination does not occur. Suppose also that AS1–AS3, (2.11) hold, and that (2.16) holds for $k \notin \mathcal{R}$. Then there exists a subsequence $\{k_j\}$ such that

$$\liminf_{\substack{j\to\infty \\ k_j\in\mathcal{C}}} \chi_{k_j} = 0 \ \text{ and } \ \liminf_{\substack{j\to\infty \\ k_j\in\mathcal{A}}} \pi_{k_j} = 0. \tag{3.41}$$

Proof. We start by observing that Lemma 11 implies that the second conclusion of (2.11) holds for k sufficiently large. Moreover, as in Lemma 11, we obtain (3.35) and therefore (3.36) for each $k \in \mathcal{S}$, $k \geq k_0$. Suppose now, for the purpose of obtaining a contradiction, that (3.12) and (3.14) hold. Assume first that $k \in \mathcal{C}$. In this case, and notice that

$$m_k(x_k) - m_k(x_k + s_k) = m_k(x_k) - m_k(x_k^{\rm N}) + m_k(x_k^{\rm N}) - m_k(x_k + s_k). \tag{3.42}$$

Moreover, note, as in Lemma 5, that

$$|m_k(x_k) - m_k(x_k^{\rm N})| \leq \kappa_{\rm ubg}\|n_k\| + \kappa_{\rm umh}\|n_k\|^2,$$

which in turn yields that

$$\lim_{k\to\infty}[m_k(x_k) - m_k(x_k^{\rm N})] = 0$$

because of Lemma 11 and the second conclusion of (2.11). This limit, together with (3.35), (3.36) and (3.42), then gives that

$$\lim_{\substack{k\to\infty \\ k\in\mathcal{S}}}[m_k(x_k^{\rm N}) - m_k(x_k + s_k)] = 0. \tag{3.43}$$

But (2.16), (3.12), AS2 and Lemma 12 together imply that, for all $k \geq k_0$

$$m_k(x_k^{\rm N}) - m_k(x_k+s_k) \geq \kappa_{\rm tmd}\chi_k \min\left[\frac{\chi_k}{\beta_k}, \Delta_k\right] \geq \kappa_{\rm tmd}\epsilon \min\left[\frac{\epsilon}{\kappa_{\rm umh}}, \Delta_{\rm min}\right], \tag{3.44}$$

immediately giving a contradiction with (3.43).

On the other hand, if $k \in \mathcal{A}$, then (3.14) and (2.18) immediately imply that

$$m_k(x_k) - m_k(x_k + s_k) \geq \kappa_{\rm tmd} \min[\epsilon, \Delta_{\rm min}] > 0,$$

which, together with (2.21) and the fact that $k \in \mathcal{S}$, contradicts the boundedness of f. Hence (3.12) and (3.14) cannot hold together and the desired result follows. □

We may summarize all of the above in our main global convergence result.

Theorem 14 Suppose that Algorithm 2.1 is applied to problem (1.1) and that finite termination does not occur. Suppose also that AS1–AS3 and (2.11) hold, and that (2.16) holds for $k \notin \mathcal{R}$. Let $\{x_k\}$ be the sequence of iterates produced by the algorithm. Then either the restoration procedure terminates unsuccessfully by converging to an infeasible first-order critical point of problem (2.17), or there is a subsequence $\{k_j\}$ for which

$$\lim_{j\to\infty} x_{k_j} = x_*$$

and x_* is a first-order critical point for problem (1.1).

Proof. Suppose that the restoration iteration always terminates successfully. From AS3, Lemmas 10, 11 and 13, we obtain that, for some subsequence $\{k_j\}$,

$$\lim_{j\to\infty} \theta_{k_j} = \lim_{\substack{j\to\infty \\ k_j\in\mathcal{C}}} \chi_{k_j} = \lim_{\substack{j\to\infty \\ k_j\in\mathcal{A}}} \pi_{k_j} = 0. \tag{3.45}$$

The conclusion then follows from Lemma 2. □

Can we dispense with AS3 to obtain this result? Firstly, this assumption ensures that the objective and constraint functions remain bounded above and below (see (3.4)). This is crucial for the rest of the analysis because the convergence of the iterates to feasibility depends on the fact that the area of the filter is finite. Thus, if AS3 does not hold, we have to verify that (3.4) holds for other reasons. The second part of this statement may be ensured quite simply by initializing the filter to $(\theta^{\max}, -\infty)$, for some $\theta^{\max} > \theta_0$, in Step 0 of the algorithm. This has the effect of putting an upper bound on the infeasibility of all iterates, which may be useful in practice. However, this does not prevent the objective function from being unbounded below in

$$\mathcal{C}(\theta^{\max}) = \{x \in \mathbb{R}^n \mid \theta(x) \le \theta^{\max}\}$$

and we cannot exclude the possibility that a sequence of infeasible iterates might both continue to improve the value of the objective function and satisfy (2.20). If $\mathcal{C}(\theta^{\max})$ is bounded, AS3 is most certainly satisfied. If this is not the case, we could assume that

$$f^{\min} \leq f(x) \leq f^{\max} \text{ and } 0 \leq \theta(x) \leq \theta^{\max} \text{ for } x \in \mathcal{C}(\theta^{\max}) \tag{3.46}$$

for some values of $f^{\min}$ and $f^{\max}$ and simply monitor that the values $f(x_k)$ are
reasonable—in view of the problem being solved—as the algorithm proceeds. To summarize, we may replace AS1 and AS3 by the following assumption.

AS4: The functions f and c are twice continuously differentiable on an open set containing $\mathcal{C}(\theta^{\max})$, their first and second derivatives are uniformly bounded on $\mathcal{C}(\theta^{\max})$, and (3.46) holds.

The reader should note that AS4 no longer ensures the existence of a limit point, but only that (3.45) holds for some subsequence $\{k_j\}$. Furthermore, the comments following the statement of (2.11) no longer apply if limit points at infinity are allowed.

4. Conclusion and Perspectives

We have introduced a hybrid trust-region SQP-filter algorithm for general nonlinear programming, that mixes composite steps with potentially cheaper alternative steps, and we have shown this algorithm to be globally convergent to first-order critical points. This hybrid algorithm has the potential of being numerically more efficient than its version that only uses composite steps, as analyzed in Fletcher et al. (1999). However, the authors are well aware that this potential must be confirmed by numerical experiments.

References

L. T. Biegler, J. Nocedal, and C. Schmid. A reduced Hessian method for large-scale constrained optimization. *SIAM Journal on Optimization*, **5**(2), 314–347, 1995.

R. H. Bielschowsky and F. A. M. Gomes. Dynamical control of infeasibility in nonlinearly constrained optimization. Presentation at the Optimization 98 Conference, Coimbra, 1998.

R. H. Byrd, J. Ch. Gilbert, and J. Nocedal. A trust region method based on interior point techniques for nonlinear programming. *Mathematical Programming, Series A*, **89**(1), 149–186, 2000*a*.

R. H. Byrd, M. E. Hribar, and J. Nocedal. An interior point algorithm for large scale nonlinear programming. *SIAM Journal on Optimization*, **9**(4), 877–900, 2000*b*.

R. H. Byrd, R. B. Schnabel, and G. A. Shultz. A trust region algorithm for nonlinearly constrained optimization. *SIAM Journal on Numerical Analysis*, **24**, 1152–1170, 1987.

A. R. Conn, N. I. M. Gould, D. Orban, and Ph. L. Toint. A primal-dual trust-region algorithm for minimizing a non-convex function subject to bound and linear equality constraints. *Mathematical Programming*, **87**(2), 215–249, 2000.

A. R. Conn, N. I. M. Gould, A. Sartenaer, and Ph. L. Toint. Global convergence of a class of trust region algorithms for optimization using inexact projections on convex constraints. *SIAM Journal on Optimization*, **3**(1), 164–221, 1993.

R. S. Dembo and U. Tulowitzki. On the minimization of quadratic functions subject to box constraints. School of Organization and Management Working paper Series B no. 71, Yale University, Yale, USA, 1983.

J. E. Dennis and L. N. Vicente. On the convergence theory of trust-region based algorithms for equality-constrained optimization. *SIAM Journal on Optimization*, **7**(4), 927–950, 1997.

J. E. Dennis, M. El-Alem, and M. C. Maciel. A global convergence theory for general trust-region based algorithms for equality constrained optimization. *SIAM Journal on Optimization*, **7**(1), 177–207, 1997.

J. E. Dennis, M. El-Alem, and K. A. Williamson. A trust-region approach to nonlinear systems of equalities and inequalities. *SIAM Journal on Optimization*, **9**(2), 291–315, 1999.

M. El-Alem. Global convergence without the assumption of linear independence for a trust-region algorithm for constrained optimization. *Journal of Optimization Theory and Applications*, **87**(3), 563–577, 1995.

M. El-Alem. A global convergence theory for a general class of trust-region-based algorithms for constrained optimization without assuming regularity. *SIAM Journal on Optimization*, **9**(4), 965–990, 1999.

M. El-Hallabi and R. A. Tapia. An inexact trust-region feasible-point algorithm for nonlinear systems of equalities and inequalities. Technical Report TR95-09, Department of Computational and Applied Mathematics, Rice University, Houston, Texas, USA, 1995.

A. V. Fiacco. *Introduction to sensitivity and stability analysis in nonlinear programming*. Academic Press, London, 1983.

R. Fletcher and S. Leyffer. Nonlinear programming without a penalty function. Numerical Analysis Report NA/171, Department of Mathematics, University of Dundee, Dundee, Scotland, 1997.

R. Fletcher and S. Leyffer. User manual for filterSQP. Numerical Analysis Report NA/181, Department of Mathematics, University of Dundee, Dundee, Scotland, 1998.

R. Fletcher, N. I. M. Gould, S. Leyffer, and Ph. L. Toint. Global convergence of trust-region SQP-filter algorithms for nonlinear programming. Technical Report 99/03, Department of Mathematics, University of Namur, Namur, Belgium, 1999.

R. Fletcher, S. Leyffer, and Ph. L. Toint. On the global convergence of an SLP-filter algorithm. Technical Report 98/13, Department of Mathematics, University of Namur, Namur, Belgium, 1998.

R. Fletcher, S. Leyffer, and Ph. L. Toint. On the global convergence of an SQP-filter algorithm. Technical Report 00/??, Department of Mathematics, University of Namur, Namur, Belgium, 2000.

N. I. M. Gould, M. E. Hribar, and J. Nocedal. On the solution of equality constrained quadratic problems arising in optimization. Technical Report RAL-TR-98-069, Rutherford Appleton Laboratory, Chilton, Oxfordshire, England, 1998.

M. Lalee, J. Nocedal, and T. D. Plantenga. On the implementation of an algorithm for large-scale equality constrained optimization. *SIAM Journal on Optimization*, **8**(3), 682–706, 1998.

X. Liu and Y. Yuan. A robust trust-region algorithm for solving general nonlinear programming problems. Presentation at the International Conference on Nonlinear Programming and Variational Inequalities, Hong Kong, 1998.

J. J. Moré and G. Toraldo. On the solution of large quadratic programming problems with bound constraints. *SIAM Journal on Optimization*, **1**(1), 93–113, 1991.

W. Murray and F. J. Prieto. A sequential quadratic programming algorithm using an incomplete solution of the subproblem. *SIAM Journal on Optimization*, **5**(3), 590–640, 1995.

E. O. Omojokun. *Trust region algorithms for optimization with nonlinear equality and inequality constraints.* PhD thesis, University of Colorado, Boulder, Colorado, USA, 1989.

A. Sartenaer. Automatic determination of an initial trust region in nonlinear programming. *SIAM Journal on Scientific Computing*, **18**(6), 1788–1803, 1997.

Ph. L. Toint. Global convergence of a class of trust region methods for nonconvex minimization in Hilbert space. *IMA Journal of Numerical Analysis*, **8**(2), 231–252, 1988.

A. Vardi. A trust region algorithm for equality constrained minimization: convergence properties and implementation. *SIAM Journal on Numerical Analysis*, **22**(3), 575–591, 1985.

Y. Yuan. Trust region algorithms for nonlinear programming. *in* Z. C. Shi, ed., 'Contemporary Mathematics', Vol. 163, pp. 205–225, Providence, Rhode-Island, USA, 1994. American Mathematical Society.

SOME ASPECTS OF NONLINEAR SEMIDEFINITE PROGRAMMING

Florian Jarre
Institut für Mathematik
Universität Düsseldorf
Universitätsstraße 1
D-40225 Düsseldorf, Germany
jarre@opt.uni-duesseldorf.de

Abstract This paper is an extended abstract of a survey talk given at the IFIP TC7 Conference in Trier, July 2001. We consider linear and nonlinear semidefinite programming problems and concentrate on selected aspects that are relevant for understanding dual barrier methods. The paper is aimed at graduate students to highlight some issues regarding smoothness, regularity, and computational complexity without going into details.

Keywords: Semidefinite programming, smoothness, regularity, interior method, local minimization.

1. Introduction

In this paper we consider nonlinear semidefinite programming problems (NLSDP's) and concentrate on some aspects relevant to a dual barrier method. Other approaches for solving NLSDP's are the program package LOQO of Vanderbei (1997) based on a primal-dual approach, or recent work of Vanderbei et.al. (2000). Also the work of Kocvara and Stingl (2001) solving large scale semidefinite programs based on a modified barrier approach seems very promising. The modified barrier approach does not require the barrier parameter to converge to zero and may thus overcome some of the problems related to ill-conditioning in traditional interior methods. Optimality conditions for NLSDP's are considered in Forsgren (2000); Shapiro and Scheinberg (2000).

Some problems considered in this paper do not satisfy any constraint qualification. For such problems primal-dual methods do not appear to be suitable. Another question addressed in this paper is the question of how to avoid "poor" local minimizers, a question that may be even more difficult to investigate for primal dual methods than it is for barrier methods.

1.1 Notation

The following notation has become standard in the literature on linear semidefinite programs. The space of symmetric $n \times n$-matrices is denoted by $\mathcal{S}^n$. The inequality

$$X \succeq 0, \qquad (X \succ 0)$$

is used to indicate that X is a symmetric positive semidefinite (positive definite) $n \times n$-matrix. By

$$\langle C, X \rangle = C \bullet X = \mathrm{trace}(C^T X) = \sum_{i,j} C_{i,j} X_{i,j}$$

we denote the standard scalar product on the space of $n \times n$-matrices inducing the Frobenius norm,

$$X \bullet X = \|X\|_F^2.$$

For given symmetric matrices $A^{(i)}$ we define a *linear* map $\mathcal{A}$ from $\mathcal{S}^n$ to $\mathbb{R}^m$ by

$$\mathcal{A}(X) = \begin{pmatrix} A^{(1)} \bullet X \\ \vdots \\ A^{(m)} \bullet X \end{pmatrix}.$$

The adjoint operator $\mathcal{A}^*$ satisfying

$$\langle \mathcal{A}^*(y), X \rangle = y^T \mathcal{A}(X) \qquad \forall\, X \in \mathcal{S}^n,\ y \in \mathbb{R}^m$$

is given by

$$\mathcal{A}^*(y) = \sum_{i=1}^{m} y_i A^{(i)}.$$

2. Linear semidefinite programs

In this section we consider a pair of primal and dual (linear) semidefinite programs in standard form,

$$(P) \qquad \text{minimize } C \bullet X \ \text{ s.t. } \ \mathcal{A}(X) = b,\ X \succeq 0$$

and

$$(D) \qquad \text{maximize } b^T y \ \text{ s.t. } \ \mathcal{A}^*(y) + S = C,\ S \succeq 0.$$

When using the notion "semidefinite program" (SDP) we always refer to a linear semidefinite program; nonlinear SDP's will be denoted by NLSDP. The data for (P) and (D) are a linear map $\mathcal{A}$, a vector $b \in I\!R^m$ and a matrix $C \in \mathcal{S}^n$. We use the convention that the infimum of (P) is $+\infty$ whenever (P) does not have any feasible solution X, and the supremum of (D) is $-\infty$ if the feasible set of (D) is empty. If there exists a matrix $X \succ 0$ (not just $X \succeq 0$) that is feasible for (P) then we call X "strictly" feasible for (P) and say that (P) satisfies Slaters condition. Likewise, if there exists a matrix $S \succ 0$ that is feasible for (D) we call (D) strictly feasible. If Slaters condition is satisfied by (P) or by (D) then the optimal values of (P) and (D) coincide, and if both problems satisfy Slaters condition, then the optimal solutions X^{opt} and y^{opt}, S^{opt} of both problems exist and satisfy the equation

$$X^{opt} S^{opt} = 0. \tag{1}$$

Conversely, any pair X and y, S of feasible points for (P) and (D) satisfying (1) is optimal for both problems, see e.g. Shapiro and Scheinberg (2000). Condition (1) implies that there exists a unitary matrix U that simultaneously diagonalizes X^{opt} and S^{opt}. Moreover, the eigenvalues of X^{opt} and S^{opt} to the same eigenvector are complementary.

The two main applications of semidefinite programs are relaxations for combinatorial optimization problems, see e.g. Alizadeh (1991); Helmberg et.al. (1996); Goemans and Williamson (1994), and semidefinite programs arising from Lyapunov functions or from the positive real lemma in control theory, see e.g. Boyd et.al. (1994); Leibfritz (2001); Scherer (1999). Next, we give two simple examples for such applications.

2.1 A first simple example

In our first example we consider the differential equation

$$\dot{x}(t) = Ax(t)$$

for some vector function $x : I\!R \to I\!R^n$. By definition, this system is called stable if for all initial values $x^{(0)} = x(0)$ the solutions $x(t)$ converge to zero when $t \to \infty$. It is well known, see e.g. Hirsch and Smale (1974), that this is the case if and only if the real part of all eigenvalues of A is negative,

$$\mathrm{Re}(\lambda_i(A)) < 0 \quad \text{for } 1 \le i \le n.$$

By Lyapunov's theorem, this is the case if and only if

$$\exists P \succ 0 : \quad -A^T P - PA \succ 0.$$

Let us now assume that the system matrix A is subject to uncertainties that can be "confined" to a polyhedron with m given vertices $A^{(i)}$, i.e.

$$A = A(t) \in \text{conv}\{A^{(1)}, \ldots, A^{(m)}\} \quad \text{for } t \geq 0.$$

In this case the existence of a Lyapunov matrix $P \succ 0$ with

$$-(A^{(i)})^T P - PA^{(i)} \succ 0 \quad \text{for } 1 \leq i \leq m \tag{2}$$

implies that

$$-A(t)^T P - PA(t) \succ 0 \quad \text{for all } t \geq 0,$$

and hence,

$$0 > x(t)^T \left(A(t)^T P + PA(t)\right) x(t) = (A(t)x(t))^T Px(t) + x(t)^T P\,(A(t)x(t))$$

$$= \frac{d}{dt}\left(x(t)^T Px(t)\right) = \frac{d}{dt}\|x(t)\|_P^2$$

whenever $x(t) \neq 0$. This implies that $\|x(t)\|_P^2 \to 0$, and hence the existence of a matrix $P \succ 0$ satisfying (2) is a sufficient condition to prove stability of the uncertain system. (The above argument only shows that $\|x(t)\|_P$ is monotonously decreasing. In order to show that $\|x(t)\|_P$ converges to zero, one can find a strictly negative bound for $\frac{d}{dt}\|x(t)\|_P^2$ using the largest real part of the eigenvalues of $(A^{(i)})^T P + PA^{(i)}$.)

There are straightforward ways to formulate the problem of finding a matrix $P \succ 0$ satisfying (2) as a linear semidefinite program, see e.g. Boyd et.al. (1994). While this simple example results in a linear semidefinite program, other problems from controller design often result in bilinear semidefinite programs that are no longer convex, see e.g. Leibfritz (2001); Scherer (1999); Freund and Jarre (2000).

2.2 A second simple example

Binary quadratic programs (also known as max-cut-problems) have few applications in VLSI layout or in spin glass models from physics. Their most important property, however, appears to be the fact that these problems are $\mathcal{NP}$-complete (and hence, there is no known polynomial time method for solving these problems). What makes these problems so appealing is that they appear to be quite easy.

Let

$$\mathcal{MC} = \text{conv}\left(\left\{xx^T \mid x_i \in \{\pm 1\} \quad \text{for } 1 \leq i \leq n\right\}\right) \subset \mathcal{S}^n$$

be the max cut polytope. Hence, $\mathcal{MC}$ is the convex hull of all rank-1 matrices generated by ± 1-vectors x. Any binary quadratic program or any max-cut problem can be written in the following form:

$$\text{minimize } C \bullet X \text{ s.t. } X \in \mathcal{MC}. \tag{3}$$

This is a standard linear program with the drawback that the feasible set $\mathcal{MC}$ is defined as convex hull of exponentially many points xx^T, rather than being defined by (a polynomial number of) linear inequalities.

Let $e = (1, \dots, 1)^T$ be the vector of all ones. It is straightforward to see that $\mathcal{MC}$ can be written in the form

$$\mathcal{MC} = \operatorname{conv}\left(\{X \succeq 0 \mid \operatorname{diag}(X) = e, \quad \operatorname{rank}(X) = 1\}\right).$$

Due to the condition $\operatorname{diag}(X) = e$ the set $\mathcal{MC}$ lies in an affine subspace of $\mathcal{S}^n$ of dimension $n(n-1)/2$. $\mathcal{MC}$ has 2^{n-1} vertices that are pairwise adjacent i.e. connected by an edge (a 1-dimensional extreme set of $\mathcal{MC}$).

Note that the constraints of this second definition of $\mathcal{MC}$ appear to be smooth constraints; a semidefiniteness constraint, a linear constraint, and a rank condition. These conditions, however, imply that there are only finitely many "discrete" elements of which the convex hull is taken. In some sense the constraints contain a hidden binary constraint allowing only certain matrices with entries ± 1. When the rank constraint is omitted, we obtain the standard SDP relaxation of the max-cut problem,

$$\mathcal{SDP} = \{X \succeq 0 \mid \operatorname{diag}(X) = e\}$$

satisfying $\mathcal{MC} \subset \mathcal{SDP}$. A relaxed version of (3) is thus given by

$$\text{minimize } C \bullet X \quad \text{s.t.} \quad X \in \mathcal{SDP}. \tag{4}$$

This problem is a linear SDP of the form (P) and can be solved efficiently using, for example, interior point methods, see e.g. Helmberg et.al. (1996). Goemans and Williamson (1994) have shown how to obtain an excellent approximation of the max-cut problem (3) using the solution X of (4).

A quite interesting inner approximation of $\mathcal{MC}$ leading to a nonlinear semidefinite program is described in Chapter 3.3.

2.3 Smoothness of semidefiniteness constraint

To understand the complexity of nonlinear semidefinite programs we briefly address the question of smoothness and regularity of the semidefinite cone. The set of positive semidefinite matrices can be characterized in several different forms,

$$\{X \mid X \succeq 0\}$$

$$
\begin{aligned}
&= \{X \mid \lambda_{min}(X) \ge 0\} \\
&= \{X \mid \lambda_i(X) \ge 0 \ \text{ for } 1 \le i \le n\} \\
&= \left\{X \mid u^T X u \ge 0 \ \text{ for all } u \in I\!R^n \ (\|u\| = 1)\right\} \\
&= \{X \mid \det(X_{\Sigma,\Sigma}) \ge 0 \ \text{ for all } \Sigma \subset \{1, \dots, n\}\} \\
&= \left\{X \mid \exists Z \in \mathcal{S}^n : \ X = Z^2\right\}.
\end{aligned}
$$

The first characterization uses the smallest eigenvalue $\lambda_{min}(X)$ of X. This is a nonsmooth representation. When ordering the eigenvalues in a suitable way, the eigenvalues $\lambda_i(X)$ used in the second representation have directional derivatives, but are not totally differentiable. The third representation is based on a semi-infinite constraint. From this representation one can easily deduce, for example, that $\{X \mid X \succeq 0\}$ is convex. The fourth representation is based on a finite (but exponential) number of smooth constraints, requiring all principal subdeterminants to be nonnegative. This representation certainly justifies the claim that $\{X \mid X \succeq 0\}$ is bounded by smooth constraints. As shown in Pataki (2000), the tangent plane to $\{X \mid X \succeq 0\}$ at a point $\hat{X}$ is given as follows. Let

$$\hat{X} = UDU^T$$

with a diagonal matrix D and a unitary matrix U. If $\hat{X}$ is a boundary point of $\{X \mid X \succeq 0\}$ we may assume without loss of generality that the first k diagonal entries of D satisfy $d_1 = \ldots = d_k = 0$ and $d_{k+1}, \ldots, d_n > 0$. Let ΔX be given by

$$\Delta X = U \begin{pmatrix} 0 & * \\ * & * \end{pmatrix} U^T$$

where the 0-block in the matrix on the right hand side is of size $k \times k$, and the entries $*$ are any entries of suitable dimension. All matrices ΔX of the above form belong to the tangent space of $\{X \mid X \succeq 0\}$ at $\hat{X}$.

The fourth representation also leads to the convex barrier function

$$\tilde{\Phi}(X) = -\log\det(X)$$

for the positive semidefinite cone. For this barrier function it is sufficient to consider $\Sigma = \{1, \ldots, n\}$, and to set $\tilde{\Phi}(X) = \infty$ whenever X is not positive definite.

The last representation is a projection of a quadratic equality constraint.

Most, if not all, of the above representations have been used numerically to enforce semidefiniteness of some unknown matrix X.

The set $\{X \mid X \succeq 0\}$ certainly satisfies Slaters condition, or, in the context of nonconvex minimization, any point $\bar{X} \in \{X \mid X \succeq 0\}$ trivially satisfies the constraint qualification by Robinson. However, the fourth representation above does not satisfy LICQ. (LICQ is a common regularity condition requiring that the active constraints at any point are linearly independent, see e.g. Wright and Nocedal (1999).) In fact, (for $n > 1$) there does not exist any representation of the positive semidefinite cone by nonlinear inequalities that do satisfy LICQ. Nevertheless the positive semidefinite cone and its surface are numerically tractable, and may be considered as a regular set with smooth constraints.

2.4 A dual barrier method

We consider problem (D) and eliminate the slack variable S to obtain the problem

$$\text{maximize } b^T y \text{ s.t. } C - \mathcal{A}^*(y) \succeq 0.$$

For $y \in I\!R^m$ with $C - \mathcal{A}^*(y) \succeq 0$ we then define a convex barrier function $\hat{\Phi}$,

$$\hat{\Phi}(y) = -\log\left(\det(C - \mathcal{A}^*(y))\right).$$

A plain dual barrier method can be stated as follows:

Dual barrier method
Start: Find $y^{(0)}$ with $C - \mathcal{A}^*(y^{(0)}) \succeq 0$.
For $k = 1, 2, 3, \ldots$

- Set $\mu_k = 10^{-k}$ and find

$$y^{(k)} \approx y(\mu_k) = \arg\min_y \; -\frac{b^T y}{\mu_k} + \hat{\Phi}(y)$$

  by Newton's method with line search starting at $y^{(k-1)}$.

Of course, this conceptual method needs many refinements such as an appropriate choice of the starting point and a somewhat more sophisticated update of μ_k. With such minor modifications, however, the above algorithm solves the semidefinite programming problem in polynomial time. (The notion of polynomiality in the context of nonlinear programming is to be taken with care; the solution of a linear semidefinite program can have an exponential "size" like an optimal value of 2^{2^n} for a semidefinite program with encoding length $O(n)$. Our reference to "polynomial time" is meant that the method reduces some primal dual gap function in polynomial time, see e.g. Nesterov and Nemirovski (1994).)

The key elements in guaranteeing the theoretical efficiency of the barrier method rest on two facts:

- The duality gap (or some linear measure of closeness to optimality) is of order μ,

- and the Hessian $\nabla^2\hat{\Phi}$ of the barrier function satisfies a local relative Lipschitz condition.

Both facts were shown by Nesterov and Nemirovski (1994) and rest on two conditions introduced in Nesterov and Nemirovski (1994). The first fact is implied by a local Lipschitz condition of $\hat{\Phi}$ with respect to the norm induced by $\nabla^2\hat{\Phi}(y)$, and the second fact is called self-concordance, and implies that Newton's method converges globally at a fixed rate. More details can be found in Nesterov and Nemirovski (1994); Jarre (1996).

The guaranteed convergence results in these references are much slower than what is observed in implementations of related methods. In fact, these theoretical results are much too slow to be relevant for practical applications. However, these results guarantee a certain independence of the method from the data of the problem. Even with exact arithmetic, the performance of the steepest descent method for unconstrained minimization, for example, depends on the condition number of the Hessian matrix at the optimal solution. Unlike the steepest descent method, the worst case bound for the barrier method only depends on the dimension n of the problem (D), but not on any condition numbers or any other parts of the data of the problem. In this respect, the theoretical analysis *is* relevant for practical applications.

The above barrier method is not suitable for practical implementations. The following simple acceleration scheme is essential for obtaining a more practical algorithm: Observe that the points $y(\mu)$ that are approximated at each iteration of the barrier method satisfy

$$-\frac{b}{\mu} + \nabla\hat{\Phi}(y(\mu)) = 0.$$

Differentiating this equation with respect to μ yields

$$\frac{b}{\mu^2} + \nabla^2\hat{\Phi}(y(\mu))\dot{y}(\mu) = 0.$$

For given values of μ and $y(\mu)$ this is a linear equation that can be solved for $\dot{y}(\mu)$. (The matrix is the same as the one that is used in the Newton step for finding $y(\mu)$.) Given this observation we can state a more efficient predictor corrector method.

Dual predictor corrector method
Start: Find $y^{(0)}$ and $\mu_0 > 0$ with $y^{(0)} \approx y(\mu_0)$.
For $k = 1, 2, 3, \ldots$

- Choose $\Delta\mu_k \in (0, \mu_{k-1})$ such that $\hat{y}^{(k)} = y^{(k-1)} - \Delta\mu_k\dot{y}(\mu_{k-1})$ satisfies $C - \mathcal{A}^*(\hat{y}^{(k)}) \succeq 0$.

- Set $\mu_k = \mu_{k-1} - \Delta\mu_k$ and find $y^{(k)} \approx y(\mu_k)$ by Newton's method with line search starting at $\hat{y}^{(k)}$.

It turns out that $\dot{y}(\mu_{k-1})$ can be computed fairly accurately even if only an approximate point $y^{(k-1)} \approx y(\mu_{k-1})$ is known. For details see e.g. Jarre and Saunders (1993). This predictor corrector method is "reasonably efficient", but primal-dual approaches are more efficient in general.

We will generalize this method to nonlinear semidefinite programs in the next section

3. Nonlinear Semidefinite Programs

In this section we consider nonlinear semidefinite programs of the form

$$\text{maximize } b^T y \;\text{ s.t. } \mathcal{A}(y) \succeq 0, \;\; f_i(y) \leq 0 \;\text{ for } 1 \leq i \leq m, \tag{5}$$

where $\mathcal{A} : I\!R^n \to \mathcal{S}^l$ is a smooth map and $f_i : I\!R^n \to I\!R$ are smooth functions. Note a slight change of notation, in this chapter $\mathcal{A}$ is a nonlinear operator, $\mathcal{A} : I\!R^n \to \mathcal{S}^l$.

We define a (possibly nonconvex) barrier function Φ,

$$\Phi(y) = -\log\det(\mathcal{A}(y)) - \sum_{i=1}^{m} \log(-f_i(y))$$

and local minimizers

$$y(\mu) = \text{local minimizer of } -\frac{b^T y}{\mu} + \Phi(y). \tag{6}$$

In slight abuse of notation we will denote any local minimizer by $y(\mu)$; this definition therefore does not characterize $y(\mu)$ uniquely.

Replacing $\hat{\Phi}$ with Φ, both, the barrier method and the predictor corrector method of Chapter 2.4 can also be applied to solve problem (5).

There are two questions regarding the efficiency of the predictor corrector method for solving (5). (The barrier method is certainly unpractical!)

- Does $\bar{y} = \lim_{k\to\infty} y^{(k)}$ exist, and if so, is $\bar{y}$ a "good" locally optimal solution of (5)?
- How quickly can $y^{(k)}$ be computed?

3.1 Issues of global convergence

As to the first question, one can show (see e.g. Jarre (2001)) that any accumulation point $\bar{y}$ of the sequence $y^{(k)}$ satisfies the Fritz-John condition, (for a definition see e.g. Borgwardt (2001))

$$\exists u \geq 0,\ u \neq 0: \quad -u_0 b + \sum_{i=1}^{m} u_i \nabla f_i(\bar{y}) + u_{m+1} \nabla \det(\mathcal{A}(\bar{y})) = 0.$$

While this condition is reasonable in the absence of a constraint qualification it is not suitable for semidefinite programs. Indeed, whenever $\mathcal{A}(\bar{y})$ has the eigenvalue zero of multiplicity more than one, then $\nabla \det(\mathcal{A}(\bar{y})) = 0$, so that one can choose $u_{m+1} = 1$ and $u_i = 0$ for all other i.

A more appropriate convergence result therefore is

$$\begin{array}{ll} \not\exists \Delta y: & b^T \Delta y \geq 0, \\ & \nabla f_i(\bar{y}) \Delta y < 0 \ \text{ for all } y \text{ with } f_i(\bar{y}) = 0 \\ & \mathcal{A}(\bar{y}) + \varepsilon D A(\bar{y})[\Delta y] \succ 0 \text{ for small } \varepsilon > 0. \end{array}$$

This result states that there does not exist any direction Δy starting at $\bar{y}$ that is strictly linearized feasible and does not increase the objective function.

Neither of the statements guarantees that $\bar{y}$ is a local minimizer. Indeed there are simple degenerate examples for which $\bar{y}$ is the global *maximizer* of (5). As shown in Jongen and Ruiz Jhones (1999), for nonlinear programs satisfying an LICQ condition and not containing "degenerate" critical points, the limit point of $y^{(k)}$ is a local minimizer. For such problems one can still construct examples, such that $\bar{y}$ is a very "poor" local minimizer. Nevertheless we believe that in many cases $\bar{y}$ is a minimizer whose objective value is "close" to the global minimum of (5). This intuition is motivated by the work Nesterov (1997). Nesterov considered the problem of minimizing a quadratic function over the ∞-norm unit cube. This problem may have very poor local minimizers (whose objective value is much closer to the global maximum value than it is to the global minimum). Nesterov shows that any local minimizer over a p-norm cube with a suitable value of $p = O(\log n)$ has much better global properties in the sense that it is at least as good as the result guaranteed by the semidefinite relaxation. Intuitively, this result is due to the fact that the p-norm cube "rounds" the vertices and edges of the ∞-norm cube. By this rounding procedure, the poor local minimizers disappear. In two dimensions the level sets of the logarithmic barrier function are almost indistinguishable from suitably scaled p-norm cubes. This leads us to believe that at least for quadratic minimization problems over the ∞-norm unit cube, a suitably implemented barrier method will also generate "good" local minimizers.

3.2 Efficiency of local minimization

Note that by definition, $y(\mu_k)$ is a local minimizer of (6), and hence, $\nabla^2\Phi(y(\mu_k)) \succeq 0$. In all of our test problems the iterates $y^{(k)} \approx y(\mu_k)$ satisfied the stronger condition $\nabla^2\Phi(y^{(k)}) \succ 0$. If this relation is satisfied the extrapolation step for computing $\hat{y}^{(k)}$ in the predictor corrector method can be carried out in the same way as in the convex case.

However, the iterates $y^{(k,i)}$ "on the way" from $\hat{y}^{(k)}$ to $y^{(k)}$ often do not satisfy $\nabla^2\Phi(y^{(k,i)}) \succ 0$. This implies that the concept of self-concordance that formed the basis of the dual barrier method and of the predictor corrector method for solving (D) is no longer applicable. While it is not yet possible to generalize the theory of self-concordance to nonconvex functions, it seems possible that the known Lipschitz continuity properties of $\nabla^2\hat{\Phi}$ carry over in some form to $\nabla^2\Phi$. The tool that was used for minimizing the barrier function involving $\hat{\Phi}$ in Section 2.2 is Newton's method. When $\nabla^2\Phi(y^{(k,i)}) \not\succ 0$, Newton's method with line search for approximating $y(\mu_k)$ is no longer applicable.

We need to find a suitable generalization of Newton's method to the nonconvex case involving Φ. For this generalization we keep the following properties in mind: The barrier subproblems that need to be solved at each step of the barrier method (or of the predictor corrector method) are systematically ill-conditioned. The condition number typically is $O(1/\mu)$, and the constant in the order notation is typically large. In addition, the computation of the Hessian matrices often is very expensive.

Possible minimization methods for approximating $y(\mu_k)$ include trust region methods with quasi-Newton updates of an approximate Hessian, see e.g. Conn et.al. (2000), continuation methods, or expensive plane search strategies as proposed in Jarre (2001).

In numerical examples it turned out that the minimization problems tend to be quite difficult and none of the minimization methods converge quickly. In particular, the barrier subproblems appear to be substantially more difficult to solve than in the convex case. We therefore address the complexity of smooth nonconvex local minimization. The next section shows that local minimization is $\mathcal{NP}$-hard in a certain sense.

3.3 Returning to the max cut problem

We return to the example in Chapter 2.2. As shown in Nesterov (1998) an inner approximation for the polyhedron $\mathcal{MC}$ is given by

$$\mathcal{NA} = \left\{ X \in \mathcal{SDP} \mid \sin\left[\frac{\pi}{2}X\right] \succeq 0 \right\}.$$

Here, the square brackets $\sin\left[\frac{\pi}{2}X\right]$ are used to indicate that the sin function is applied *componentwise* to each of the matrix entries of $\frac{\pi}{2}X$. The set $\mathcal{NA}$ is formed from $\mathcal{SDP}$ using the function $c : [-1, 1] \to [-1, 1]$ with $c(t) = \sin\left(\frac{\pi}{2}t\right)$. This function is a nonlinear "contraction" in the sense that $|c(t)| \leq |t|$.

It is somewhat surprising to find out that $\text{conv}(\mathcal{NA}) = \mathcal{MC}$, i.e.

$$\mathcal{NA} \subset \text{conv}(\mathcal{NA}) = \mathcal{MC} \subset \mathcal{SDP}.$$

see Nesterov (1998).

A simple picture can explain the relationship of $\mathcal{MC}$, $\mathcal{SDP}$, and $\mathcal{NA}$.

The set $\mathcal{MC}$ is a polytope whose precise description is not known in spite of its simple structure. (More precisely, there does not exist any known polynomial time algorithm which, given a point X, either returns a certificate proving that $X \in \mathcal{MC}$ or returns a separating hyperplane.)

The set $\mathcal{SDP}$ is obtained by "inflating" the set $\mathcal{MC}$ while keeping all faces of dimension $\leq n-2$ fixed. Like a balloon we "pump up" the hull of $\mathcal{MC}$ while keeping certain low-dimensional boundary manifolds fixed. (Note that $\mathcal{MC}$ has dimension $n(n-1)/2$.) The set $\mathcal{SDP}$ is convex and is "efficiently representable", i.e. there exist efficient numerical algorithms for minimizing convex functions over $\mathcal{SDP}$.

The set $\mathcal{NA}$ is obtained by shrinking $\mathcal{SDP}$ in a certain nonlinear fashion. This shrinkage is done in a certain optimal way such that all boundary manifolds of dimensions 1 and 2 of $\mathcal{MC}$ are contained in $\mathcal{NA}$. In particular, for $n = 3$ we have $\mathcal{MC} = \mathcal{NA}$, see Hirschfeld and Jarre (2001).

The set $\mathcal{NA}$ is bounded by two smooth constraints, is star shaped, contains a ball of radius 1, and is contained in a ball of radius n. By our previous considerations,
any locally optimal vertex of

$$\textit{minimize } C \bullet X \quad \textit{s.t.} \quad X \in \mathcal{NA} \tag{7}$$

solves the max cut problem (3).

Hence, in spite of the nice properties of $\mathcal{NA}$, it must be very difficult to find a local optimal vertex of (7) or to check whether a given vertex is a *local* minimum.

Note that (7) is a nonlinear semidefinite program. The difficulty of the local minimization of (7) is due to the fact that problem (7) suffers from a systematic violation of any constraint qualification. It contains many "peaks" similar to the one in

$$\left\{x \in I\!R^2 \mid x \geq 0, \; x_2 \leq x_1^3\right\}.$$

In higher dimensions such peaks become untractable.

3.4 Finding an ϵ-KKT-point

In a second example, see Hirschfeld and Jarre (2001), the so-called chained Rosenbrock function $f : \mathbb{R}^n \to \mathbb{R}$

$$f(x) = (x_1 - 1)^2 + 100 \sum_{i=2}^{n} (x_i - x_{i-1}^2)^2$$

(see also Toint (1978)) has been tested. This function has only one local minimizer which is also the global minimizer, $x = (1, \ldots, 1)^T$. Applying various trust region methods for minimizing f starting at $x^{(0)} = (-1, 1, \ldots, 1)^T$ results in running times that appear to be exponential in n. (These running times are purely experimental, and due to time limitations could only be tested for small values of n.)

At first sight this result seems to contradict a statement by Vavasis. In the paper Vavasis (1993) the following result is shown.

Consider the problem

$$\text{minimize } f(x) \quad \text{s.t.} \quad -1 \leq x_i \leq 1 \quad \text{for } 1 \leq i \leq n. \tag{8}$$

Vavasis assumes that the gradient ∇f is Lipschitz continuous with Lipschitz constant M and considers the problem of finding an ε-KKT point for (8). He presents an algorithm that takes at most $O(\frac{nM}{\varepsilon})$ gradient evaluations to find an ε-KKT point. This bound is exponential with respect to the number of digits of the required accuracy, i.e. with respect to "$-\log \varepsilon$", but linear with respect to n.

He also presents a class of functions of two variables for which *any* algorithm has a worst case complexity of at least $O(\sqrt{\frac{M}{\varepsilon}})$ gradient evaluations to find an ε-KKT point.

The conditions of Vavasis' paper apply to the Rosenbrock example as well. All points at which this function is evaluated by the trust region algorithms lie in the box $-1 \leq x_i \leq 1$, and moreover, Rosenbrocks function possesses moderately bounded norms of $\nabla^2 f$ at these points implying that M is consistently small. The reason for the observed exponential growth of the number of iterations lies in the fact that the norms of the gradients do become small very quickly (as predicted by Vavasis even for a steepest descent method), but for large n, the norm of ∇f needs to be extremely small to guarantee that the iterate is close to a local minimizer. Thus the exponential growth with respect to the number of variables is due to the fact that the ε-KKT condition is a poor condition for large n. (We don't know of any better condition though!) More results on local minimization issues are discussed in the forthcoming paper Hirschfeld and Jarre (2001).

4. Conclusion

We have highlighted some issues of nonlinear semidefinite programming related to a dual barrier method. In particular we have raised the questions of smoothness, regularity, and computational complexity related to semidefinite programs. As preliminary numerical results in Jarre (2001) indicate, variants of the predictor corrector method of the present paper are reasonably fast for medium size problems (up to 500 unknowns). The numerical results were also compared with the ones in Fukuda and Kojima (2001). In all examples it turned out that the method proposed in this paper converged to the global minimizer. This gives some further weak evidence that the method is indeed unlikely to be "trapped" near poor local minimizers. We also indicated that the local convergence of solving the barrier subproblems in the predictor corrector method is slow; improvements of this convergence behavior are the subject of future research.

References

F. Alizadeh, "Combinatorial Optimization with Interior Point Methods and Semidefinite Matrices" PhD Thesis, University of Minnesota (1991).

Optimierung, Operations Research, Spieltheorie Birkhäuser Verlag (2001).

S. Boyd, L. El Ghaoui, E. Feron, and V. Balakrishnan, *Linear Matrix Inequalities in Systems and Control Theory*, Volume 15 of *Studies in Applied Mathematics* SIAM Philadelphia, PA, (1994).

A.R. Conn, N.I.M. Gould and Ph.L. Toint, *Trust-Region Methods* MPS/SIAM Series on Optimization, SIAM, Philadelphia, (2000).

A. Forsgren, "Optimality conditions for nonconvex semidefinite programming", Mathematical Programming 88 (2000), 105-128.

R.W. Freund and F. Jarre, "An Extension of the Positive Real Lemma to Descriptor Systems" Report 00/3-09, Scientific Computing Interest Group, Bell Labs, Lucent Technologies, (2000)

M. Fukuda and M. Kojima, "Branch-and-Cut Algorithms for the Bilinear Matrix Inequality Eigenvalue Problem", *Computational Optimization and Applications* Vol.19, No.1, pp.79–105, (2001).

M.X. Goemans and D.P. Williamson, ".878-Approximation Algorithm for MAX CUT and MAX 2SAT", in *ACM Symposium on Theory of Computing (STOC)*, (1994).

C. Helmberg, F. Rendl, R.J. Vanderbei, "An Interior-Point Method for Semidefinite Programming" *SIAM J. Optim.* 6(2):342–361 (1996).

M.W. Hirsch and S. Smale *Differential equations, dynamical systems, and linear algebra* Acad. Press, New York, (1974).

B.W. Hirschfeld and F. Jarre, "Complexity Issues of Smooth Local Minimization", Technical Report, Universität Düsseldorf, in preparation (2001).

"Interior-Point Algorithms for Classes of Convex Programs" in T. Terlaky ed. *Interior Methods of Mathematical Programming* Kluwer (1996).

F. Jarre, "A QQP-Minimization Method for Semidefinite and Smooth Nonconvex Programs", Technical Report, University of Düsseldorf, Germany, to appear in revised form in *Optimization and Engineering* (2001).

F. Jarre and M.A. Saunders, "A Practical Interior-Point Method for Convex Programming", *SIAM J. Optim.* 5(1) pp.149–171 (1995).

H.T. Jongen and A. Ruiz Jhones, "Nonlinear Optimization: On the Min-Max Digraph and Global Smoothing", in A. Ioffe, S. Reich, I. Shafrir eds *Calculus of Variations and Differential Equations* Chapman & Hall / CRC Research Notes in Mathematics Series, Vol 410, CRC Press, (UK) LLC, pp.119–135, (1999).

M. Kocvara and M. Stingl, "Augmented Lagrangian Method for Semidefinite Programming" forthcoming report, Institute of Applied Mathematics, University of Erlangen-Nuremberg (2001).

F. Leibfritz, "A LMI-based algorithm for designing suboptimal static / output feedback controllers", *SIAM Journal on Control and Optimization*, Vol. 39, No. 6, pp. 1711 - 1735, (2001).

Y.E. Nesterov, Talk given at the Conference on Semidefinite Optimization, ZIB Berlin, (1997).

Y.E. Nesterov, "Semidefinite Relaxation and Nonconvex Quadratic Optimization", *Optimization Methods and Software* 9, pp.141–160, (1998).

Y.E. Nesterov and A.S. Nemirovski, *Interior Point Polynomial Algorithms in Convex Programming*, SIAM Publications, SIAM Philadelphia, PA, (1994).

G. Pataki, "The Geometry of Semidefinite Programming", in H. Wolkowicz, R. Saigal, L. Vandenberghe eds *Handbook of Semidefinite Programming: Theory, Algorithms and Applications* Kluwer's International Series (2000).

C. Scherer, "Lower bounds in multi-objective H_2/H_∞ problems", *Proc. 38th IEEE Conf. Decision and Control*, Arizona, Phoenix (1999).

A. Shapiro and K. Scheinberg, "Duality and Optimality Conditions", in H. Wolkowicz, R. Saigal, L. Vandenberghe eds *Handbook of Semidefinite Programming: Theory, Algorithms and Applications* Kluwer's International Series (2000).

Ph.L. Toint, "Some Numerical Result Using a Sparse Matrix Updating Formula in Unconstrained Optimization", *Mathematics of Computation*, vol. 32(143), pp. 839-851, (1978).

R.J. Vanderbei, "LOQO User's Manual – Version 3.10" . Report SOR 97-08, Princeton University, Princeton, NJ 08544, (1997, revised 10/06/98).

R.J. Vanderbei, H. Benson and D. Shanno, "Interior-Point Methods for Nonconvex Nonlinear Programming: Filter Methods and Merit Functions", Report ORFE 00-06, Princeton University, Princeton, NJ 08544, (2000).

S.A. Vavasis, "Black-Box Complexity of Local Minimization", *SIAM J. Optim.* 3(1) pp.60–80, (1993).

S.J. Wright and J. Nocedal, *Numerical Optimization*, Springer Verlag, (1999).

IMPLICIT FILTERING AND NONLINEAR LEAST SQUARES PROBLEMS

C. T. Kelley
North Carolina State University
Center for Research in Scientific Computation
Department of Mathematics
Box 8205, Raleigh, N. C. 27695-8205, USA
Tim_Kelley@ncsu.edu

Abstract In this paper we motivate and analyze a version of the implicit filtering algorithm by viewing it as an extension of coordinate search. We then show how implicit filtering can be combined with the damped Gauss-Newton method to solve noisy nonlinear least squares problems.

Keywords: noisy optimization, implicit filtering, damped Gauss-Newton iteration, nonlinear least squares problems

1. Introduction

The purposes of this paper are to show how a version of the implicit filtering algorithm [24, 17, 16] can be motivated and analyzed by viewing it as an elaboration of coordinate search, and to describe and analyze a implicit filtering Gauss-Newton method for nonlinear least squares problems.

Our approach to nonlinear least squares problems is based on a finite-difference form of the damped Gauss-Newton method [11, 24, 32], but differs from that in the MINPACK [30] routine `lmdif.f`. That code uses forward difference Jacobians with a user-defined difference increment, but that increment is set only once. Implicit filtering uses a central difference not only to compute more accurate Jacobians, but more importantly to avoid local minima and to decide when to reduce the difference increment.

Implicit filtering, which we describe in § 2, is a deterministic stencil-based sampling method. In general terms, implicit filtering is a finite-difference quasi-Newton method in which the size of the difference stencil decreases as the optimization progresses. In this way one hopes to "filter" low-amplitude, high-frequency noise in the objective function.

Sampling methods do not use derivatives, but rather sample the objective function on a stencil or pattern to determine the progress of the iteration and whether or not to change the size, but not the shape, of the stencil. Many of these methods, like implicit filtering, the Hooke-Jeeves [20] method, and multidirectional search [38, 39], reduce the size of the stencil in the course of the optimization. The stencil-size reduction policy leads to a convergence theory [24, 5, 39].

The best-known sampling method is the Nelder-Mead [31] algorithm. This method uses an irregular pattern that changes as the optimization progresses, and hence is not stencil-based in the sense of this paper. Analytical results for the Nelder-Mead algorithm are limited [24, 5, 26]. Theoretical developments are at also a very early stage for more aggressive sampling methods, like the DIRECT [22] algorithm, [14, 15].

Sampling methods, for the most part, need many iterations to obtain a high-precision result. Therefore, when gradient information is available and the optimization landscape is relatively smooth, conventional gradient-based algorithms usually perform far better. Sampling methods do well for problems with complex optimization landscapes like the ones in Figure 1, where nonsmoothness and nonconvexity can defeat most gradient based methods.

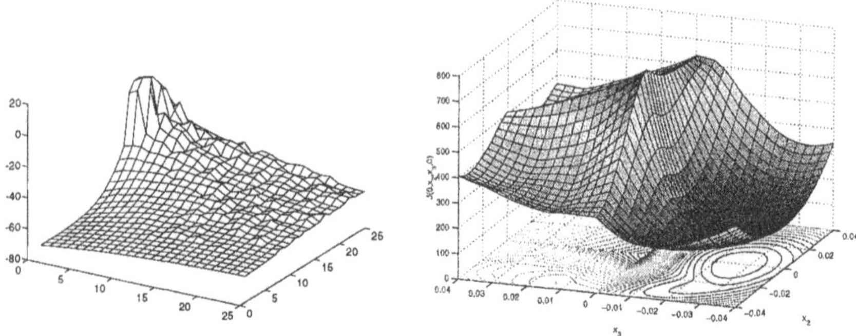

Figure 1. Optimization Landscapes

We caution the reader that sampling methods are not designed to be true global optimization algorithms. Problems with violently oscillatory optimization landscapes are candidates for genetic algorithms [19, 35], simulated annealing [25, 41], or the DIRECT algorithm [22, 21].

The paper is organized as follows. In § 2 we briefly describe the implicit filtering method and some of the convergence results. We describe the new algorithm in § 3 and prove a local convergence result. In § 4 we illustrate the ideas with a parameter identification problem.

2. Implicit Filtering

In this section we introduce implicit filtering. We show how the method can be viewed as an enhanced form of a simple coordinate search method. Convergence analysis for methods of this type is typically done in a setting far simpler than one sees in practice. Many results require smooth objective functions [28, 26, 12, 39, 8, 9] or objective functions that are small perturbations of smooth functions [29, 17, 23, 5, 24, 7, 44]. The main results in this paper make the latter assumption. We will also assume that the noise decays near an optimal point. Such decay has been observed in practice [36, 10, 42, 43, 37, 4] and methods designed with this decay in mind can perform well even when the noise does not decay to zero as optimality is approached.

2.1 Coordinate Search

We begin with a discussion of a coordinate search algorithm, the simplest of all sampling methods, and consider the unconstrained problem

$$\min_{x \in R^N} f(x). \tag{1}$$

From a current point x_c and stencil radius or **scale** h_c we sample f at the $2N$ points

$$S(x_c, h_c) = \{x_c \pm h_c e_j\}, \tag{2}$$

where e_j is the unit vector in the jth coordinate direction. Then either x_c or h_c is changed.

- If

$$f(x_c) \leq \min_{x \in S(x_c, h_c)} f(x) \tag{3}$$

  then we replace h_c by $h_+ = h_c/2$ and set $x_+ = x_c$.

- Otherwise, we replace x_c by any point in $x_+ \in S$ such that

$$f(x_+) = \min_{x \in S(x,h)} f(x)$$

  and let $h_+ = h_c$.

We refer to (3) as **stencil failure**. If f is Lipschitz continuously differentiable, then [24, 5] stencil failure implies that

$$\|\nabla f(x_c)\| = O(h_c). \tag{4}$$

Now, if f has bounded level sets, h will be reduced infinitely many times because there are only finitely many points on the grid with function values smaller than $f(x_c)$ [38]. Hence, by (4), the gradient of f will be driven to zero, giving subsequential convergence to a point that satisfies the necessary conditions for optimality.

One-sided stencils [24, 36] and more general stencils with $< 2N$ directions have also been used [1, 2, 27] and have similar theoretical properties. Our experience has been that a full centered-difference stencil is better in practice.

Sampling methods do more than solve smooth problems. Consider an objective which is the sum of a smooth function f_x and a non-smooth function ϕ, which we will refer to as the noise.

$$f(x) = f_s(x) + \phi(x) \tag{5}$$

We assume that ϕ is uniformly bounded and small relative to f_s, but make no smoothness or even continuity assumptions beyond that. High-frequency oscillations in ϕ could result in local minima of f which would trap a conventional gradient-based method far from a minimizer of f_s. If ϕ decays sufficiently rapidly near a minimizer of f, then the coordinate search method responds to f_s and, in a sense, "does not see" ϕ.

To quantify the claim above, we return to the concept of stencil failure. Define

$$\|\phi\|_{S(x,h)} = \max_{z \in S(x,h)} |\phi(x)|.$$

If (3) holds and f satisfies (5), then [24, 5]

$$\|\nabla f_s(x_c)\| = O\left(h_c + \frac{\|\phi\|_{S(x_c,h_c)}}{h_c}\right). \tag{6}$$

Now, let $\{x_n\}$ be the sequence of coordinate search iterations and $\{h_n\}$ be the sequence of stencil radii, which we will refer to as **scales**. If f has bounded level sets, then the set of possible iterations for a given scale h is finite, as they lie on a grid [39], hence $h_n \to 0$. If, moreover, the noise decays rapidly enough so that

$$\lim_{n \to \infty} \frac{\|\phi\|_{S(x_n,h_n)}}{h_n} = 0, \tag{7}$$

then $\nabla f_s(x_n) \to 0$, by (6).

This asymptotic result does not address an important practical issue. The number of times that h will be reduced during the optimization needs to be specified when the optimization begins or a limit on the number of calls to f must be imposed. Most implementations of sampling methods use one or both of these as termination criteria.

In the simple case where f_s is a convex quadratic, for example, coordinate search, therefore, "jumps over" oscillations in ϕ early in the iteration, when h is large, and, after finding a neighborhood of the minimizer, increases the resolution (*i. e.* decreases the scale) and converges.

2.2 Implicit Filtering

The version of implicit filtering which we discuss in this paper accelerates coordinate search with a quasi-Newton method. We use the sample values to construct a centered difference gradient $\nabla_h f(x_c)$. We then try to take a quasi-Newton step

$$x_+ = x_c - H_c^{-1} \nabla_{h_c} f(x_c) \tag{8}$$

where H_c is a quasi-Newton model Hessian. We find that the BFGS [6, 18, 13, 34] works well for unconstrained problems. We reduce the scale when either the norm of the difference gradient is sufficiently small or stencil failure occurs.

We formally describe implicit filtering below as a sequence of calls to a finite-difference quasi-Newton algorithm (`fdquasi`) followed by a reduction in the difference increment. The quasi-Newton iteration is terminated on entry if stencil failure is detected. The other termination criteria of the quasi-Newton iteration reflect the truncation error in the difference gradient. The tolerance for the gradient

$$\|\nabla_h f(x)\| < \tau h \tag{9}$$

is motivated both by the heuristic that the step should be at least of the same order as the scale, by the implication (6) of stencil failure, and by the error estimate [24] [24]

$$\|\nabla f_s(x) - \nabla_h f(x)\| = O\left(h^2 + \frac{\|\phi\|_{S(x,h)}}{h}\right). \tag{10}$$

The performance of implicit filtering can be sensitive to the choice of the parameter τ if, as was the case for the earliest implementations of implicit filtering [36, 17, 10], the test for stencil failure is not incorporated into the algorithm.

The line search is not guaranteed to succeed because the gradient is not exact, therefore we allow only a few reductions in the step length before exiting the quasi-Newton iteration. If the line search fails, then sufficient decrease condition

$$f(x_c - \lambda \nabla_{h_c} f(x_c)) - f(x_c) < -\alpha\lambda \|\nabla_{h_c} f(x_c)\|^2 \tag{11}$$

has been violated. Here, as is standard, [11, 24], α is a small parameter, typically 10^{-4}. If both (9) and (11) fail, then one can show in some cases [17] that the noise is sufficiently larger that the scale to justify terminating the entire optimization. This leads to the question of selection of the smallest scale, which is open. In some special cases, [17] failure of the line search can be related to the size of noise, motivating termination of the entire optimization because the assumption that $\|\phi\|$ is much smaller than h is no longer valid.

Algorithm 1 fdquasi$(x, f, pmax, \tau, h, amax)$

$p = 1$
while $p \leq pmax$ and $\|\nabla_h f(x)\| \geq \tau h$ **do**
  compute f and $\nabla_h f$
  if (2) holds **then**
    terminate and report **stencil failure**
  end if
  update the model Hessian H if appropriate; solve $Hd = -\nabla_h f(x)$
  use a backtracking line search, with at most $amax$ backtracks, to find a step length λ
  if $amax$ backtracks have been taken **then**
    terminate and report **line search failure**
  end if
  $x \leftarrow x + \lambda d$
  $p \leftarrow p + 1$
end while
if $p > pmax$ report **iteration count failure**

Implicit filtering is a sequence of calls to **fdquasi** with the difference increments or **scales** reduced after each return from **fdquasi.**

Algorithm 2 imfilter$(x, f, pmax, \tau, \{h_n\}, amax)$

for $k = 0, \ldots$ **do**
  fdquasi$(x, f, pmax, \tau, h_n, amax)$
end for

Our analysis of coordinate search depended on the fact that

$$\|\nabla f_s(x_n)\| = O\left(h_n + \frac{\|\phi\|_{S(x_n,h_n)}}{h_n}\right) \tag{12}$$

when stencil failure occurred and that h was reduced when that happened. Since stencil failure directly implies success, as do (6) and (9) together, the convergence result for coordinate search will hold for implicit filtering provided the line search only fails finitely often and the quasi-Newton iteration terminates because of stencil failure or satisfaction of (9), We summarize these observations in a theorem from [24].

Theorem 1 *Let f satisfy* (5) *and let ∇f_s be Lipschitz continuous. Let $h_n \to 0$ and let $\{x_n\}$ be the implicit filtering sequence. Assume that either* (3) *or* (11) *hold after each call to* `fdquasi` *(*i. e. *there is no line search failure or iteration count failure) for all but finitely many k. Then if*

$$\lim_{k \to \infty} (h_n + h_n^{-1} \|\phi\|_{S(x,h_n)}) = 0 \tag{13}$$

then any limit point of the sequence $\{x_n\}$ is a critical point of f_s.

Theorem 1 does not explain the performance of implicit filtering in practice. In fact, other methods, such as coordinate search, Hooke-Jeeves, and MDS, also satisfy the conclusion of Theorem 1 if (13) holds, [24, 40]. Implicit filtering performs well only if a quasi-Newton model Hessian is used. The reasons for the efficacy of the quasi-Newton methods are not fully understood. A step toward such an understanding is in [7], where a superlinear convergence result is presented. That result is somewhat like the one we give in § 3 and we will summarize it here.

Assumptions on the rate of decrease of $\{h_n\}$ and of the size of ϕ must be made to prove convergence rates. Landscapes like those in Figure 1 motivated the qualitative decay assumption (13). To obtain superlinear convergence one must ask for much more and demand that h and ϕ satisfy

$$\|\nabla_h \phi(x)\| = O(\|x - x^*\|^{1+p}) \tag{14}$$

for some $p > 0$. Here x^* is a local minimizer of f_s. Satisfaction of (14) is possible in practice if both ϕ and the scales h decrease near x^*. As an example, suppose that f_s has a local minimizer x^*, $\nabla^2 f_s$ is Lipschitz continuous in a neighborhood of x^*, $\nabla^2 f_s(x^*)$ is positive definite, and for x sufficiently near x^*,

$$|\phi(x)| = O(\|x - x^*\|^{2+2p}), \tag{15}$$

for some $p > 0$. In that case, if one sets

$$h_{n+1} = \|\nabla_{h_n} f(x_{n+1})\|^{1+p}, \tag{16}$$

and other technical assumptions hold, then one can show that the implicit filtering iteration, with the BFGS update, is locally superlinearly convergent to x^*.

3. Gauss-Newton Iteration and Implicit Filtering

For the remainder of this paper we focus on nonlinear least squares objective functions

$$f(x) = \frac{1}{2}\sum_{i=1}^{M} \|r_i(x)\|_2^2 = \frac{1}{2}R(x)^T R(x). \tag{17}$$

We assume that

$$R(x) = R_s(x) + \Phi(x) \tag{18}$$

where $R_s : R^N \to R^M$ is Lipschitz continuously differentiable. Here the noise Φ in the residual does not correspond to noise in any data in the problem, but rather noise in the computation of R. As an example, if one is doing a nonlinear fit to data, R might have the form $R = M(x) - d$, where d is a vector of data and x are the model parameters. The noise we have in mind is in the computation of M, not in d.

The noise Φ in R can be related to the noise ϕ in f by

$$\phi(x) = R(x)^T \Phi(x) + \Phi(x)^T \Phi(x)/2. \tag{19}$$

3.1 Implicit Filtering Gauss-Newton (IFGN) Algorithm

Our implementation of implicit filtering for nonlinear least squares differs from the one described in § 2 in two ways:

- The Jacobian of the residual, not the gradient of the objective function, is approximated by finite differences.
- The Gauss-Newton model Hessian is used instead of a quasi-Newton model Hessian.

We let $\nabla_h R(x)$ be the centered difference gradient of R based on the stencil $S(x, h)$. Our finite difference Gauss-Newton iteration Algorithm `fdgauss`, must be prepared for stencil failure and failure of the line search. The sufficient decrease condition is now

$$f(x_c - \lambda d) - f(x_c) < -\alpha\lambda((\nabla_h R(x_c))^T R(x_c))^T d. \tag{20}$$

where

$$d = -(\nabla_h R(x_c)^T \nabla_h R(x_c))^{-1} \nabla_h R(x_c)^T R(x_c)$$

is the IFGN direction.

Algorithm 3 fdgauss$(x, R, pmax, \tau, h, amax)$

$p = 1$
while $p \leq pmax$ and $\|(\nabla_h R(x))^T R(x)\| \geq \tau h$ **do**
  compute $f = R(x)^T R(x)/2$ and $\nabla_h R$
  if (2) holds **then**
    terminate and report **stencil failure**
  end if
  set $H = (\nabla_h R(x))^T (\nabla_h R(x))$; solve $Hd = -\nabla_h R(x)^T R(x)$.
  use a backtracking line search, with at most $amax$ backtracks, to find a step length λ
  if $amax$ backtracks have been taken **then**
    terminate and report **line search failure**
  end if
  $x \leftarrow x + \lambda d$
  $p \leftarrow p + 1$
end while
if $p > pmax$ report **iteration count failure**

The implicit filtering form of the damped Gauss-Newton method, (Algorithm `IFGN`) calls `fdgauss` repeatedly, reducing the scale with each iteration.

Algorithm 4 IFGN$(x, R, pmax, \tau, \{h_n\}, amax)$

for $k = 0, \ldots$ **do**
  `fdgauss`$(x, R, pmax, \tau, h_n, amax)$
end for

3.2 Convergence Analysis

We will make a distinction between the central difference gradient of $f = R^T R/2$ and the difference gradient computed via $(\nabla_h R)^T R$, since the two approximate gradients have different errors, especially in the small residual case.

For any function $\psi : R^N \to R^L$ (here $L = 1$ or $L = M$), define

$$\|\psi\|_{S(x,h)} = \max_{z \in S(x,h)} \|\psi(x)\|.$$

and

$$E(x, h, \psi) = h^2 + \frac{\|\psi\|_{S(x,h)}}{h}.$$

We can rewrite (10) as

$$\|\nabla f_s(x) - \nabla_h f(x)\| = O(E(x,h,\phi)). \tag{21}$$

Lemma 3.1 gives the analog of (21) for nonlinear least squares problems in (24) and refines (21) in (23). The error in $(\nabla_h R(x))^T R(x)$ is scaled by the residual norm, a fact we exploit for zero residual problems in Lemma 3.3.

Lemma 3.1 *Let R be given by* (18). *Assume that there is $K > 0$ such that*

$$\|\Phi\|_{S(x,h)} \le K\|R_s(x)\|. \tag{22}$$

Then

$$\|\nabla f_s(x) - \nabla_h f(x)\| = O\left(h^2 + \frac{\|R_s\|_{S(x,h)}\|\Phi\|_{S(x,h)}}{h}\right), \tag{23}$$

$$\|\nabla f_s(x) - (\nabla_h R(x))^T R(x)\| = O(\|R_s(x)\|E(x,h,\Phi)), \tag{24}$$

and

$$\|R_s'(x)^T R_s'(x) - (\nabla_h R(x))^T \nabla_h R(x)\| = O(E(x,h,\Phi)). \tag{25}$$

The constants in the O-terms depend on the norm and the Lipschitz constant of R'.

Proof. The estimate (23) follows from (10) and (19).

We now prove (24). By definition,

$$\begin{aligned}
(\nabla_h R(x))^T R(x) &= (\nabla_h (R_s(x) + \Phi(x)))^T (R_s(x) + \Phi(x)) \\
&= (\nabla_h R_s(x))^T R_s(x) + O\left(\|R(x)\|\frac{\|\Phi\|_{S(x,h)}}{h}\right) \\
&= \nabla f_s(x) + O\left(\|R_s(x)\|E(x,h,\Phi) + \frac{\|\Phi\|^2_{S(x,h)}}{h}\right) \\
&= \nabla f_s(x) + O\left(\|R_s(x)\|E(x,h,\Phi)\right).
\end{aligned}$$

as asserted.

The proof of (25) is similar. ☐

Lemma 3.1 leads directly to a simple convergence result. which, for zero residual problems with only a few stencil failures, requires only that $E(x_n, h_n, \Phi)$ be bounded, a weaker condition than (7).

Theorem 2 *Let R satisfy* (18) *and assume that R' is Lipschitz continuous. Let $h_n \to 0$ and let $\{x_n\}$ be the implicit filtering sequence. Assume that all but finitely many calls to* `fdgauss` *return with stencil failure or*

$$\|(\nabla_{h_n} R(x_n))^T R(x_n)\| < \tau h_n, \tag{26}$$

that the model Hessians $R(x_n)^T R(x_n)$ are nonsingular, and that the model Hessians and their inverses are uniformly bounded. Then if

$$\lim_{n\to\infty} E(x_n, h_n, \Phi) = 0 \tag{27}$$

then any limit point of x_n is a critical point of f. If, moreover, all but finitely many calls to `fdgauss` *return with* (26), *then* (27) *can be replaced by*

$$\lim_{n\to\infty} \|R_s(x_n)\| E(x_n, h_n, \Phi) = 0 \tag{28}$$

Proof. The convergence assumption (27) requires that

$$\|\Phi\|_{S(x_n,h_n)}/h_n \to 0.$$

In view of (19), this is equivalent to (7) if (22) holds. Hence the first assertion of the theorem is simply a restatement of Theorem 1.

If the finite-difference Gauss-Newton iteration terminates all but finitely many times with (26), then

$$\|\nabla f_s(x_n)\| \leq \tau h_n + O(\|R_s(x_n)\| E(x_n, h_n, \Phi))$$

by (24). This completes the proof. □

3.3 Local Convergence

To analyze the local convergence behavior of the IFGN iteration, we must assume that the model Hessians are well conditioned and bounded. Let x^* be a local minimizer of $f_s(x) = R_s^T(x) R_s(x)$ for which the standard assumptions for convergence of the Gauss-Newton iteration

$$x_+^{GN} = x_c - ((\nabla_h R_s(x_c))^T \nabla_h R_s(x_c))^{-1} R_s'(x_c) R_s(x_c),$$

hold (smoothness, nonsingularity of the model Hessian, sufficiently small residual).

To quantify this we will assume:

Assumption 3.1 *There is $\rho_0 > 0$ such that*

- *R_s is Lipschitz continuously differentiable in the set*

$$\mathcal{D} = \{x \mid \|x - x^*\| \leq \rho_0\},$$

- *the Gauss-Newton model Hessian* $R_s'(x)^T R_s'(x)$ *and its inverse are uniformly bounded in* $\mathcal{D}$, *and*
- *there are* $r_{GN} \in (0,1)$ *and* $C_{GN} > 1$ *such that for all* $x_c \in \mathcal{D}$,

$$\|e_+^{GN}\| \leq C_{GN}(\|e_c\|^2 + \|R_s(x^*)\|\|e_c\|) \leq r_{GN}\|e_c\|. \qquad (29)$$

As is standard, we let $e = x - x^*$ for $x \in R^N$, with the iteration index for e being inherited from the one for x.

Lemma 3.2 *Let* R *be given by* (18). *Let* (22) *and Assumption 3.1 hold and let* $x_c \in \mathcal{D}$. *Then if*

$$\sup_{x \in \mathcal{D}} E(x, h, \Phi)$$

is sufficiently small, the IFGN model Hessian $(\nabla_h R(x_c))^T \nabla_h R(x_c)$ *is nonsingular. Moreover, if*

$$x_+ = x_c - ((\nabla_h R(x_c))^T \nabla_h R(x_c))^{-1} \nabla_h R(x_c)^T R(x_c)$$

then

$$\|e_+\| = \|e_+^{GN}\| + O(\|R_s(x_c)\| E(x_c, h, \Phi)). \qquad (30)$$

Proof. Let $x_c \in \mathcal{D}$. Assumption 3.1 and (25) imply that

$$\|(R_s'(x_c)^T R_s'(x_c))^{-1} - ((\nabla_h R(x_c))^T \nabla_h R(x_c))^{-1}\| = O(E(x_c, h, \Phi)). \qquad (31)$$

Now,

$$x_+ = x_+^{GN} + E_H \nabla f_s(x_c) + (R'(x_c)^T R'(x_c))^{-1} E_g$$

where

$$E_H = (R'(x_c)^T R'(x_c))^{-1} - ((\nabla_h R(x_c))^T \nabla_h R(x_c))^{-1}$$

and

$$E_g = \nabla f_s(x) - (\nabla_h R(x))^T R(x).$$

Since $\nabla f_s(x_c) = O(\|e_c\|)$, we apply (31) to obtain

$$E_H \nabla f_s(x_c) = O(\|R_s(x_c)\| E(x_c, h, \Phi)).$$

The conclusion now follows from (22) and (24). □

Theorem 3 *Let* R *be given by* (18). *Let* (22) *and Assumption 3.1 hold. Let* $x_0 \in \mathcal{D}$. *Let* $h_n \to 0$. *Assume that the implicit filtering sequence* $\{x_n\} \subset \mathcal{D}$ *and that the line search fails only finitely many times. Then if* (27) *holds then* $x_n \to x^*$.

3.4 Rates of Convergence

To obtain rates of convergence we must make stronger assumptions on Φ, on the scales, and on the convergence rates of the Gauss-Newton iteration for the smooth problem. We must augment (29) with a lower bound that states that the Gauss-Newton iteration for R_s converges no faster than the standard Gauss-Newton convergence rate. This latter assumption is a nondegeneracy condition on R'' and is needed for the superlinear convergence results.

Assumption 3.2 *There are $p \in (0,1]$ and $C_p > 0$ such that*

$$\|\Phi(x_c)\| \leq C_p \|e_c\|^{2+2p} \tag{32}$$

for all $x_c \in \mathcal{D}$. In addition to (29),

$$C_{GN}^{-1}(\|e_c\|^2 + \|R_s(x^*)\|\|e_c\|) \leq \|e_+^{GN}\| \tag{33}$$

for all $x_c \in \mathcal{D}$.

Lemma 3.3 *Let Assumptions 3.1 and 3.2 hold. Then if x_c is sufficiently near x^* and*

$$C_h^{-1}\|e_c\|^{1+p} \leq h_c \leq C_h \|e_c\|^{(1+p)/2} \tag{34}$$

then there are $r_{GN} < r < 1$ and $C > 1$ such that

$$C^{-1}\|e_+^{GN}\| \leq \|e_+\| \leq C\|e_+^{GN}\| \leq r\|e_c\|, \tag{35}$$

Proof. We will show that

$$\|R_s(x_c)\|E(x_c, h_c, \Phi) = o(\|e_+^{GN}\|) \tag{36}$$

for x_c near x^*. The result will follow from Lemma 3.2 for x_c sufficiently near x^*.

Lemma 3.3 and (32) imply that

$$E(x_c, h, \Phi) = O(\|e_c\|^{1+p}).$$

We consider two cases. If the smooth problem is a zero residual problem ($R_s(x^*) = 0$), then

$$\|R_s(x_c)\|E(x_c, h_c, \Phi) = O(\|e_c\|^{2+p}).$$

In this case, (33) implies (36).

If $R_s(x^*) \neq 0$, then

$$\|R_s(x_c)\| E(x_c, h_c, \Phi) = O(\|e_c\|^{1+p}).$$

However, in that case (33) implies that

$$\|e_+^{GN}\| \geq C_{GN}^{-1} \|R_s(x^*)\| \|e_c\|$$

and (36) holds. This completes the proof. □

In order to apply Lemma 3.3 we need to make sure that (34) holds throughout the iteration. The most direct way to do this is to update h_n with an analog of (16)

$$h_{n+1} = \|(\nabla_{h_n} R(x_{n+1}))^T R(x_{n+1})\|^{1+p}. \tag{37}$$

Theorem 4 *Let Assumptions 3.1 and 3.2 hold. The if x_0 is sufficiently near x^*,*

$$\|\nabla f_s(x_0)\|^{1+p}/2 \leq h_0 \leq 2\|\nabla f_s(x_0)\|^{(1+p)/2}, \tag{38}$$

and the implicit filtering sequence is defined by Algorithm `IFGN` *and* (37), *then $x_n \to x^*$ and*

$$C^{-1}\|e_{n+1}^{GN}\| \leq \|e_{n+1}\| \leq C\|e_{n+1}^{GN}\| \leq r\|e_n\|, \tag{39}$$

for all $n \geq 0$.

Proof. Our assumptions imply that (38) is equivalent to (34) with, for example,

$$C_h = \sup_{x \in \mathcal{D}} \|\nabla^2 f_x(x)\|.$$

Hence, proceeding by induction, we need only show that

$$\|\nabla f_s(x_n)\|^{1+p}/2 \leq h_n \leq 2\|\nabla f_s(x_n)\|^{(1+p)/2} \tag{40}$$

for $n > 0$.

By (24), if h_n satisfies (40), then

$$\begin{aligned} h_{n+1} &= (\|\nabla f_s(x_{n+1})\| + \|R_s(x_{n+1})\| E(x_{n+1}, h_n, \Phi))^{1+p} \\ &= \left(\|\nabla f_s(x_{n+1})\| + o(\|e_{n+1}^{GN}\|)\right)^{1+p} \\ &= \|\nabla f_s(x_{n+1})\|^{1+p} + o(\|\nabla R_s(x_{n+1})\|^{1+p}). \end{aligned}$$

Hence h_{n+1} satisfies (40) for x_0 sufficiently near x^*. □

Remark: Theorem 4 says that the local convergence of IFGN is as good asymptotically as Gauss-Newton, if one counts only nonlinear iterations. For zero residual problems, one need not reduce the scales as rapidly. If we replace (34) by

$$C_h^{-1}\|e_c\|^{1+p} \leq h_c \tag{41}$$

then (35) becomes

$$\|e_+\| \leq C\|e_+^{GN}\| + O(\|R_s(x_c)\|(h_c^2 + \|e_c\|^{1+p})). \tag{42}$$

This will imply superlinear convergence for zero residual problems for which (22) and (32) hold if $h_n \to 0$. The computations in § 4 illustrate this.

4. Numerical Example

We report on the performance of IFGN on a parameter identification problem taken from [24, 7, 3]. Here $N = 2$ and $M = 100$. The problem is to identify the stiffness k and damping c in a harmonic oscillator so that the numerical solution of

$$u'' + cu' + ku = 0; u(0) = u_0, u'(0) = 0$$

best fits the data in the least squares sense.

For this example the data are values of the exact solution at $t_i = i/100$ for $1 \leq i \leq 100$. The numerical solution was computed with the MATLAB ODE15s integrator [33].

We compare three variations of implicit filtering, IFGN with a fixed sequence of scales and an adaptive sequence that attempts to satisfy (37), and a version of the implicit filtering/BFGS algorithm from [24, 7] that has been modified to use adaptive scales. In all three we limit the optimization to a budget of 100 calls to the function. This does not mean that an iteration is terminated before completion, rather we monitor the number of function evaluations after a call to the finite difference optimizer returns and stop the optimization if the number of function evaluations has exceeded the budget after the completion of the iteration.

For all the computations the initial iterate is $(c, k) = (2, 3)$. The sequence of scales used in the examples is

$$h_n^{(1)} = 2^{-n}, \; n = 4, \ldots, 13. \tag{43}$$

Following [7], we implement adaptive scales based on a scaled and safeguarded form of (37),

$$h_{n+1}^{(2)} = \max\left(\min\left[h_{n+1}^{(1)}, \left(\frac{\|(\nabla_{h_n} R(x_{n+1})^T R(x_{n+1})\|}{\|(\nabla_{h_0} R(x_0)^T R(x_0)\|}\right)^{1+p}\right], h_{min}\right) \tag{44}$$

where $p = 1/2$ and $h_{min} = 10^{-5}$. h_{min} is roughly the cube root of machine roundoff and is the optimal choice of h for a central difference.

In the examples the line search strategy is to reduce the step by half if the sufficient decrease condition (either (11) for implicit filtering or (20) for IFGN) fails. Within both algorithms **fdquasi** and **fdguass**, $amax = 10$ and $pmax = 100$.

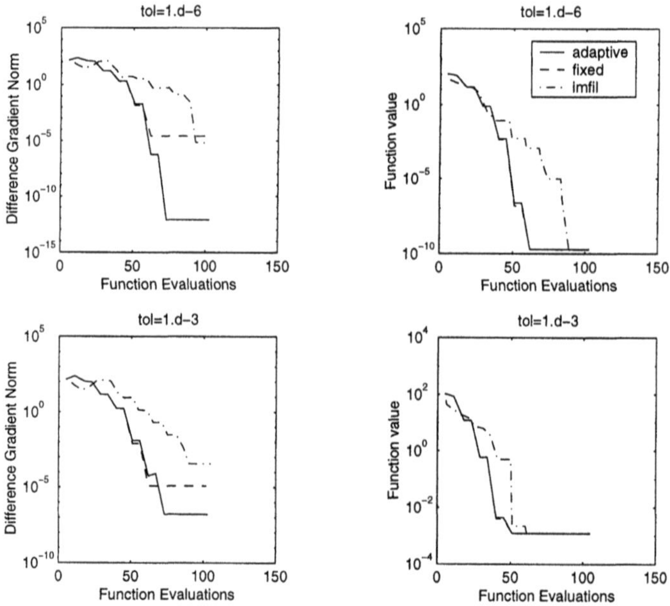

Figure 2. Parameter Identification Example

In Figure 4 we plot the norm of the difference gradient and the size of the function for the three variations of implicit filtering and two values, 10^{-6} and 10^{-3}, of the tolerance given to ODE15s. One can see that the two variations of IFGN did substantially better than an implementation of implicit filtering that did not exploit the least squares structure. A more subtle difference, explained by the remark at the end of § 3, is that while the use adaptive scales made no visible difference in IFGN's ability to reduce the residual (the curves overlap, indicating that the rate of convergence for both methods is equally fast, *i. e.* superlinear), it did make the difference gradient a much better indicator of the progress of the optimization (the scales that are reduced most rapidly produce more accurate gradients).

We see similar behavior for a small, but non-zero, residual problem. In Figure 2 we show the results from the parameter ID problem with uniformly distributed random numbers in the interval $[0, 10^{-4}]$ added to the data. The gradients behave in the same way as in the experiment with exact data, while the limiting function values reflect the non-zero residual in the high-accuracy simulation. In the low-accuracy simulation, the tolerances given to the integrator are smaller than the noise in the data, so the figures are almost identical to the one for the noise-free case.

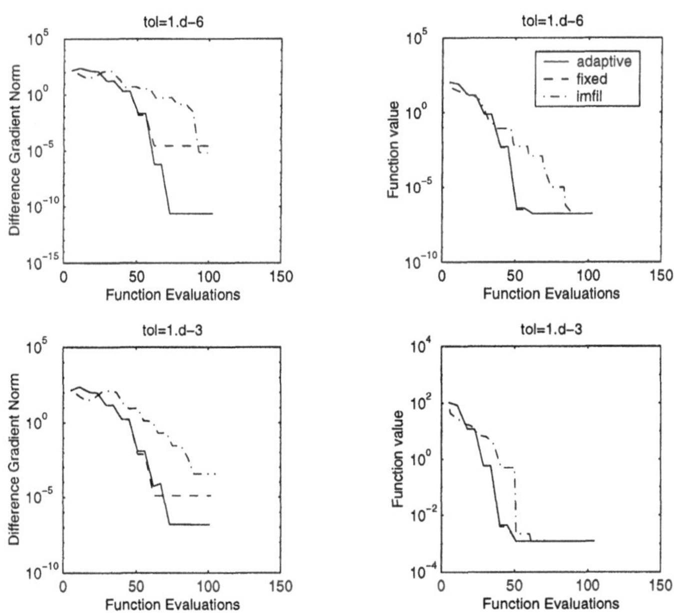

Figure 3. Parameter Identification Example; Random Noise in Data

References

[1] C. AUDET AND J. E. DENNIS, *Analysis of generalized pattern searches.* submitted for publication, 2000.

[2] ———, *A pattern search filter method for nonlinear programming without derivatives.* submitted for publication, 2000.

[3] H. T. BANKS AND H. T. TRAN, *Mathematical and experimental modeling of physical processes.* Department of Mathematics, North Carolina State University, unpublished lecture notes for Mathematics 573-4, 1997.

[4] A. Battermann, J. M. Gablonsky, A. Patrick, C. T. Kelley, T. Coffey, K. Kavanagh, and C. T. Miller, *Solution of a groundwater control problem with implicit filtering*, Optimization and Engineering, 3 (2002), pp. 189–199.

[5] D. M. Bortz and C. T. Kelley, *The simplex gradient and noisy optimization problems*, in Computational Methods in Optimal Design and Control, J. T. Borggaard, J. Burns, E. Cliff, and S. Schreck, eds., vol. 24 of Progress in Systems and Control Theory, Birkhäuser, Boston, 1998, pp. 77–90.

[6] C. G. Broyden, *A new double-rank minimization algorithm*, AMS Notices, 16 (1969), p. 670.

[7] T. D. Choi and C. T. Kelley, *Superlinear convergence and implicit filtering*, SIAM J. Optim., 10 (2000), pp. 1149–1162.

[8] A. R. Conn, K. Scheinberg, and P. L. Toint, *On the convergence of derivative-free methods for unconstrained optimization*, in Approximation Theory and Optimization: Tributes to M. J. D. Powell, A. Iserles and M. Buhmann, eds., Cambridge, U.K., 1997, Cambridge University Press, pp. 83–108.

[9] ———, *Recent progress in unconstrained optimization without derivatives*, Math. Prog. Ser. B, 79 (1997), pp. 397–414.

[10] J. W. David, C. T. Kelley, and C. Y. Cheng, *Use of an implicit filtering algorithm for mechanical system parameter identification*, 1996. SAE Paper 960358, 1996 SAE International Congress and Exposition Conference Proceedings, Modeling of CI and SI Engines, pp. 189–194, Society of Automotive Engineers, Washington, DC.

[11] J. E. Dennis and R. B. Schnabel, *Numerical Methods for Unconstrained Optimization and Nonlinear Equations*, no. 16 in Classics in Applied Mathematics, SIAM, Philadelphia, 1996.

[12] J. E. Dennis and V. Torczon, *Direct search methods on parallel machines*, SIAM J. Optim., 1 (1991), pp. 448 - 474.

[13] R. Fletcher, *A new approach to variable metric methods*, Comput. J., 13 (1970), pp. 317–322.

[14] J. M. Gablonsky, *Modifications of the DIRECT Algorithm*, PhD thesis, North Carolina State University, Raleigh, North Carolina, 2001.

[15] J. M. Gablonsky and C. T. Kelley, *A locally-biased form of the DIRECT algorithm*, Journal of Global Optimization, 21 (2001), pp. 27–37.

[16] P. Gilmore, *An Algorithm for Optimizing Functions with Multiple Minima*, PhD thesis, North Carolina State University, Raleigh, North Carolina, 1993.

[17] P. Gilmore and C. T. Kelley, *An implicit filtering algorithm for optimization of functions with many local minima*, SIAM J. Optim., 5 (1995), pp. 269–285.

[18] D. Goldfarb, *A family of variable metric methods derived by variational means*, Math. Comp., 24 (1970), pp. 23–26.

[19] J. H. Holland, *Genetic algorithms and the optimal allocation of trials*, SIAM J. Comput., 2 (1973).

[20] R. Hooke and T. A. Jeeves, *'Direct search' solution of numerical and statistical problems*, Journal of the Association for Computing Machinery, 8 (1961), pp. 212–229.

[21] D. R. Jones, *The DIRECT global optimization algorithm.* to appear in the Encylopedia of Optimization, 1999.

[22] D. R. JONES, C. C. PERTTUNEN, AND B. E. STUCKMAN, *Lipschitzian optimization without the Lipschitz constant*, J. Optim. Theory Appl., 79 (1993), pp. 157–181.

[23] C. T. KELLEY, *Detection and remediation of stagnation in the Nelder-Mead algorithm using a sufficient decrease condition*, SIAM J. Optim., 10 (1999), pp. 43–55.

[24] ——, *Iterative Methods for Optimization*, no. 18 in Frontiers in Applied Mathematics, SIAM, Philadelphia, 1999.

[25] S. KIRKPATRICK, C. D. GEDDAT, AND M. P. VECCHI, *Optimization by simulated annealing*, Science, 220 (1983), pp. 671–680.

[26] J. C. LAGARIAS, J. A. REEDS, M. H. WRIGHT, AND P. E. WRIGHT, *Convergence properties of the Nelder-Mead simplex algorithm in low dimensions*, SIAM J. Optim., 9 (1998), pp. 112–147.

[27] R. M. LEWIS AND V. TORCZON, *Rank ordering and positive bases in pattern search algorithms*, Tech. Rep. 96-71, Institute for Computer Applications in Science and Engineering, December 1996.

[28] S. LUCIDI AND M. SCIANDRONE, *On the global convergence of derivative free methods for unconstrained optimization.* Reprint, Università di Roma "La Sapienza", Dipartimento di Informatica e Sistemistica, 1997.

[29] ——, *A derivative-free algorithm for bound constrained optimization.* Reprint, Instituto di Analisi dei Sistemi ed Informatica, Consiglio Nazionale delle Richerche, 1999.

[30] J. J. MORÉ, B. S. GARBOW, AND K. E. HILLSTROM, *User guide for MINPACK-1*, Tech. Rep. ANL-80-74, Argonne National Laboratory, 1980.

[31] J. A. NELDER AND R. MEAD, *A simplex method for function minimization*, Comput. J., 7 (1965), pp. 308–313.

[32] J. NOCEDAL AND S. J. WRIGHT, *Numerical Optimization*, Springer, New York, 1999.

[33] L. F. SHAMPINE AND M. W. REICHELT, *The MATLAB ODE suite*, SIAM J. Sci. Comput., 18 (1997), pp. 1–22.

[34] D. F. SHANNO, *Conditioning of quasi-Newton methods for function minimization*, Math. Comp., 24 (1970), pp. 647–657.

[35] M. SRINIVAS AND L. M. PATNAIK, *Genetic algorithms: a survey*, Computer, 27 (1994), pp. 17–27.

[36] D. STONEKING, G. BILBRO, R. TREW, P. GILMORE, AND C. T. KELLEY, *Yield optimization using a GaAs process simulator coupled to a physical device model*, IEEE Transactions on Microwave Theory and Techniques, 40 (1992), pp. 1353–1363.

[37] D. E. STONEKING, G. L. BILBRO, R. J. TREW, P. GILMORE, AND C. T. KELLEY, *Yield optimization using a GaAs process simulator coupled to a physical device model*, in Proceedings IEEE/Cornell Conference on Advanced Concepts in High Speed Devices and Circuits, IEEE, 1991, pp. 374–383.

[38] V. TORCZON, *Multidirectional Search*, PhD thesis, Rice University, Houston, Texas, 1989.

[39] ——, *On the convergence of the multidimensional direct search*, SIAM J. Optim., 1 (1991), pp. 123–145.

[40] ——, *On the convergence of pattern search algorithms*, SIAM J. Optim., 7 (1997), pp. 1–25.

[41] P. van Laarhoven and E. Aarts, *Simulated annealing, theory and practice*, Kluwer, Dordrecht, 1987.

[42] T. A. Winslow, R. J. Trew, P. Gilmore, and C. T. Kelley, *Doping profiles for optimum class B performance of GaAs mesfet amplifiers*, in Proceedings IEEE/Cornell Conference on Advanced Concepts in High Speed Devices and Circuits, IEEE, 1991, pp. 188–197.

[43] ——, *Simulated performance optimization of GaAs MESFET amplifiers*, in Proceedings IEEE/Cornell Conference on Advanced Concepts in High Speed Devices and Circuits, IEEE, 1991, pp. 393–402.

[44] S. K. Zavriev, *On the global optimization properties of finite-difference local descent algorithms*, J. Global Optimization, 3 (1993), pp. 67–78.

DATA MINING VIA SUPPORT VECTOR MACHINES

O. L. Mangasarian
Computer Sciences Department
University of Wisconsin
1210 West Dayton Street
Madison, WI 53706 *
olvi@cs.wisc.edu

Abstract Support vector machines (SVMs) have played a key role in broad classes of problems arising in various fields. Much more recently, SVMs have become the tool of choice for problems arising in data classification and mining. This paper emphasizes some recent developments that the author and his colleagues have contributed to such as: generalized SVMs (a very general mathematical programming framework for SVMs), smooth SVMs (a smooth nonlinear equation representation of SVMs solvable by a fast Newton method), Lagrangian SVMs (an unconstrained Lagrangian representation of SVMs leading to an extremely simple iterative scheme capable of solving classification problems with millions of points) and reduced SVMs (a rectangular kernel classifier that utilizes as little as 1% of the data).

1. Introduction

This paper describes four recent developments, one theoretical, three algorithmic, all centered on support vector machines (SVMs). SVMs have become the tool of choice for the fundamental classification problem of machine learning and data mining. We briefly outline these four developments now.

In Section 2 new formulations for SVMs are given as convex mathematical programs which are often quadratic or linear programs. By setting apart the two functions of a support vector machine: separation of points by a nonlinear surface in the original space of patterns, and maximizing the distance between separating planes in a higher di-

*Work supported by grant CICYT TAP99-1075-C02-02.

mensional space, we are able to define indefinite, possibly discontinuous, kernels, not necessarily inner product ones, that generate highly nonlinear separating surfaces. Maximizing the distance between the separating planes in the higher dimensional space is surrogated by support vector suppression, which is achieved by minimizing any desired norm of support vector multipliers. The norm may be one induced by the separation kernel if it happens to be positive definite, or a Euclidean or a polyhedral norm. The latter norm leads to a linear program whereas the former norms lead to convex quadratic programs, all with an arbitrary separation kernel. A standard support vector machine can be recovered by using the same kernel for separation and support vector suppression.

In Section 3 we apply smoothing methods, extensively used for solving important mathematical programming problems and applications, to generate and solve an unconstrained smooth reformulation of the support vector machine for pattern classification using a completely arbitrary kernel. We term such reformulation a smooth support vector machine (SSVM). A fast Newton-Armijo algorithm for solving the SSVM converges globally and quadratically. Numerical results and comparisons demonstrate the effectiveness and speed of the algorithm. For example, on six publicly available datasets, tenfold cross validation correctness of SSVM was the highest compared with four other methods as well as the fastest.

In Section 4 an implicit Lagrangian for the dual of a simple reformulation of the standard quadratic program of a linear support vector machine is proposed. This leads to the minimization of an unconstrained differentiable convex function in a space of dimensionality equal to the number of classified points. This problem is solvable by an extremely simple linearly convergent Lagrangian support vector machine (LSVM) algorithm. LSVM requires the inversion at the outset of a single matrix of the order of the much smaller dimensionality of the original input space plus one. The full algorithm is given in this paper in 11 lines of MATLAB code without any special optimization tools such as linear or quadratic programming solvers. This LSVM code can be used "as is" to solve classification problems with millions of points.

In Section 5 an algorithm is proposed which generates a nonlinear kernel-based separating surface that requires as little as 1% of a large dataset for its explicit evaluation. To generate this nonlinear surface, the *entire* dataset is used as a constraint in an optimization problem with very few variables corresponding to the 1% of the data kept. The remainder of the data can be thrown away after solving the optimization problem. This is achieved by making use of a *rectangular* $m \times \bar{m}$ kernel $K(A, \bar{A}')$ that greatly reduces the size of the quadratic program to be solved and simplifies the characterization of the nonlinear separating surface. Here, the m rows of A represent the original m data points while the $\bar{m}$ rows of $\bar{A}$ represent a greatly reduced $\bar{m}$ data points. Computational results indicate that test set correctness for the reduced support vector machine (RSVM), with a nonlinear separating surface that depends on a small randomly selected portion of the dataset, is better than that of a conventional support vector machine (SVM) with a nonlinear surface that explicitly depends on the entire dataset, and much better than a conventional SVM using a small random sample of the data. Computational times, as well as memory usage, are much smaller for RSVM than that of a conventional SVM using the entire dataset.

A word about our notation. All vectors will be column vectors unless transposed to a row vector by a prime superscript $'$. For a vector x in the n-dimensional real space R^n, the plus function x_+ is defined as $(x_+)_i = \max\ \{0, x_i\}$, $i = 1, \ldots, n$, while x_* denotes the step function defined as $(x_*)_i = 1$ if $x_i > 0$ and $(x_*)_i = 0$ if $x_i \leq 0$, $i = 1, \ldots, n$. The scalar (inner) product of two vectors x and y in the n-dimensional real space R^n will be denoted by $x'y$ and the p-norm of x will be denoted by $\|x\|_p$. If $x'y = 0$, we than write $x \perp y$. For a matrix $A \in R^{m \times n}$, A_i is the ith row of A which is a row vector in R^n. A column vector of ones of arbitrary dimension will be denoted by e. For $A \in R^{m \times n}$ and $B \in R^{n \times l}$, the kernel $K(A, B)$ maps $R^{m \times n} \times R^{n \times l}$ into $R^{m \times l}$. In particular, if x and y are column vectors in R^n then, $K(x', y)$ is a real number, $K(x', A')$ is a row vector in R^m and $K(A, A')$ is an $m \times m$ matrix. If f is a real valued function defined on the n-dimensional real space R^n, the gradient of f at x is denoted by $\nabla f(x)$ which is a row vector in R^n and the $n \times n$ Hessian matrix of second partial derivatives of f at x is denoted by $\nabla^2 f(x)$. The base of the natural logarithm will be denoted by ε.

2. The Generalized Support Vector Machine (GSVM) [25]

We consider the problem of classifying m points in the n-dimensional real space R^n, represented by the $m \times n$ matrix A, according to membership of each point A_i in the classes +1 or -1 as specified by a given $m \times m$ diagonal matrix D with ones or minus ones along its diagonal. For this problem the standard support vector machine with a linear kernel AA'

[38, 11] is given by the following for some $\nu > 0$:

$$\begin{array}{rrcl} \min\limits_{(w,\gamma,y)\in R^{n+1+m}} & \nu e'y + \frac{1}{2}w'w & & \\ \text{s.t.} & D(Aw - e\gamma) + y & \geq & e \\ & y & \geq & 0. \end{array} \tag{1}$$

Here w is the normal to the bounding planes:

$$\begin{array}{rcrcr} x'w & - & \gamma & = & +1 \\ x'w & - & \gamma & = & -1, \end{array} \tag{2}$$

and γ determines their location relative to the origin. The first plane above bounds the class +1 points and the second plane bounds the class -1 points when the two classes are strictly linearly separable, that is when the slack variable $y = 0$. The linear separating surface is the plane

$$x'w = \gamma, \tag{3}$$

midway between the bounding planes (2). See Figure 1. If the classes are linearly inseparable then the two planes bound the two classes with a "soft margin" determined by a nonnegative slack variable y, that is:

$$\begin{array}{rcrcrcrl} x'w & - & \gamma & + & y_i & \geq & +1, & \text{for } x' = A_i \text{ and } D_{ii} = +1, \\ x'w & - & \gamma & - & y_i & \leq & -1, & \text{for } x' = A_i \text{ and } D_{ii} = -1. \end{array} \tag{4}$$

The 1-norm of the slack variable y is minimized with weight ν in (1). The quadratic term in (1), which is twice the reciprocal of the square of the 2-norm distance $\frac{2}{\|w\|_2}$ between the two bounding planes of (2) in the n-dimensional space of $w \in R^n$ for a *fixed* γ, maximizes that distance, often called the "margin". Figure 1 depicts the points represented by A, the bounding planes (2) with margin $\frac{2}{\|w\|_2}$, and the separating plane (3) which separates $A+$, the points represented by rows of A with $D_{ii} = +1$, from $A-$, the points represented by rows of A with $D_{ii} = -1$.

In the GSVM formulation we attempt to discriminate between the classes +1 and -1 by a *nonlinear separating surface* which subsumes the linear separating surface (3), and is induced by some kernel $K(A, A')$, as follows:

$$K(x', A')Du = \gamma, \tag{5}$$

where $K(x', A') \in R^m$, *e.g.* $K(x', A') = x'A$ for the linear separating surface (3) and $w = A'Du$. The parameters $u \in R^m$ and $\gamma \in R$ are determined by solving a mathematical program, typically quadratic or linear. In special cases, such as the standard SVM (13) below, u can be

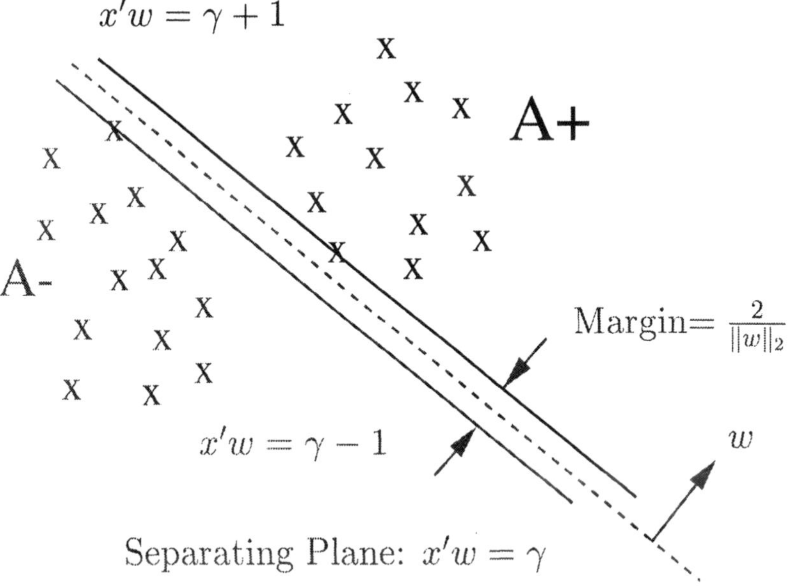

Figure 1. The bounding planes (2) with margin $\frac{2}{\|w\|_2}$, and the plane (3) separating $A+$, the points represented by rows of A with $D_{ii} = +1$, from $A-$, the points represented by rows of A with $D_{ii} = -1$.

interpreted as a dual variable. A point $x \in R^n$ is classified in class +1 or -1 according to whether the *decision function*

$$(K(x', A')Du - \gamma)_*, \tag{6}$$

yields 1 or 0 respectively. Here $(\cdot)_*$ denotes the step function defined in the Introduction. The kernel function $K(x', A')$ implicitly defines a nonlinear map from $x \in R^n$ to some other space $z \in R^k$ where k may be much larger than n. In particular if the kernel K is an inner product kernel under Mercer's condition [13, pp 138-140],[38, 11, 5] (an assumption that we will not make in this paper) then for x and y in R^n:

$$K(x, y) = h(x)'h(y), \tag{7}$$

and the separating surface (5) becomes:

$$h(x)'h(A')Du = \gamma, \tag{8}$$

where h is a function, not easily computable, from R^n to R^k, and $h(A') \in R^{k \times m}$ results from applying h to the m columns of A'. The difficulty in computing h and the possible high dimensionality of R^k have been important factors in using a kernel K as a generator of an implicit nonlinear separating surface in the original feature space R^n but which is linear in the high dimensional space R^k. Our separating surface (5) written in terms of a kernel function retains this advantage and is linear in its parameters, u, γ. We now state a mathematical program that generates such a surface for a general kernel K as follows:

$$\begin{array}{lrcl} \displaystyle\min_{u,\gamma,y} & \nu e'y + f(u) & & \\ \text{s.t.} & D(K(A,A')Du - e\gamma) + y & \geq & e \\ & y & \geq & 0. \end{array} \tag{9}$$

Here f is some convex function on R^m, typically some norm or semi-norm, and ν is some positive parameter that weights the separation error $e'y$ versus suppression of the separating surface parameter u. Suppression of u can be interpreted in one of two ways. We interpret it here as minimizing the number of support vectors, i.e. constraints of (9) with positive multipliers. A more conventional interpretation is that of maximizing some measure of the distance or margin between the bounding parallel planes in R^k, under appropriate assumptions, such as f being a quadratic function induced by a positive definite kernel K as in (13) below. As is well known, this leads to improved generalization by minimizing an upper bound on the VC dimension [38, 35].

We term a solution of the mathematical program (9) and the resulting decision function (6) a *generalized support vector machine*, GSVM. In what follows derive a number of special cases, including the standard support vector machine.

We consider first support vector machines that include the standard ones [38, 11, 5] and which are obtained by setting f of (9) to be a convex quadratic function $f(u) = \frac{1}{2}u'Hu$, where $H \in R^{m\times m}$ is some symmetric positive definite matrix. The mathematical program (9) becomes the following convex quadratic program:

$$\begin{array}{lrcl} \displaystyle\min_{u,\gamma,y} & \nu e'y + \frac{1}{2}u'Hu & & \\ \text{s.t.} & D(K(A,A')Du - e\gamma) + y & \geq & e \\ & y & \geq & 0. \end{array} \tag{10}$$

The Wolfe dual [39, 22] of this convex quadratic program is:

$$\begin{array}{lrcl} \displaystyle\min_{r\in R^m} & \frac{1}{2}r'DK(A,A')DH^{-1}DK(A,A')'Dr - e'r & & \\ \text{s.t.} & e'Dr & = & 0 \\ & 0 \leq r & \leq & \nu e. \end{array} \tag{11}$$

Furthermore, the primal variable u is related to the dual variable r by:

$$u = H^{-1}DK(A,A')'Dr. \tag{12}$$

If we assume that the kernel $K(A,A')$ is symmetric positive definite and let $H = DK(A,A')D$, then our dual problem (11) degenerates to the dual problem of the standard support vector machine [38, 11, 5] with $u = r$:

$$\begin{array}{lrcl} \min\limits_{u \in R^m} & \frac{1}{2}u'DK(A,A')Du - e'u & & \\ \text{s.t.} & e'Du & = & 0 \\ & 0 \leq u & \leq & \nu e. \end{array} \tag{13}$$

The positive definiteness assumption on $K(A,A')$ in (13) can be relaxed to positive *semi*definiteness while maintaining the convex quadratic program (10), with $H = DK(A,A')D$, as the direct dual of (13) without utilizing (11) and (12). The symmetry and positive semidefiniteness of the kernel $K(A,A')$ for this version of a support vector machine is consistent with the support vector machine literature. The fact that $r = u$ in the dual formulation (13), shows that the variable u appearing in the original formulation (10) is also the dual multiplier vector for the first set of constraints of (10). Hence the quadratic term in the objective function of (10) can be thought of as suppressing as many multipliers of support vectors as possible and thus minimizing the number of such support vectors. This is another (nonstandard) interpretation of the standard support vector machine that is usually interpreted as maximizing the margin or distance between parallel separating planes.

This leads to the idea of using other values for the matrix H other than $DK(A,A')D$ that will also suppress u. One particular choice is interesting because it puts no restrictions on K: no symmetry, no positive definiteness or semidefiniteness and not even continuity. This is the choice $H = I$ in (10) which leads to a dual problem (11) with $H = I$ and $u = DK(A,A')'Dr$ as follows:

$$\begin{array}{lrcl} \min\limits_{r \in R^m} & \frac{1}{2}r'DK(A,A')K(A,A')'Dr - e'r & & \\ \text{s.t.} & e'Dr & = & 0 \\ & 0 \leq r & \leq & \nu e. \end{array} \tag{14}$$

We note immediately that $K(A,A')K(A,A')'$ is positive semidefinite with no assumptions on $K(A,A')$, and hence the above problem is an always solvable convex quadratic program for any kernel $K(A,A')$. In fact by the Frank-Wolfe existence theorem [15], the quadratic program (10) is solvable for *any* symmetric positive definite matrix H because its objective function is bounded below by zero. Hence by quadratic programming duality its dual problem (11) is also solvable. Any solution of (10) can be used to generate a nonlinear decision function (6). Thus we are free to choose any symmetric positive definite matrix H to generate a support vector machine. Experimentation will be needed to determine what are the most appropriate choices for H.

By using the 1-norm instead of the 2-norm a linear programming formulation for the GSVM can be obtained. We refer the interested reader to [25].

We turn our attention now to an efficient method for generating SVMs based on smoothing ideas that have already been effectively used to solve various mathematical programs [7, 8, 6, 9, 10, 16, 37, 12].

3. SSVM: Smooth Support Vector Machines [21]

In our smooth approach, the square of 2-norm of the slack variable y is minimized with weight $\frac{\nu}{2}$ instead of the 1-norm of y as in (1). In addition the distance between the planes (2) is measured in the $(n+1)$-dimensional space of $(w, \gamma) \in R^{n+1}$, that is $\frac{2}{\|(w,\gamma)\|_2}$. Measuring the margin in this $(n+1)$-dimensional space instead of R^n induces strong convexity and has little or no effect on the problem as was shown in [26, 27, 21, 20]. Thus using twice the reciprocal squared of this margin instead, yields our modified SVM problem as follows:

$$\begin{array}{ll} \min\limits_{w,\gamma,y} & \frac{\nu}{2}y'y + \frac{1}{2}(w'w + \gamma^2) \\ \text{s.t.} & D(Aw - e\gamma) + y \geq e \\ & y \geq 0. \end{array} \tag{15}$$

At the solution of problem (15), y is given by

$$y = (e - D(Aw - e\gamma))_+, \tag{16}$$

where, as defined in the Introduction, $(\cdot)_+$ replaces negative components of a vector by zeros. Thus, we can replace y in (15) by $(e - D(Aw - e\gamma))_+$ and convert the SVM problem (15) into an equivalent SVM which is an unconstrained optimization problem as follows:

$$\min_{w,\gamma} \frac{\nu}{2}\|(e - D(Aw - e\gamma))_+\|_2^2 + \frac{1}{2}(w'w + \gamma^2). \tag{17}$$

This problem is a strongly convex minimization problem without any constraints. It is easy to show that it has a unique solution. However, the objective function in (17) is not twice differentiable which precludes the use of a fast Newton method. We thus apply the smoothing techniques of [7, 8] and replace x_+ by a very accurate smooth approximation [21, Lemma 2.1] that is given by $p(x, \alpha)$, the integral of the sigmoid function $\frac{1}{1+\varepsilon^{-\alpha x}}$ of neural networks [23], that is

$$p(x, \alpha) = x + \frac{1}{\alpha}\log(1 + \varepsilon^{-\alpha x}), \ \alpha > 0. \tag{18}$$

This p function with a smoothing parameter α is used here to replace the plus function of (17) to obtain a smooth support vector machine **(SSVM)**:

$$\begin{aligned} \min_{(w,\gamma)\in R^{n+1}} \Phi_\alpha(w,\gamma) &:= \\ = \min_{(w,\gamma)\in R^{n+1}} &\frac{\nu}{2}\|p(e - D(Aw - e\gamma), \alpha)\|_2^2 + \frac{1}{2}(w'w + \gamma^2). \end{aligned} \tag{19}$$

It can be shown [21, Theorem 2.2] that the solution of problem (15) is obtained by solving problem (19) with α approaching infinity. Advantage can be taken of the twice differentiable property of the objective function of (19) to utilize a quadratically convergent algorithm for solving the smooth support vector machine (19) as follows.

Algorithm 3.1 Newton-Armijo Algorithm for SSVM (19)
Start with any $(w^0, \gamma^0) \in R^{n+1}$. Having (w^i, γ^i), stop if the gradient of the objective function of (19) is zero, that is $\nabla\Phi_\alpha(w^i, \gamma^i) = 0$. Else compute (w^{i+1}, γ^{i+1}) as follows:

(i) **Newton Direction**: Determine direction $d^i \in R^{n+1}$ by setting equal to zero the linearization of $\nabla\Phi_\alpha(w, \gamma)$ around (w^i, γ^i) which gives $n+1$ linear equations in $n+1$ variables:

$$\nabla^2\Phi_\alpha(w^i, \gamma^i)d^i = -\nabla\Phi_\alpha(w^i, \gamma^i)'. \tag{20}$$

(ii) **Armijo Stepsize** [1]: Choose a stepsize $\lambda_i \in R$ such that:

$$(w^{i+1}, \gamma^{i+1}) = (w^i, \gamma^i) + \lambda_i d^i \tag{21}$$

where $\lambda_i = \max\{1, \frac{1}{2}, \frac{1}{4}, \ldots\}$ such that :

$$\Phi_\alpha(w^i, \gamma^i) - \Phi_\alpha((w^i, \gamma^i) + \lambda_i d^i) \geq -\delta\lambda_i \nabla\Phi_\alpha(w^i, \gamma^i)d^i \tag{22}$$

where $\delta \in (0, \frac{1}{2})$.

Note that a key difference between our smoothing approach and that of the classical SVM [38, 11] is that we are solving here a linear system of equations (20) instead of solving a quadratic program as is the case with the classical SVM. Furthermore, it can be shown [21, Theorem 3.2] that the smoothing algorithm above converges quadratically from any starting point.

To obtain a nonlinear SSVM we consider the GSVM formulation (9) with a 2-norm squared error term on y instead of the 1-norm, and instead of the convex term $f(u)$ that suppresses u we use a 2-norm squared of $\begin{bmatrix} u \\ \gamma \end{bmatrix}$ to suppress both u and γ. We obtain then:

$$\begin{array}{lrcl} \min\limits_{u,\gamma,y} & \frac{\nu}{2}y'y + \frac{1}{2}(u'u+\gamma^2) & & \\ \text{s.t.} & D(K(A,A')Du - e\gamma) + y & \geq & e \\ & y & \geq & 0. \end{array} \tag{23}$$

We repeat the same arguments as above, in going from (15) to (19), to obtain the SSVM with a nonlinear kernel $K(A, A')$:

$$\min_{u,\gamma} \ \frac{\nu}{2}\|p(e - D(K(A,A')Du - e\gamma), \alpha)\|_2^2 + \frac{1}{2}(u'u+\gamma^2), \tag{24}$$

where $K(A, A')$ is a kernel map from $R^{m\times n} \times R^{n\times m}$ to $R^{m\times m}$. We note that this problem, which is capable of generating highly nonlinear separating surfaces, still retains the strong convexity and differentiability properties for any arbitrary kernel. All of the convergence results for a linear kernel hold here for a nonlinear kernel [21].

The effectiveness and speed of the smooth support vector machine (SSVM) approach can be demonstrated by comparing it numerically with other methods. In order to evaluate how well each algorithm generalizes to future data, tenfold cross-validation is performed on each dataset [36]. To evaluate the efficacy of SSVM, computational times of SSVM were compared with robust linear program (RLP) algorithm [2], the feature selection concave minimization (FSV) algorithm, the support vector machine using the 1-norm approach ($\text{SVM}_{\|\cdot\|_1}$) and the classical support vector machine ($\text{SVM}_{\|\cdot\|_2^2}$) [3, 38, 11]. All tests were run on six publicly available datasets: the Wisconsin Prognostic Breast Cancer Database [31] and four datasets from the Irvine Machine Learning Database Repository [34]. It turned out that tenfold testing correctness of the SSVM was the highest for these five methods on all datasets tested as well as the computational speed. Detailed numerical results are given in [21].

As a test of effectiveness of the SSVM in generating a highly nonlinear separating surface, we tested it on the 1000-point checkerboard dataset of [19] depicted in Figure 2. We used the following a Gaussian kernel in the SSVM formulation (24):

Gaussian Kernel : $\quad \varepsilon^{-\mu\|A_i - A_j\|_2^2}, \ i,j = 1,2,3\ldots m.$

The value of the parameter μ used as well as values of the parameters ν and α of the nonlinear SSVM (24) are all given in Figure 3 which depicts the separation obtained. Note that the boundaries of the checkerboard are as sharp as those of [26], obtained by a linear programming solution, and considerably sharper than those of [19], obtained by a Newton approach applied to a quadratic programming formulation.

We turn now to an extremely simple iterative algorithm for SVMs that requires neither a quadratic program nor a linear program to be solved.

4. LSVM: Lagrangian Support Vector Machines [28]

We propose here an algorithm based on an implicit Lagrangian of the dual of a simple reformulation of the standard quadratic program of a linear support vector machine. This leads to the minimization of an unconstrained differentiable convex function in a space of dimensionality equal to the number of classified points. This problem is solvable by an extremely simple linearly convergent Lagrangian support vector machine (LSVM) algorithm. LSVM requires the inversion at the outset of a single matrix of the order of the much smaller dimensionality of the original input space plus one. The full algorithm is given in this paper in 11 lines of MATLAB code without any special optimization tools such as linear or quadratic programming solvers. This LSVM code can be used "as is" to solve classification problems with millions of points. For example, 2 million points in 10 dimensional input space were classified by a linear surface in 6.7 minutes on a 250-MHz UltraSPARC II [28].

The starting point for LSVM is the primal quadratic formulation (15) of the SVM problem. Taking the dual [24] of this problem gives:

$$\min_{0 \leq u \in R^m} \frac{1}{2} u'(\frac{I}{\nu} + D(AA' + ee')D)u - e'u. \tag{25}$$

The variables (w, γ) of the primal problem which determine the separating surface $x'w = \gamma$ are recovered directly from the solution of the dual (25) above by the relations:

$$w = A'Du, \;\; y = \frac{u}{\nu}, \;\; \gamma = -e'Du. \tag{26}$$

We immediately note that the matrix appearing in the dual objective function is positive definite and that there is no equality constraint and no upper bound on the dual variable u. The only constraint present is a nonnegativity one. These facts lead us to our simple iterative Lagrangian SVM Algorithm which requires the inversion of a positive definite $(n+1) \times (n+1)$ matrix, at the beginning of the algorithm followed by a straightforward linearly convergent iterative scheme that requires no optimization package.

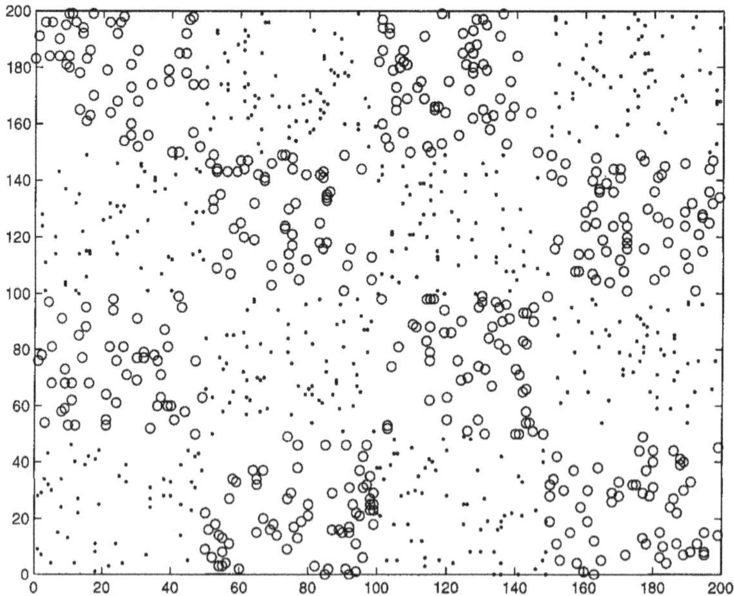

Figure 2. Checkerboard training dataset

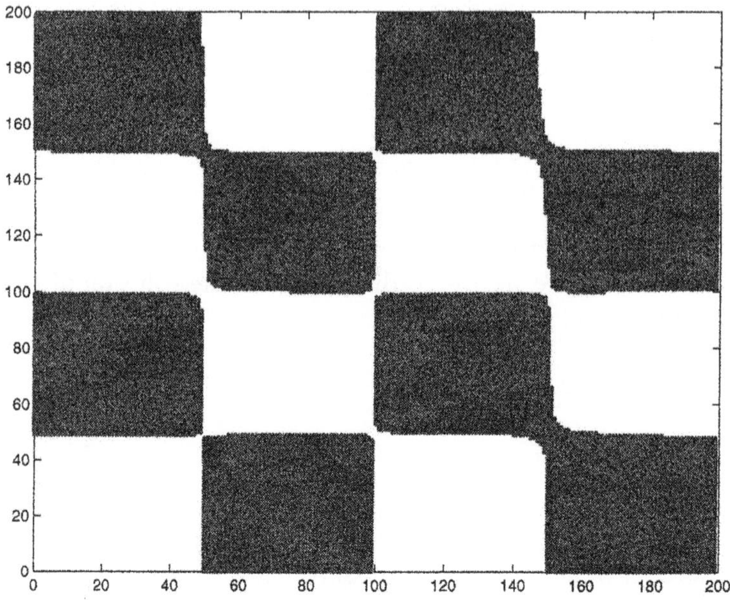

Figure 3. Gaussian kernel separation of checkerboard dataset ($\nu = 10$, $\alpha = 5$, $\mu = 2$)

Before stating our algorithm we define two matrices to simplify notation as follows:

$$H = D[A \quad -e], \qquad Q = \frac{I}{\nu} + HH'. \tag{27}$$

With these definitions the dual problem (25) becomes

$$\min_{0 \leq u \in R^m} f(u) := \frac{1}{2} u'Qu - e'u. \tag{28}$$

It will be understood that within the LSVM Algorithm, the single time that Q^{-1} is computed at the outset of the algorithm, the SMW identity [17] will be used:

$$(\frac{I}{\nu} + HH')^{-1} = \nu(I - H(\frac{I}{\nu} + H'H)^{-1}H'), \tag{29}$$

where ν is a positive number and H is an arbitrary $m \times k$ matrix. Hence only an $(n+1) \times (n+1)$ matrix is inverted.

The LSVM Algorithm is based directly on the Karush-Kuhn-Tucker necessary and sufficient optimality conditions [24, KTP 7.2.4, page 94] for the dual problem (28) which are the following:

$$0 \leq u \perp Qu - e \geq 0. \tag{30}$$

By using the easily established identity between any two real numbers (or vectors) a and b:

$$0 \leq a \perp b \geq 0 \iff a = (a - \alpha b)_+, \ \alpha > 0, \tag{31}$$

the optimality condition (30) can be written in the following equivalent form for any positive α:

$$Qu - e = ((Qu - e) - \alpha u)_+. \tag{32}$$

These optimality conditions lead to the following very simple iterative scheme which constitutes our LSVM Algorithm:

$$u^{i+1} = Q^{-1}(e + ((Qu^i - e) - \alpha u^i)_+), \ i = 0, 1, \ldots, \tag{33}$$

for which we will establish global linear convergence from any starting point under the easily satisfiable condition:

$$0 < \alpha < \frac{2}{\nu}. \tag{34}$$

We implement this condition as $\alpha = 1.9/\nu$ in all our experiments, where ν is the parameter of our SVM formulation (25). It turns out, and this is the way that led us to this iterative scheme, that the optimality condition (32), is also the necessary and sufficient condition for the unconstrained minimum of the implicit Lagrangian [30] associated with the dual problem (28):

$$\min_{u \in R^m} L(u, \alpha) =$$
$$= \min_{u \in R^m} \frac{1}{2}u'Qu - e'u + \frac{1}{2\alpha}(\|(-\alpha u + Qu - e)_+\|^2 - \|Qu - e\|^2). \quad (35)$$

Setting the gradient with respect to u of this convex and differentiable Lagrangian to zero gives

$$(Qu - e) + \frac{1}{\alpha}(Q - \alpha I)((Q - \alpha I)u - e)_+ - \frac{1}{\alpha}Q(Qu - e) = 0, \quad (36)$$

or equivalently:

$$(\alpha I - Q)((Qu - e) - ((Q - \alpha I)u - e)_+) = 0, \quad (37)$$

which is equivalent to the optimality condition (32) under the assumption that α is positive and not an eigenvalue of Q.

In [28] global linear convergence of the iteration (33) under condition (34) is established as follows.

Algorithm 4.1 LSVM Algorithm & Its Global Convergence *[28] Let $Q \in R^{m \times m}$ be the symmetric positive definite matrix defined by (27) and let (34) hold. Starting with an arbitrary $u^0 \in R^m$, the iterates u^i of (33) converge to the unique solution $\bar{u}$ of (28) at the linear rate:*

$$\|Qu^{i+1} - Q\bar{u}\| \leq \|I - \alpha Q^{-1}\| \cdot \|Qu^i - Q\bar{u}\|. \quad (38)$$

A complete MATLAB [32] code of LSVM which is capable of solving problems with millions of points using only native MATLAB commands is given below in Code 4.2. The input parameters, besides A, D and ν of (27), which define the problem, are: itmax, the maximum number of iterations and tol, the tolerated nonzero error in $\|u^{i+1} - u^i\|$ at termination which can be shown [28] to constitute a bound on the distance to the unique solution of the problem from the current iterate.

Code 4.2 LSVM MATLAB Code

```
function [it, opt, w, gamma] = svml(A,D,nu,itmax,tol)
% lsvm with SMW for min 1/2*u'*Q*u-e'*u s.t. u=>0,
% Q=I/nu+H*H', H=D[A -e]
% Input: A, D, nu, itmax, tol; Output: it, opt, w, gamma
% [it, opt, w, gamma] = svml(A,D,nu,itmax,tol);
  [m,n]=size(A);alpha=1.9/nu;e=ones(m,1);H=D*[A -e];it=0;
  S=H*inv((speye(n+1)/nu+H'*H));
  u=nu*(1-S*(H'*e));oldu=u+1;
  while it<itmax & norm(oldu-u)>tol
    z=(1+pl(((u/nu+H*(H'*u))-alpha*u)-1));
    oldu=u;
    u=nu*(z-S*(H'*z));
    it=it+1;
  end;
  opt=norm(u-oldu);w=A'*D*u;gamma=-e'*D*u;

function pl = pl(x); pl = (abs(x)+x)/2;
```

Using this MATLAB code, 2 million random points in 10-dimensional space were classified in 6.7 minutes in 6 iterations to $e-5$ accuracy using a 250-MHz UltraSPARC II with 2 gigabyte memory. In contrast a linear programming formulation using CPLEX [14] ran out of memory. Other favorable numerical comparisons with other methods are contained in [28].

We turn now to our final topic of extracting very effective classifiers from a minimal portion of a large dataset.

5. RSVM: Reduced Support Vector Machines [20]

In this section we describe an algorithm that generates a nonlinear kernel-based separating surface which requires as little as 1% of a large dataset for its explicit evaluation. To generate this nonlinear surface, the *entire* dataset is used as a constraint in an optimization problem with very few variables corresponding to the 1% of the data kept. The remainder of the data can be thrown away after solving the optimization problem. This is achieved by making use of a *rectangular* $m \times \bar{m}$ kernel $K(A, \bar{A}')$ that greatly reduces the size of the quadratic program to be solved and simplifies the characterization of the nonlinear separating surface. Here as before, the m rows of A represent the original m data points while the $\bar{m}$ rows of $\bar{A}$ represent a greatly reduced $\bar{m}$ data points. Computational results indicate that test set correctness for the reduced support vector machine (RSVM), with a nonlinear separating surface that depends on a small randomly selected portion of the dataset, is better than that of a conventional support vector machine (SVM) with a nonlinear surface that explicitly depends on the entire dataset, and much better than a conventional SVM using a small random sample of the data. Computational times, as well as memory usage, are much smaller for RSVM than that of a conventional SVM using the entire dataset.

The motivation for RSVM comes from the practical objective of generating a nonlinear separating surface (5) for a large dataset which uses only a small portion of the dataset for its characterization. The difficulty in using nonlinear kernels on large datasets is twofold. First, there is the computational difficulty in solving the the potentially huge unconstrained optimization problem (24) which involves the kernel function $K(A, A')$ that typically leads to the computer running out of memory even before beginning the solution process. For example, for the Adult dataset with 32562 points, which is actually solved with RSVM [20], this would mean a matrix with over one billion entries for a conventional SVM. The second difficulty comes from utilizing the formula (5) for the separating surface on a new unseen point x. The formula dictates that we store and utilize the entire data set represented by the 32562×123 matrix A which may be prohibitively expensive storage-wise and computing-time-wise. For example for the Adult dataset just mentioned which has an input space of 123 dimensions, this would mean that the nonlinear surface (5) requires a storage capacity for 4,005,126 numbers. To avoid all these difficulties and based on experience with chunking methods [4, 29], we hit upon the idea of using a very small random subset of the dataset given by $\bar{m}$ points of the original m data points with $\bar{m} << m$, that we call $\bar{A}$ and use $\bar{A}'$ in place of A' in *both* the unconstrained optimization problem (24), to cut problem size and computation time, and for the same purposes in evaluating the nonlinear surface (5). Note that the matrix A is left intact in $K(A, \bar{A}')$, whereas $\bar{A}'$ has replaced A'. Computational testing results show a standard deviation of 0.002 or less of test set correctness over 50 random choices for $\bar{A}$. By contrast if *both* A and A' are replaced by $\bar{A}$ and $\bar{A}'$ respectively, then test set correctness declines substantially compared to RSVM, while the standard deviation, of test set correctness over 50 cases, increases more than tenfold over that of RSVM.

The justification for our proposed approach is this. We use a small random $\bar{A}$ sample of our dataset as a representative sample with respect to the *entire* dataset A both in solving the optimization problem (24) and in evaluating the the nonlinear separating surface (5). We interpret this as a possible instance-based learning [33, Chapter 8] where the small sample $\bar{A}$ is learning from the much larger training set A by forming the appropriate rectangular kernel relationship $K(A, \bar{A}')$ between the original and reduced sets. This formulation works extremely well computationally as evidenced by the computational results of [20].

By using the formulations described in Section 3 for the full dataset $A \in R^{m\times n}$ with a square kernel $K(A, A') \in R^{m\times m}$, and modifying these formulations for the reduced dataset $\bar{A} \in R^{\bar{m}\times n}$ with corresponding diagonal matrix $\bar{D}$ and rectangular kernel $K(A, \bar{A}') \in R^{m\times \bar{m}}$, we obtain our RSVM Algorithm below. This algorithm solves, by smoothing, the RSVM quadratic program obtained from (23) by replacing A' with $\bar{A}'$ as follows:

$$\begin{array}{lrcl} \displaystyle\min_{(\bar{u},\gamma,y)\in R^{\bar{m}+1+m}} & \frac{\nu}{2}y'y + \frac{1}{2}(\bar{u}'\bar{u} + \gamma^2) & & \\ \text{s.t.} & D(K(A, \bar{A}')\bar{D}\bar{u} - e\gamma) + y & \geq & e \\ & y & \geq & 0. \end{array} \tag{39}$$

Algorithm 5.1 *RSVM Algorithm*

(i) Choose a random subset matrix $\bar{A} \in R^{\bar{m}\times n}$ of the original data matrix $A \in R^{m\times n}$. Typically $\bar{m}$ is 1% to 10% of m.

(ii) Solve the following modified version of the SSVM (24) where A' **only** *is replaced by $\bar{A}'$ with corresponding $\bar{D} \subset D$:*

$$\min_{(\bar{u},\gamma)\in R^{\bar{m}+1}} \frac{\nu}{2}\|p(e - D(K(A, \bar{A}')\bar{D}\bar{u} - e\gamma), \alpha)\|_2^2 + \frac{1}{2}(\bar{u}'\bar{u} + \gamma^2), \tag{40}$$

which is equivalent to solving (23) with A' **only** *replaced by $\bar{A}'$.*

(iii) The separating surface is given by (5) with A' replaced by $\bar{A}'$ as follows:

$$K(x', \bar{A}')\bar{D}\bar{u} = \gamma, \tag{41}$$

where $(\bar{u}, \gamma) \in R^{\bar{m}+1}$ is the unique solution of (40), and $x \in R^n$ is a free input space variable of a new point.

(iv) A new input point $x \in R^n$ is classified into class $+1$ or -1 depending on whether the step function:

$$(K(x', \bar{A}')\bar{D}\bar{u} - \gamma)_*, \tag{42}$$

is $+1$ or zero, respectively.

As stated earlier, this algorithm is quite insensitive as to which submatrix $\bar{A}$ is chosen for (40)-(41), as far as tenfold cross-validation correctness is concerned. In fact, another choice for $\bar{A}$ is to choose it randomly but only keep rows that are more than a certain minimal distance apart. This leads to a slight improvement in testing correctness but increases computational time somewhat. Replacing *both* A and A' in a conventional SVM by a randomly chosen reduced matrix $\bar{A}$ and its transpose $\bar{A}'$ gives poor testing set results that vary significantly with the choice of $\bar{A}$. This fact can be demonstrated graphically as follows.

The checkerboard dataset [18, 19] already used earlier, consists of 1000 points in R^2 of black and white points taken from sixteen black and white squares of a checkerboard. This dataset is chosen in order to depict graphically the effectiveness of RSVM using a random 5% of the given 1000-point training dataset compared to the very poor performance of a conventional SVM on the same 5% randomly chosen subset. Figure 4 shows the poor pattern approximating a checkerboard obtained by a conventional SVM using a Gaussian kernel, that is solving (23) with *both* A and A' replaced by the randomly chosen $\bar{A}$ and its transpose $\bar{A}'$ respectively. Test set correctness of this conventional SVM using the reduced $\bar{A}$ and $\bar{A}'$ averaged, over 15 cases, 43.60% for the 50-point dataset, on a test set of 39601 points. In contrast, using our RSVM Algorithm 4.1 on the *same* randomly chosen submatrices $\bar{A}'$, yields the much more accurate representations of the checkerboard depicted in Figures 5 with corresponding average test set correctness of 96.70% on the same test set.

6. Conclusion and Extensions

We have described the important role of support vector machines in solving the key problem of classification that arises in data mining and machine learning. In particular we have described a general framework for support vector machines and given three highly effective algorithms for generating linear and nonlinear classifiers. In all our results mathematical programming plays key theoretical and algorithmic roles. Some extensions of the these ideas include multicategory classification, classification based on criteria other than belonging to a halfspace, incremental classification of massive streaming datasets, concurrent feature and data selection for optimal classification, classification based on minimal data subsets and multiple instance classification.

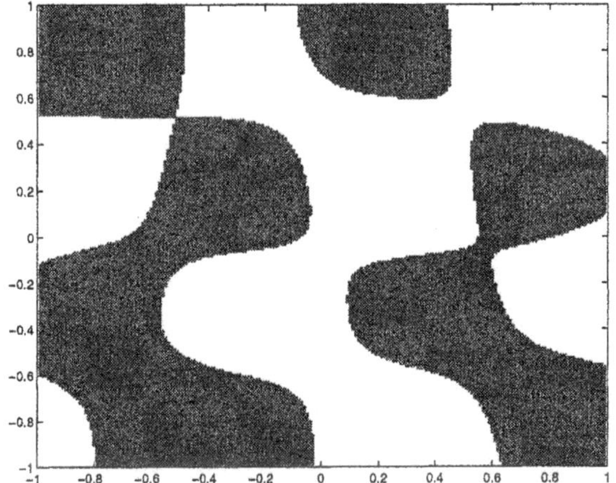

Figure 4. SVM: Checkerboard resulting from a randomly selected 50 points, out of a 1000-point dataset, and used in a conventional Gaussian kernel SVM (23). The resulting nonlinear surface, separating white and black areas, generated using the 50 random points only, depends explicitly on those points only. Correctness on a 39601-point test set averaged 43.60% on 15 randomly chosen 50-point sets, with a standard deviation of 0.0895 and best correctness of 61.03% depicted above.

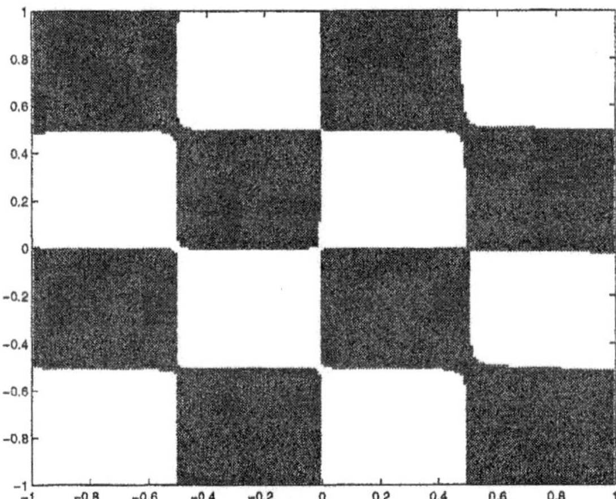

Figure 5. RSVM: Checkerboard resulting from randomly selected 50 points and used in a **reduced** Gaussian kernel SVM (39). The resulting nonlinear surface, separating white and black areas, generated using the entire 1000-point dataset, depends explicitly on the 50 points **only**. The remaining 950 points can be thrown away once the separating surface has been generated. Correctness on a 39601-point test set averaged 96.7% on 15 randomly chosen 50-point sets, with a standard deviation of 0.0082 and best correctness of 98.04% depicted above.

Acknowledgements

The research described in this Data Mining Institute Report 01-05, May 2001, was supported by National Science Foundation Grants CCR-9729842 and CDA-9623632, by Air Force Office of Scientific Research Grant F49620-00-1-0085 and by the Microsoft Corporation.

References

[1] L. Armijo. Minimization of functions having Lipschitz-continuous first partial derivatives. *Pacific Journal of Mathematics*, 16:1–3, 1966.

[2] K. P. Bennett and O. L. Mangasarian. Robust linear programming discrimination of two linearly inseparable sets. *Optimization Methods and Software*, 1:23–34, 1992.

[3] P. S. Bradley and O. L. Mangasarian. Feature selection via concave minimization and support vector machines. In J. Shavlik, editor, *Machine Learning Proceedings of the Fifteenth International Conference(ICML '98)*, pages 82–90, San Francisco, California, 1998. Morgan Kaufmann. ftp://ftp.cs.wisc.edu/math-prog/tech-reports/98-03.ps.

[4] P. S. Bradley and O. L. Mangasarian. Massive data discrimination via linear support vector machines. *Optimization Methods and Software*, 13:1–10, 2000. ftp://ftp.cs.wisc.edu/math-prog/tech-reports/98-03.ps.

[5] C. J. C. Burges. A tutorial on support vector machines for pattern recognition. *Data Mining and Knowledge Discovery*, 2(2):121–167, 1998.

[6] B. Chen and P. T. Harker. Smooth approximations to nonlinear complementarity problems. *SIAM Journal of Optimization*, 7:403–420, 1997.

[7] Chunhui Chen and O. L. Mangasarian. Smoothing methods for convex inequalities and linear complementarity problems. *Mathematical Programming*, 71(1):51–69, 1995.

[8] Chunhui Chen and O. L. Mangasarian. A class of smoothing functions for nonlinear and mixed complementarity problems. *Computational Optimization and Applications*, 5(2):97–138, 1996.

[9] X. Chen, L. Qi, and D. Sun. Global and superlinear convergence of the smoothing Newton method and its application to general box constrained variational inequalities. *Mathematics of Computation*, 67:519–540, 1998.

[10] X. Chen and Y. Ye. On homotopy-smoothing methods for variational inequalities. *SIAM Journal on Control and Optimization*, 37:589–616, 1999.

[11] V. Cherkassky and F. Mulier. *Learning from Data - Concepts, Theory and Methods.* John Wiley & Sons, New York, 1998.

[12] P. W. Christensen and J.-S. Pang. Frictional contact algorithms based on semismooth Newton methods. In *Reformulation: Nonsmooth, Piecewise Smooth, Semismooth and Smoothing Methods, M. Fukushima and L. Qi, (editors)*, pages 81–116, Dordrecht, Netherlands, 1999. Kluwer Academic Publishers.

[13] R. Courant and D. Hilbert. *Methods of Mathematical Physics.* Interscience Publishers, New York, 1953.

[14] CPLEX Optimization Inc., Incline Village, Nevada. *Using the CPLEX(TM) Linear Optimizer and CPLEX(TM) Mixed Integer Optimizer (Version 2.0)*, 1992.

[15] M. Frank and P. Wolfe. An algorithm for quadratic programming. *Naval Research Logistics Quarterly*, 3:95–110, 1956.

[16] M. Fukushima and L. Qi. *Reformulation: Nonsmooth, Piecewise Smooth, Semismooth and Smoothing Methods.* Kluwer Academic Publishers, Dordrecht, The Netherlands, 1999.

[17] G. H. Golub and C. F. Van Loan. *Matrix Computations.* The John Hopkins University Press, Baltimore, Maryland, 3rd edition, 1996.

[18] T. K. Ho and E. M. Kleinberg. Building projectable classifiers of arbitrary complexity. In *Proceedings of the 13th International Conference on Pattern Recognition*, pages 880–885, Vienna, Austria, 1996. http://cm.bell-labs.com/who/tkh/pubs.html. Checker dataset at: ftp://ftp.cs.wisc.edu/math-prog/cpo-dataset/machine-learn/checker.

[19] L. Kaufman. Solving the quadratic programming problem arising in support vector classification. In B. Schölkopf, C. J. C. Burges, and A. J. Smola, editors, *Advances in Kernel Methods - Support Vector Learning*, pages 147–167. MIT Press, 1999.

[20] Y.-J. Lee and O. L. Mangasarian. RSVM: Reduced support vector machines. Technical Report 00-07, Data Mining Institute, Computer Sciences Department, University of Wisconsin, Madison, Wisconsin, July 2000. Proceedings of the First SIAM International Conference on Data Mining, Chicago, April 5-7, 2001, CD-ROM Proceedings. ftp://ftp.cs.wisc.edu/pub/dmi/tech-reports/00-07.ps.

[21] Yuh-Jye Lee and O. L. Mangasarian. SSVM: A smooth support vector machine. *Computational Optimization and Applications*, 20:5–22, 2001. Data Mining Institute, University of Wisconsin, Technical Report 99-03. ftp://ftp.cs.wisc.edu/pub/dmi/tech-reports/99-03.ps.

[22] O. L. Mangasarian. *Nonlinear Programming.* McGraw–Hill, New York, 1969. Reprint: SIAM Classic in Applied Mathematics 10, 1994, Philadelphia.

[23] O. L. Mangasarian. Mathematical programming in neural networks. *ORSA Journal on Computing*, 5(4):349–360, 1993.

[24] O. L. Mangasarian. *Nonlinear Programming.* SIAM, Philadelphia, PA, 1994.

[25] O. L. Mangasarian. Generalized support vector machines. In A. Smola, P. Bartlett, B. Schölkopf, and D. Schuurmans, editors, *Advances in Large Margin Classifiers*, pages 135–146, Cambridge, MA, 2000. MIT Press. ftp://ftp.cs.wisc.edu/math-prog/tech-reports/98-14.ps.

[26] O. L. Mangasarian and D. R. Musicant. Successive overrelaxation for support vector machines. *IEEE Transactions on Neural Networks*, 10:1032–1037, 1999. ftp://ftp.cs.wisc.edu/math-prog/tech-reports/98-18.ps.

[27] O. L. Mangasarian and D. R. Musicant. Data discrimination via nonlinear generalized support vector machines. In M. C. Ferris, O. L. Mangasarian, and J.-S. Pang, editors, *Complementarity: Applications, Algorithms and Extensions*, pages 233–251, Dordrecht, January 2001. Kluwer Academic Publishers. ftp://ftp.cs.wisc.edu/math-prog/tech-reports/99-03.ps.

[28] O. L. Mangasarian and D. R. Musicant. Lagrangian support vector machines. *Journal of Machine Learning Research*, 1:161–177, 2001. ftp://ftp.cs.wisc.edu/pub/dmi/tech-reports/00-06.ps.

[29] O. L. Mangasarian and D. R. Musicant. Large scale kernel regression via linear programming. *Machine Learning*, 46:255–269, 2002. ftp://ftp.cs.wisc.edu/pub/dmi/tech-reports/99-02.ps.

[30] O. L. Mangasarian and M. V. Solodov. Nonlinear complementarity as unconstrained and constrained minimization. *Mathematical Programming, Series B*, 62:277–297, 1993.

[31] O. L. Mangasarian, W. N. Street, and W. H. Wolberg. Breast cancer diagnosis and prognosis via linear programming. *Operations Research*, 43(4):570–577, July-August 1995.

[32] MATLAB. *User's Guide.* The MathWorks, Inc., Natick, MA 01760, 1994-2001. http://www.mathworks.com.

[33] T. M. Mitchell. *Machine Learning.* McGraw-Hill, Boston, 1997.

[34] P. M. Murphy and D. W. Aha. UCI machine learning repository, 1992. www.ics.uci.edu/~mlearn/MLRepository.html.

[35] B. Schölkopf. *Support Vector Learning.* R. Oldenbourg Verlag, Munich, 1997.

[36] M. Stone. Cross-validatory choice and assessment of statistical predictions. *Journal of the Royal Statistical Society*, 36:111–147, 1974.

[37] P. Tseng. Analysis of a non-interior continuation method based on Chen-Mangasarian smoothing functions for complementarity problems. In *Reformulation: Nonsmooth, Piecewise Smooth, Semismooth and Smoothing Methods, M. Fukushima and L. Qi, (editors)*, pages 381–404, Dordrecht, Netherlands, 1999. Kluwer Academic Publishers.

[38] V. N. Vapnik. *The Nature of Statistical Learning Theory.* Springer, New York, 1995.

[39] P. Wolfe. A duality theorem for nonlinear programming. *Quarterly of Applied Mathematics*, 19:239–244, 1961.

PROPERTIES OF OLIGOPOLISTIC MARKET EQUILIBRIA IN LINEARIZED DC POWER NETWORKS WITH ARBITRAGE AND SUPPLY FUNCTION CONJECTURES

Jong-Shi Pang
Department of Mathematical Sciences
The Johns Hopkins University
Baltimore, Maryland 21218-2682, U.S.A.
jsp@vicp1.mts.jhu.edu

Benjamin F. Hobbs
Department of Geography and Environmental Engineering
The Johns Hopkins University
Baltimore, Maryland 21218-2682, U.S.A.
bhobbs@jhu.edu

Christopher J. Day
Enron Europe Limited
40 Grosvenor Place
London SW1X 7EN, United Kingdom
christopher.j.day@enron.com

Abstract We present mathematical models for a power market on a linearized DC network with affine demand. The models represent the conjecture that each power generating company may hold regarding how rival firms will change their outputs if prices change. The classic Cournot model is a special case of this conjecture. The models differ in how arbitrage is handled, and their formulations give rise to nonlinear mixed complementarity problems. In the Stackelberg version, the generators anticipate how arbitrage would affect prices at different locations, and therefore treat the arbitrage amounts as decision variables in their profit maximization problems. In the other version, arbitrage is exogenous to the firms. We show that solutions to the latter model are also solutions to the Stackelberg model. We also demonstrate existence and uniqueness properties for the exogenous arbitrage model.

1. Introduction

In restructured power markets, electric power generators have been privatized or freed of regulatory constraints on prices. The intent of restructuring is to provide incentives for innovation and more efficient production and consumption of electricity [5]. However, because of market failures, these benefits may not be fully realized. A market failure that has been of particular concern to regulators and the public is market power [12]. Market power is defined as the ability of a market participant to unilaterally alter prices in its own favor, and to sustain those price changes. Transmission capacity limits that restrict power imports and exports are an important source of market power for generating companies, as they allow firms within an isolated region to raise prices above competitive levels [2].

The potential for market power to be exercised within a given power system can be studied through laboratory experiments, empirical analysis, and modeling. There are many models of strategic interaction in transmission constrained systems (for reviews, see [7] or [8]). Models can be used to unveil unanticipated ways in which market power might be exercised on networks, to identify locations where prices can be manipulated, to assess the effects of adding transmission capacity upon prices, and to examine the competitive effects of company mergers or divestments. The most common oligopolistic modeling frameworks employed in power market analyses are based on the ideas of Cournot games and Supply Function Equilibria (SFE), defined below.

The purpose of this paper is to analyze the existence and uniqueness properties of solutions of a new model of oligopolistic power generators. The model represents the power network using a linearized "DC" load flow model [13], and includes a flexible representation of interactions of competing generating firms. We term this representation the "conjectured supply functions" (CSFs) approach. A CSF is a function representing the beliefs of a firm concerning how total supply from rival firms will react to price. Two versions of a linear CSF have been proposed: one in which the slope of conjectured supply response is constant and the intercept is to be solved for, and another in which the intercept is given but the slope is to be determined. The former CSP yields a linear mixed complementarity problem (MCP) for the market equilibrium, while the latter gives a nonlinear MCP.

The CSF model can be viewed as a generalization of the Cournot models of [7] and [14] in that the amount that rival firms are anticipated to adjust prices in response to a price change is not restricted to zero (the Cournot assumption). Instead, each generating company is allowed to conjecture that rival firms will react to price increases or decreases. By making different assumptions about the assumed supply response, different degrees of competitive intensity can be modeled, ranging from pure competition (infinitely large positive response by rivals to price increases) to oligopolistic Cournot competition (no response). Positive sloped CSFs represent a competitive intensity between the Cournot and pure competition extremes. A detailed justification of the CSF approach to modeling competition on transmission networks is given in [4], along with an application to the United Kingdom power system.

It should be noted that the CSF modeling approach is distinct from the widely used supply function equilibrium (SFE) approach to market modeling [1, 9]. The SFE is a Nash game in bid functions, in which suppliers provide a function to a central auctioneer that relates their willingness to supply to the price. The SFE approach also yields prices intermediate between the pure competition and Cournot extremes, but is plagued by computational challenges along with problems of nonuniqueness and, in some cases, nonexistence of solutions [2]. The fundamental difference between the SFE and CSF approaches is that the anticipated supply response of competitors is endogenous in SFE models and is consistent with the competitor's actual bid function, while in the CSF approach, the conjectured supply response of competing firms is instead based on an assumed parameter (slope or intercept). It is this difference that allows SFEs to be formulated as mixed complementarity problems that are relatively easy to solve and yield solutions whose existence and uniqueness properties can be demonstrated.

Questions concerning the existence and uniqueness of equilibrium solutions to market models are important for two reasons. First, public policy is in part based on policy analyses using market models; if unique solutions cannot be assured, then the question arises as to whether the conclusions of an analysis depend on which of several possible solutions is selected. Second, if a solution exists and is unique, then computational procedures do not need to check for multiple solutions, and are therefore simpler. This paper focuses on the existence and uniqueness properties of the solution of the nonlinear MCP (fixed intercept model), as those properties for the linear MCP (fixed slope) are readily established using the results of [11].

The paper begins by defining notation and the profit maximizing problems that are common to all the models presented in this paper

(Section 2). Those common problems include the profit maximization problems for the independent system operator (ISO) who allocates scarce transmission capacity, and the arbitrager who eliminates any noncost-based price differences among nodes in the network. Consumers are represented by downward sloping demand curves. The various models introduced in the paper differ in terms of their representation of the profit maximization problem for the oligopolistic power producer. The first model, Model I, is introduced in Section 3. There, the power producer makes production and sales decisions recognizing that demand responds to price, that rival producers will react to price changes (according to the assumed CSF), and that noncost-based price differences will be arbitraged away. Inclusion of arbitrage means that the arbitrager's equilibrium conditions are introduced as constraints in the producer's constraint set. After introducing the producer profit maximization problem, we obtain the nonlinear MCP that represents the market equilibrium. Section 4 presents Model II which differs from Model I in that the arbitrager's equilibrium conditions are kept outside of the producer's problem, resulting in a model which can be analyzed more fully than Model I. In Section 5, relevant theory of monotone linear complementarity problems is introduced which will be the basis for the demonstrations of the model properties. This theory is used in Section 6 to establish the existence of solutions to Models I and II, the conditions under which solutions to Model II exist and certain of the variables (prices, total generation, sales, and profits) are unique.

2. The ISO and Arbitrage Models

In this and the next two sections, we present the mixed NCP formulations of the market equilibrium with conjectured supply functions. The resulting models become the respective linear complementarity models considered in [11] when the intercepts tend to minus infinity. In what follows, we present the NCP models, establish the existence of solutions and analyze their properties.

2.1 Notation

Before presenting the mathematical formulations for the models, we summarize the notation.

Parameters

$\mathcal{N}$: set of nodes, excluding the hub
$\mathcal{A}$: set of transmission elements in the full network
$\mathcal{F}$: set of firms
α_i : fixed intercept of supply function at node i
c_{fi} : cost per unit generation at node i by firm f

P_i^0 : price intercept of supply function at node i
Q_i^0 : quantity intercept of supply function at node i
T_k^+ : capacity on transmission element k
T_k^- : capacity in the reverse direction of transmission element k
CAP_{fi} : production capacity at node i for firm f
PDF_{ik} : power distribution factor for node i on element k, describing the megawatt (MW) increase in flow resulting from 1 MW of power injection at i and 1 MW of withdrawal at a hub node.

Variables

s_{fi} : amount of sales at node i by firm f
g_{fi} : generation at node i by firm f
p_{fi} : price at node i anticipated by firm f
y_i : amount of transmission service from hub H to node i
w_i : transmission price from hub H to node i
p_f : price at the hub node, anticipated by firm f
a_{fi} : amount that arbitragers sell at node i, anticipated by firm f
$\lambda_k^{\pm}$: dual variables of transmission capacity constraints in ISO's problem
γ_{fi} : dual variable of production capacity constraint in firm f's problem
φ_f : dual variable of balance equation between supply and generation in firm f's problem
π_i : market price at node i

Vectors & Matrices

$\mathbf{1}$: vector of ones of appropriate size
$\boldsymbol{I}$: identity matrix of appropriate order
$\boldsymbol{E}$: square matrix of ones of appropriate order
Π: $|\mathcal{N}| \times |\mathcal{A}|$ matrix of PDF_{ik}, $i \in \mathcal{N}$ and $k \in \mathcal{A}$
$\boldsymbol{s}$: $(|\mathcal{N}| \times |\mathcal{F}|)$-vector of s_{fi}, $i \in \mathcal{N}$ and $f \in \mathcal{F}$
$\boldsymbol{g}$: $(|\mathcal{N}| \times |\mathcal{F}|)$-vector of g_{fi}, $i \in \mathcal{N}$ and $f \in \mathcal{F}$
$\boldsymbol{\pi}$: $|\mathcal{N}|$-vector of equilibrium prices π_i, $i \in \mathcal{N}$
$\boldsymbol{\gamma}$: $(|\mathcal{N}| \times |\mathcal{F}|)$-vector of γ_{fi}, $i \in \mathcal{N}$ and $f \in \mathcal{F}$
$\lambda^{\pm}$: $|\mathcal{A}|$-vectors of $\lambda_k^{\pm}$, $k \in \mathcal{A}$
c: $|\mathcal{N}| \times |\mathcal{F}|)$-vector of c_{fi}, $i \in \mathcal{N}$ and $f \in \mathcal{F}$
CAP: $(|\mathcal{N}| \times |\mathcal{F}|)$-vector of CAP_{fi}, $i \in \mathcal{N}$ and $f \in \mathcal{F}$.

The components of the vectors $\boldsymbol{s}$, $\boldsymbol{g}$, c, and CAP are grouped by firms; that is

$$\boldsymbol{s} \equiv (s_1, \ldots, s_{|\mathcal{F}|})^T,$$

where each s_f is the $|\mathcal{N}|$-vector with components s_{fi}, $i \in \mathcal{N}$. The other three vectors $\boldsymbol{g}$, c, and CAP are similarly arranged. Except for the supply intercepts and some power distribution factors, all parameters of the models are positive.

2.2 The ISO's problem

The ISO's problem is the following linear program (LP). Given the transmission prices w_i, $i \in \mathcal{N}$, compute y_i, $i \in \mathcal{N}$ in order to

$$\begin{array}{lllr} \text{maximize} & \sum_{i \in \mathcal{N}} w_i \, y_i & & \\ \text{subject to} & \sum_{i \in \mathcal{N}} y_i \, = \, 0, & & (\eta) \\ & \sum_{i \in \mathcal{N}} \mathrm{PDF}_{ik} \, y_i \, \leq \, T_k^+, & \forall \, k \, \in \, \mathcal{A}, & (\lambda_k^+) \\ & \sum_{i \in \mathcal{N}} \mathrm{PDF}_{ik} \, y_i \, \geq \, -T_k^-, & \forall \, k \, \in \, \mathcal{A}, & (\lambda_k^-), \end{array} \tag{1}$$

where we write the dual variables in parentheses next to the corresponding constraints. Note that the variables y_i are unrestricted in sign. A positive (negative) y_i means that there is a net flow into (out of) node i. It is trivial to note that $y = 0$ is always a feasible solution to the above LP, because the $T_k^{\pm}$ are positive scalars. The optimality conditions of the LP can be written as a mixed LCP in the variables y_i for $i \in \mathcal{N}$, $\lambda_k^{\pm}$ for $k \in \mathcal{A}$ and η, parameterized by the transmission fees w_i, $i \in \mathcal{N}$:

$$\begin{array}{l} 0 \leq \lambda_k^- \perp T_k^- + \sum_{i \in \mathcal{N}} \mathrm{PDF}_{ik} \, y_i \geq 0, \quad k \in \mathcal{A}, \\ 0 \leq \lambda_k^+ \perp T_k^+ - \sum_{i \in \mathcal{N}} \mathrm{PDF}_{ik} \, y_i \geq 0, \quad k \in \mathcal{A}, \\ 0 = \sum_{i \in \mathcal{N}} y_i, \\ 0 = w_i + \sum_{k \in \mathcal{A}} \mathrm{PDF}_{ik} \, (\lambda_k^- - \lambda_k^+) + \eta, \quad i \in \mathcal{N}. \end{array} \tag{2}$$

2.3 The arbitrager's problem

The arbitrager maximizes its profit by buying and selling power in the market, given the prices at the nodes in the network. With a_i denoting the arbitrage amount sold at node i, the arbitrager's profit maximization problem is very simple: for fixed prices p_i and costs w_i, compute a_i, $i \in \mathcal{N}$ in order to

$$\begin{array}{ll} \text{maximize} & \sum_{i \in \mathcal{N}} (p_i - w_i) \, a_i \\ \text{subject to} & \sum_{i \in \mathcal{N}} a_i \, = \, 0,; \end{array} \tag{3}$$

the transmission fee at node i is included in the objective function because the arbitrager must also pay this cost. The arbitrage amounts are measured as the net sales at a node; thus the sum of all the arbitrage amounts must equal to zero. Note that a_i is unrestricted in sign. A positive a_i represents the amount sold by the arbitrager at node i; in this case, the arbitrager is receiving p_i for each unit sold but is paying w_i for the transmission. If a_i is negative, then $|a_i|$ is the quantity that the arbitrager bought from node i; in this case, the arbitrager is paying p_i per unit and paying $-w_i$ per unit to ship out of i.

The problem (3) is trivially solvable. In particular, this problem is equivalent to the two equations:

$$\begin{aligned} p_i - w_i - p_H &= 0, \quad \forall i \in \mathcal{N} \\ \sum_{i \in \mathcal{N}} a_i &= 0. \end{aligned} \tag{4}$$

In turn, the first equation implies

$$p_i - p_j = w_i - w_j, \quad \forall i, j \in \mathcal{N},$$

which says that the difference in prices at two distinct nodes is exactly the difference between the transmission fees at those two nodes.

3. Model I

In this model, each firm that produces power anticipates the arbitrage amounts by including the variables a_{fi} and a supply function conjecture with fixed intercept in its profit maximization problem. The constraints that these variables satisfy are basically (4), where p_i is determined by the price function:

$$p_{fi} \equiv P_i^0 - \frac{P_i^0}{Q_i^0}\left(\sum_{t \in \mathcal{F}} s_{ti} + a_{fi}\right).$$

(Note the addition of the subscript i in p_{fi} as this is now the price at i anticipated by f.) The supply function conjecture is expressed by the equation

$$s_{-fi} \equiv \sum_{t \neq f} s_{ti} = \frac{p_{fi} - \alpha_i}{\pi_i - \alpha_i} s^*_{-fi}.$$

Note that π_i is a base price at which $s_{-fi} = s^*_{-fi}$ and is exogenous to the firms. Substituting s_{-fi} into the former equation and simplifying, we obtain

$$p_{fi} = \left(Q_i^0 - s_{fi} - a_{fi} + \frac{\alpha_i}{\pi_i - \alpha_i} s^*_{-fi}\right) \Big/ \left(\frac{Q_i^0}{P_i^0} + \frac{s^*_{-fi}}{\pi_i - \alpha_i}\right).$$

Letting p_f be the firm's anticipated price at the hub, firm f's problem is: with s^*_{-fi}, π_i, and w_i, $i \in \mathcal{N}$ fixed, find s_{fi}, g_{fi}, a_{fi}, p_{fi} for $i \in \mathcal{N}$, and p_f in order to

$$\text{maximize} \quad \sum_{i\in\mathcal{N}} (p_{fi} - w_i) s_{fi} - \sum_{i\in\mathcal{N}} (c_{fi} - w_i) g_{fi}$$

$$\text{subject to} \quad g_{fi} \leq \text{CAP}_{fi}, \quad \forall i \in \mathcal{N}$$

$$\sum_{i\in\mathcal{N}} (s_{fi} - g_{fi}) = 0,$$

$$p_{fi} = \frac{Q_i^0 - s_{fi} - a_{fi} + \dfrac{\alpha_i}{\pi_i - \alpha_i} s^*_{-fi}}{\dfrac{Q_i^0}{P_i^0} + \dfrac{s^*_{-fi}}{\pi_i - \alpha_i}}, \quad \forall i \in \mathcal{N}$$

$$p_{fi} = p_f + w_i, \quad \forall i \in \mathcal{N}$$

$$\sum_{i\in\mathcal{N}} a_{fi} = 0,$$

$$s_{fi}, g_{fi} \geq 0, \quad \forall i \in \mathcal{N}.$$

The three equations

$$p_{fi} = \left(Q_i^0 - s_{fi} - a_{fi} + \frac{\alpha_i}{\pi_i - \alpha_i} s^*_{-fi} \right) \Big/ \left(\frac{Q_i^0}{P_i^0} + \frac{s^*_{-fi}}{\pi_i - \alpha_i} \right), \quad \forall i \in \mathcal{N}$$

$$p_{fi} = p_f + w_i, \quad \forall i \in \mathcal{N}$$

$$\sum_{i\in\mathcal{N}} a_{fi} = 0,$$

uniquely determine p_{fi} and a_{fi} for $i \in \mathcal{N}$ and p_f in terms of s_{tj}, w_j and π_j for all $t \in \mathcal{F}$ and $j \in \mathcal{N}$. For the purpose of restating firm f's maximization problem, it suffices to solve for p_f, obtaining,

$$p_f = \frac{\sum_{i\in\mathcal{N}} Q_i^0 - \sum_{i\in\mathcal{N}} w_i \left(\dfrac{Q_i^0}{P_i^0} + \dfrac{s^*_{-fi}}{\pi_i - \alpha_i} \right) - \sum_{i\in\mathcal{N}} s_{fi} + \sum_{i\in\mathcal{N}} \dfrac{\alpha_i}{\pi_i - \alpha_i} s^*_{-fi}}{\sum_{i\in\mathcal{N}} \left(\dfrac{Q_i^0}{P_i^0} + \dfrac{s^*_{-fi}}{\pi_i - \alpha_i} \right)}.$$

Let $p_f(\boldsymbol{s}, \boldsymbol{\pi}, \boldsymbol{w})$ denote the fraction on the right-hand side as a function of the vectors $\boldsymbol{s}$, $\boldsymbol{\pi}$, and $\boldsymbol{w}$. This function depends on the intercepts α_i; but since these are parameters of the model, we do not write them as the arguments of p_f. The function p_f also depends on s^*_{-fi} and π_i; at equilibrium, the latter variables will be equated with s_{-fi} and p_{fi}, respectively. Observe that p_f is a linear function of s_f, with the other arguments fixed.

We can now restate firm f's problem in the simplified form: with s_{ti} $(t \neq f)$, π_i, and w_i, $i \in \mathcal{N}$ fixed, find s_{fi} and g_{fi} for $i \in \mathcal{N}$ in order to

$$\begin{aligned}
\text{maximize} \quad & \sum_{i \in \mathcal{N}} p_f(\boldsymbol{s}, \boldsymbol{\pi}, \boldsymbol{w})\, s_{fi} - \sum_{i \in \mathcal{N}} (\, c_{fi} - w_i \,)\, g_{fi} \\
\text{subject to} \quad & g_{fi} \leq \text{CAP}_{fi}, \quad \forall\, i \in \mathcal{N}, \quad (\, \gamma_{fi} \,) \\
& \sum_{i \in \mathcal{N}} (\, s_{fi} - g_{fi} \,) = 0, \quad (\, \varphi_f \,) \\
& s_{fi},\, g_{fi} \geq 0, \quad \forall\, i \in \mathcal{N}.
\end{aligned}$$

The above problem is a quadratic concave maximization problem in the variables s_{fi} and g_{fi} for $i \in \mathcal{N}$, parameterized by s_{ti} for $t \neq f$ and π_i and w_i for $i \in \mathcal{N}$. We can write the optimality conditions for the problem as follows:

$$\begin{aligned}
& 0 \leq s_{fi} \perp -p_f(\boldsymbol{s}, \boldsymbol{\pi}, \boldsymbol{w}) + \frac{\displaystyle\sum_{j \in \mathcal{N}} s_{fj}}{\displaystyle\sum_{j \in \mathcal{N}} \left(\frac{Q_j^0}{P_j^0} + \frac{s^*_{-fj}}{\pi_j - \alpha_j} \right)} + \varphi_f \geq 0, \quad i \in \mathcal{N}, \\
& 0 \leq g_{fi} \perp c_{fi} - w_i + \gamma_{fi} - \varphi_f \geq 0, \qquad i \in \mathcal{N}, \\
& 0 \leq \gamma_{fi} \perp \text{CAP}_{fi} - g_{fi} \geq 0, \qquad i \in \mathcal{N}, \\
& 0 = \sum_{i \in \mathcal{N}} (\, s_{fi} - g_{fi} \,).
\end{aligned}$$

To complete the description of the model, we need to relate the ISO's problem to the firms' problems. This is accomplished via the market clearing condition, which is simply a flow balancing equation:

$$y_i = \sum_{t \in \mathcal{F}} (\, s_{ti} - g_{ti} \,) + a_{fi}, \quad \forall\, (\, f, i \,) \in \mathcal{F} \times \mathcal{N}. \tag{5}$$

In addition, we stipulate that

$$p_f(\boldsymbol{s}, \boldsymbol{\pi}, \boldsymbol{w}) + w_i = \pi_i, \quad \forall\, (\, f, i \,) \in \mathcal{F} \times \mathcal{N} \tag{6}$$

and $s^*_{-fi} = s_{-fi}$ for all $(f,i) \in \mathcal{F}\times\mathcal{N}$. From the definition of $p_f(\boldsymbol{s},\boldsymbol{\pi},\boldsymbol{w})$, the last two conditions yield

$$S \equiv \sum_{f\in\mathcal{F}}\sum_{i\in\mathcal{N}} s_{fi} = \sum_{i\in\mathcal{N}} Q_i^0 - \sum_{i\in\mathcal{N}} \pi_i \frac{Q_i^0}{P_i^0},$$

which expresses the total sales S of all firms in all markets in terms of the market prices π_i. Substituting (6) into the last equation, we obtain

$$p_f(\boldsymbol{s},\boldsymbol{\pi},\boldsymbol{w}) = \left(\sum_{i\in\mathcal{N}} Q_i^0 - \sum_{i\in\mathcal{N}} w_i \frac{Q_i^0}{P_i^0} - S\right) \Big/ \left(\sum_{i\in\mathcal{N}} \frac{Q_i^0}{P_i^0},\right)$$

which shows among other things that $p_f(\boldsymbol{s},\boldsymbol{\pi},\boldsymbol{w})$ is the same for all firms and

$$p_{fi} = p_f(\boldsymbol{s},\boldsymbol{\pi},\boldsymbol{w}) + w_i = \pi_i;$$

thus the price p_{fi} is independent of the firms. Substituting the equality $p_{fi} = \pi_i$ into the expression

$$p_{fi} = \left(Q_i^0 - s_{fi} - a_{fi} + \frac{\alpha_i}{\pi_i - \alpha_i} s_{-fi}\right) \Big/ \left(\frac{Q_i^0}{P_i^0} + \frac{s_{-fi}}{\pi_i - \alpha_i}\right) \tag{7}$$

yields

$$a_{fi} = Q_i^0 - \pi_i \frac{Q_i^0}{P_i^0} - \sum_{t\in\mathcal{F}} s_{ti},$$

which shows that the arbitrage amounts anticipated by the firms depend only on the market i. Substituting the expression

$$\pi_i = \left(\sum_{j\in\mathcal{N}} Q_j^0 - \sum_{j\in\mathcal{N}} w_j \frac{Q_j^0}{P_j^0} - S\right) \Big/ \left(\sum_{j\in\mathcal{N}} \frac{Q_j^0}{P_j^0}\right) + w_i$$

$$= \sigma + \sum_{j\in\mathcal{N}} (\delta_{ij} - \rho_i)\, w_j - \omega S$$

into the above expression for a_{fi}, we obtain

$$a_{fi} = R_i^0 - \sum_{t\in\mathcal{F}} s_{ti} + \rho_i S - \sum_{j\in\mathcal{N}} \zeta_{ij} w_j,$$

where

$$\omega \equiv \frac{1}{\displaystyle\sum_{i\in\mathcal{N}} \frac{Q_i^0}{P_i^0}}, \qquad \sigma \equiv \omega \sum_{i\in\mathcal{N}} Q_i^0,$$

$$\rho_i \equiv \omega \frac{Q_i^0}{P_i^0}, \qquad R_i^0 \equiv Q_i^0 - \frac{Q_i^0}{P_i^0}\sigma, \quad i \in \mathcal{N},$$

and with δ_{ij} denoting the Kronecker delta; i.e., δ_{ij} is equal to one if $i = j$ and equal to zero if $i \neq j$,

$$\zeta_{ij} \equiv \frac{Q_i^0}{P_i^0} (\delta_{ij} - \rho_j), \quad i, j \in \mathcal{N}$$

$$= \begin{cases} \dfrac{Q_i^0}{P_i^0} \left(1 - \omega \dfrac{Q_i^0}{P_i^0} \right) & \text{if } i = j \\ -\omega \dfrac{Q_i^0}{P_i^0} \dfrac{Q_j^0}{P_j^0} & \text{if } i \neq j. \end{cases}$$

Substituting the above expression for a_{fi} into (5), we obtain

$$y_i = R_i^0 - \sum_{t \in \mathcal{F}} g_{ti} + \rho_i S - \sum_{j \in \mathcal{N}} \zeta_{ij} w_j. \tag{8}$$

Proceeding as in [11], we can show that the resulting Model I is to compute

$$\{ \lambda_k^{\pm} : k \in \mathcal{A} \}, \; \{ s_{fi}, g_{fi}, \gamma_{fi} : i \in \mathcal{N}, \, f \in \mathcal{F} \}, \text{ and } \{ \varphi_f' : f \in \mathcal{F} \}$$

such that

$$0 \leq \lambda_k^- \perp q_k^- + \sum_{\ell \in \mathcal{A}} \left[\sum_{i,j \in \mathcal{N}} \text{PDF}_{ik} \, \zeta_{ij} \, \text{PDF}_{j\ell} \right] (\lambda_\ell^- - \lambda_\ell^+) +$$

$$\left(\sum_{i \in \mathcal{N}} \rho_i \, \text{PDF}_{ik} \right) S - \sum_{i \in \mathcal{N}} \sum_{f \in \mathcal{F}} \text{PDF}_{ik} \, g_{fi} \geq 0, \quad \forall \, k \in \mathcal{A},$$

$$0 \leq \lambda_k^+ \perp q_k^+ + \sum_{\ell \in \mathcal{A}} \left[\sum_{i,j \in \mathcal{N}} \text{PDF}_{ik} \, \zeta_{ij} \, \text{PDF}_{j\ell} \right] (\lambda_\ell^+ - \lambda_\ell^-) -$$

$$\left(\sum_{i \in \mathcal{N}} \rho_i \, \text{PDF}_{ik} \right) S + \sum_{i \in \mathcal{N}} \sum_{f \in \mathcal{F}} \text{PDF}_{ik} \, g_{fi} \geq 0, \quad \forall \, k \in \mathcal{A},$$

$$0 \leq s_{fi} \perp -\sigma + \omega \, S + \frac{\displaystyle\sum_{j \in \mathcal{N}} s_{fj}}{\displaystyle\sum_{j \in \mathcal{N}} \left(\frac{Q_j^0}{P_j^0} + \frac{s_{-fj}}{\pi_j - \alpha_j} \right)} + \varphi_f' +$$

$$\sum_{k \in \mathcal{A}} \left[\sum_{j \in \mathcal{N}} \rho_j \, \text{PDF}_{jk} \right] (\lambda_k^+ - \lambda_k^-) \geq 0, \; \forall \, (f, i) \in \mathcal{F} \times \mathcal{N},$$

$$
\begin{aligned}
0 \le g_{fi} \quad \perp \quad & c_{fi} + \sum_{k \in \mathcal{A}} \mathrm{PDF}_{ik} \, (\lambda_k^- - \lambda_k^+) + \\
& \gamma_{if} - \varphi_f' \ge 0, \quad \forall (f, i) \in \mathcal{F} \times \mathcal{N},
\end{aligned}
$$

$$
\begin{aligned}
0 \le \gamma_{fi} \quad \perp \quad & \mathrm{CAP}_{fi} - g_{fi} \ge 0, \quad \forall (f, i) \in \mathcal{F} \times \mathcal{N}, \\
0 = \quad & \sum_{i \in \mathcal{N}} \, (s_{fi} - g_{fi}), \quad \forall f \in \mathcal{F},
\end{aligned}
$$

where

$$
q_k^\pm \equiv T_k^\pm \mp \sum_{i \in \mathcal{N}} R_i^0 \, \mathrm{PDF}_{ik}, \quad \forall k \in \mathcal{A}.
$$

In vector form, we have,

$$
q^\pm = T^\pm \mp \Pi^T R^0 = T^\pm \mp \Pi^T \Xi \, P^0.
$$

We observe that

$$
\begin{aligned}
\pi_i &= \sigma - \omega \, S + \sum_{j \in \mathcal{N}} \sum_{k \in \mathcal{A}} (\delta_{ij} - \rho_j) \, \mathrm{PDF}_{jk} \, (\lambda_k^+ - \lambda_k^-) \\
&= \sigma - \omega \, S + \frac{P_i^0}{Q_i^0} \sum_{j \in \mathcal{N}} \sum_{k \in \mathcal{A}} \zeta_{ij} \, \mathrm{PDF}_{jk} \, (\lambda_k^+ - \lambda_k^-),
\end{aligned} \tag{9}
$$

which expresses the regional prices π_i in terms of the total sales S and the dual variables of the transmission capacity constraints. Subsequently, we show that π_i is uniquely determined by the total sales only.

Let $h_{\mathrm{I}} : \Re^{|\mathcal{F}| \times |\mathcal{N}| + 2|\mathcal{A}|} \to \Re^{|\mathcal{F}| \times |\mathcal{N}|}$ be defined by

$$
h_{\mathrm{I},f}(\boldsymbol{s}, \lambda^\pm) \equiv \frac{\displaystyle\sum_{j \in \mathcal{N}} s_{fj}}{\displaystyle\sum_{j \in \mathcal{N}} \left(\frac{Q_j^0}{P_j^0} + \frac{s_{-fj}}{\pi_j - \alpha_j} \right)} \mathbf{1}_{|\mathcal{N}|}, \quad \forall f \in \mathcal{F},
$$

where π_i is given by (9). We can now write Model I in vector-matrix form. First we assemble the variables of the system in two vectors:

$$
\boldsymbol{x} \equiv \begin{pmatrix} \lambda^- \\ \lambda^+ \\ \boldsymbol{s} \\ \boldsymbol{g} \\ \gamma \end{pmatrix} \in \Re^{(2|\mathcal{A}| + 3|\mathcal{N}| \times |\mathcal{F}|)} \quad \text{and} \quad \varphi \in \Re^{|\mathcal{F}|}.
$$

(For notational simplification, we drop the $'$ in the variable φ). Next, we define the $|\mathcal{A}| \times |\mathcal{A}|$ symmetric positive semidefinite matrix

$$\Lambda \equiv \Pi^T \Xi \Pi;$$

also define two matrices in partitioned form:

$$\boldsymbol{M}_{\text{oli}} \equiv \begin{bmatrix} \boldsymbol{M}_\lambda & \boldsymbol{M}_{\lambda s} & \boldsymbol{M}_{\lambda g} & 0 \\ -(\boldsymbol{M}_{\lambda s})^T & \boldsymbol{M}_s & 0 & 0 \\ -(\boldsymbol{M}_{\lambda g})^T & 0 & 0 & \boldsymbol{I} \\ 0 & 0 & -\boldsymbol{I} & 0 \end{bmatrix}, \quad \boldsymbol{N} \equiv \begin{bmatrix} 0 \\ \boldsymbol{J} \\ -\boldsymbol{J} \\ 0 \end{bmatrix},$$

where

$$\boldsymbol{M}_\lambda \equiv \begin{bmatrix} \Lambda & -\Lambda \\ -\Lambda & \Lambda \end{bmatrix} \in \Re^{2|\mathcal{A}| \times 2|\mathcal{A}|},$$

$$\boldsymbol{M}_{\lambda s} \equiv \begin{bmatrix} \Pi^T \rho \mathbf{1}^T_{|\mathcal{N}|} & \cdots & \Pi^T \rho \mathbf{1}^T_{|\mathcal{N}|} \\ -\Pi^T \rho \mathbf{1}^T_{|\mathcal{N}|} & \cdots & -\Pi^T \rho \mathbf{1}^T_{|\mathcal{N}|} \end{bmatrix} \in \Re^{2|\mathcal{A}| \times (|\mathcal{N}| \times |\mathcal{F}|)},$$

$$= \begin{bmatrix} \Pi^T \rho \\ -\Pi^T \rho \end{bmatrix} \begin{bmatrix} \mathbf{1}^T_{|\mathcal{N}|} & \cdots & \mathbf{1}^T_{|\mathcal{N}|} \end{bmatrix}$$

$$\boldsymbol{M}_{\lambda g} \equiv \begin{bmatrix} -\Pi^T & \cdots & -\Pi^T \\ \Pi^T & \cdots & \Pi^T \end{bmatrix} \in \Re^{2|\mathcal{A}| \times (|\mathcal{N}| \times |\mathcal{F}|)},$$

$$\boldsymbol{M}_s \equiv \omega \begin{bmatrix} \boldsymbol{E} & \boldsymbol{E} & \cdots & \boldsymbol{E} \\ \boldsymbol{E} & \boldsymbol{E} & \cdots & \boldsymbol{E} \\ \vdots & \vdots & \ddots & \vdots \\ \boldsymbol{E} & \boldsymbol{E} & \cdots & \boldsymbol{E} \end{bmatrix} \in \Re^{(|\mathcal{N}| \times |\mathcal{F}|) \times (|\mathcal{N}| \times |\mathcal{F}|)},$$

(with each $\boldsymbol{E}$ being the $|\mathcal{N}| \times |\mathcal{N}|$ matrix of all ones)

$$\boldsymbol{J} \equiv \begin{bmatrix} \mathbf{1}_{|\mathcal{N}|} & 0 & \cdots & 0 \\ 0 & \mathbf{1}_{|\mathcal{N}|} & \cdots & 0 \\ \vdots & \vdots & \ddots & \vdots \\ 0 & 0 & \cdots & \mathbf{1}_{|\mathcal{N}|} \end{bmatrix} \in \Re^{(|\mathcal{N}| \times |\mathcal{F}|) \times |\mathcal{F}|}.$$

The matrix $\boldsymbol{M}_{\text{oli}}$ is square and of order $(2|\mathcal{A}|+3|\mathcal{N}|\times|\mathcal{F}|)$; the matrix $\boldsymbol{N}$ is rectangular and of order $(2|\mathcal{A}|+3|\mathcal{N}|\times|\mathcal{F}|)$ by $|\mathcal{F}|$. Define a constant vector partitioned in accordance with $\boldsymbol{M}_{\text{oli}}$:

$$\boldsymbol{q}_{\text{oli}} \equiv \begin{pmatrix} q^- \\ q^+ \\ -\sigma\,\mathbf{1}_{|\mathcal{N}|\times|\mathcal{F}|} \\ c \\ \text{CAP} \end{pmatrix} \in \Re^{(2|\mathcal{A}|+3|\mathcal{N}|\times|\mathcal{F}|)}.$$

With the above vectors and matrices, Model I can now be stated as the following mixed NCP:

$$\begin{aligned} 0 &\leq \boldsymbol{x} \perp \boldsymbol{q}_{\text{oli}} + \boldsymbol{M}_{\text{oli}}\boldsymbol{x} + \boldsymbol{N}\varphi + \begin{pmatrix} 0 \\ 0 \\ h_{\text{I}}(\boldsymbol{s},\lambda^{\pm}) \\ 0 \\ 0 \end{pmatrix} \geq 0 \\ 0 &= \boldsymbol{N}^T\boldsymbol{x}. \end{aligned} \tag{10}$$

If not for the nonlinear function $h_{\text{I}}(\boldsymbol{s},\lambda^{\pm})$, Model I would be a linear complementarity problem (LCP), which is exactly the one treated in [11]. The existence of a solution to (10) relies on bounding the components $h_{\text{I},f}(\boldsymbol{s},\lambda^{\pm})$ for all $f \in \mathcal{F}$. In turn, this relies on bounding the prices π_i for $i \in \mathcal{N}$. In Section 6, we show how to obtain the necessary bounds via LCP theory.

4. Model II

In this model, each firm takes the arbitrage amounts as input parameters in its profit maximization problem. Specifically, with the price p_i given by (7), firm f's problem is: with $s^*_{-fi}(= s_{-fi})$, a_i and π_i fixed for all $i \in \mathcal{N}$, find s_{fi} and g_{fi} for all $i \in \mathcal{N}$ in order to

$$\begin{array}{ll} \text{maximize} & \displaystyle\sum_{i \in \mathcal{N}} \left(\frac{Q_i^0 - s_{fi} - a_i + \dfrac{\alpha_i}{\pi_i - \alpha_i} s_{-fi}^*}{\dfrac{Q_i^0}{P_i^0} + \dfrac{s_{-fi}^*}{\pi_i - \alpha_i}} - w_i \right) s_{fi} \\ & \displaystyle - \sum_{i \in \mathcal{N}} (c_{fi} - w_i)\, g_{fi} \\ \text{subject to} & \displaystyle\sum_{i \in \mathcal{N}} (s_{fi} - g_{fi}) = 0 \\ \text{and} & s_{fi} \geq 0, \quad 0 \leq g_{fi} \leq \mathrm{CAP}_{fi}, \ \forall (f,i) \in \mathcal{F} \times \mathcal{N}. \end{array} \tag{11}$$

Model II is complete with the inclusion of the ISO's problem plus the arbitrage constraint:

$$\frac{Q_i^0 - s_{fi} - a_i + \dfrac{\alpha_i}{\pi_i - \alpha_i} s_{-fi}^*}{\dfrac{Q_i^0}{P_i^0} + \dfrac{s_{-fi}^*}{\pi_i - \alpha_i}} - w_i - p_H = 0, \quad \forall i \in \mathcal{N},$$

$$\sum_{i \in \mathcal{N}} a_i = 0,$$

the flow balancing equation (5):

$$y_i = \sum_{t \in \mathcal{F}} (s_{ti} - g_{ti}) + a_{fi}, \quad \forall (f,i) \in \mathcal{F} \times \mathcal{N},$$

and the price equation

$$\pi_i = w_i + p_f, \quad \forall i \in \mathcal{N}.$$

Following a similar derivation as before, we can show that Model II can be formulated as the following NCP:

$$\begin{aligned} 0 \leq \boldsymbol{x} \perp \boldsymbol{q}_{\mathrm{oli}} + \boldsymbol{M}_{\mathrm{oli}}\boldsymbol{x} + \boldsymbol{N}\varphi + \begin{pmatrix} 0 \\ 0 \\ h_{\mathrm{II}}(\boldsymbol{s}, \lambda^{\pm}) \\ 0 \\ 0 \end{pmatrix} &\geq 0 \\ 0 = \boldsymbol{N}^T \boldsymbol{x}, & \end{aligned} \tag{12}$$

where with π_i given by (9), $h_{\text{II}} : \Re^{|\mathcal{F}|\times|\mathcal{N}|+2|\mathcal{A}|} \to \Re^{|\mathcal{F}|\times|\mathcal{N}|}$ is given by

$$h_{\text{II},fi}(\boldsymbol{s},\lambda^{\pm}) \equiv \frac{s_{fi}}{\dfrac{Q_i^0}{P_i^0} + \dfrac{s_{-fi}}{\pi_i - \alpha_i}}, \quad \forall (f,i) \in \mathcal{F} \times \mathcal{N},$$

where π_i is given by (9). The two NCPs (10) and (12) differ in the two closely related nonlinear functions h_{I} and h_{II}.

5. Complementarity Theory

The key to the analysis of Models I and II is the theory of monotone LCPs. This theory in turn yields an existence result of a special variational inequality that is the cornerstone for the existence of solutions to the supply-function based market models. In this section, we present the prerequisite LCP theory and its implications.

We begin by recalling that the LCP range of a matrix $\boldsymbol{M} \in \Re^{n\times n}$, denoted $\mathcal{R}(\boldsymbol{M})$, is the set of all vectors $\boldsymbol{q} \in \Re^n$ for which the LCP $(\boldsymbol{q}, \boldsymbol{M})$ has a solution. Our first result pertains to the solutions of an LCP defined by a symmetric positive semidefinite matrix. Although part (a) of this result is known and parts (b) and (c) hold in more general contexts (see [6]) we give a full treatment of the following theorem because it is the basis for the entire subsequent development.

Theorem 1 Let $\boldsymbol{M} \in \Re^{n\times n}$ be a symmetric positive semidefinite matrix.

(a) For every $\boldsymbol{q} \in \mathcal{R}(\boldsymbol{M})$, the solutions of the LCP $(\boldsymbol{q}, \boldsymbol{M})$ are w-unique; that is, if z^1 and z^2 are any two solutions of the LCP $(\boldsymbol{q}, \boldsymbol{M})$, then $\boldsymbol{M}z^1 = \boldsymbol{M}z^2$. Let $w(\boldsymbol{q})$ denote the common vector $\boldsymbol{q} + \boldsymbol{M}z$ for any solution z of the LCP $(\boldsymbol{q}, \boldsymbol{M})$.

(b) There exists a constant $c > 0$ such that

$$\| w(\boldsymbol{q}) \| \le c \| \boldsymbol{q} \|, \quad \forall \boldsymbol{q} \in \mathcal{R}(\boldsymbol{M}).$$

(c) The function $w : \mathcal{R}(\boldsymbol{M}) \to \Re^n$ is continuous.

Proof. Statement (a) is a well-known result in LCP theory. We next prove (b) by contradiction. Suppose no constant c satisfying (b) exists. There exists a sequence of vectors $\{\boldsymbol{q}^k\} \subset \mathcal{R}(\boldsymbol{M})$ satisfying

$$\| w(\boldsymbol{q}^k) \| > k \| \boldsymbol{q}^k \|$$

for every k. We have $w(\boldsymbol{q}^k) \neq 0$ for every k and

$$\lim_{k\to\infty} \frac{\boldsymbol{q}^k}{\| w(\boldsymbol{q}^k) \|} = 0.$$

Without loss of generality, we may assume that

$$\lim_{k \to \infty} \frac{w(\boldsymbol{q}^k)}{\| w(\boldsymbol{q}^k) \|} = v^\infty$$

for some vector v^∞, which must be nonzero. We may further assume that

$$\operatorname{supp}(w(\boldsymbol{q}^k)) \equiv \{ i : w_i(\boldsymbol{q}^k) > 0 \}$$

is the same for all k, which we denote α. With $\bar{\alpha}$ denoting the complement of α in $\{1, \ldots, n\}$, we have, for every k,

$$\begin{aligned} 0 &= \boldsymbol{q}^k_{\bar{\alpha}} + \boldsymbol{M}_{\bar{\alpha}\bar{\alpha}} z^k_{\bar{\alpha}} \\ w_\alpha(\boldsymbol{q}^k) &= \boldsymbol{q}^k_\alpha + \boldsymbol{M}_{\alpha\bar{\alpha}} z^k_{\bar{\alpha}} > 0 \end{aligned}$$

for some vector $z^k_{\bar{\alpha}} \geq 0$. Dividing by $\|w(\boldsymbol{q}^k)\|$, we deduce the existence of a nonnegative vector $z^\infty_{\bar{\alpha}}$, which is not necessarily related to the sequence $\{z^k_{\bar{\alpha}}\}$, such that

$$\begin{aligned} 0 &= \boldsymbol{M}_{\bar{\alpha}\bar{\alpha}} z^\infty_{\bar{\alpha}} \\ v^\infty_\alpha &= \boldsymbol{M}_{\alpha\bar{\alpha}} z^\infty_{\bar{\alpha}}. \end{aligned}$$

Since $\boldsymbol{M}$ is symmetric positive semidefinite, the above implies $v^\infty_{\bar{\alpha}}$, thus v^∞, is equal to zero. This contradiction establishes part (b).

To prove part (c), let $\{q^k\} \subset \mathcal{R}(\boldsymbol{M})$ converge to a limit vector q^∞ which must necessarily belong to $\mathcal{R}(\boldsymbol{M})$ because the LCP range is a closed cone. For each k, let $z^k \in \mathrm{SOL}(q^k, \boldsymbol{M})$ be such that $w(q^k) = q^k + \boldsymbol{M} z^k$. The sequence $\{w(q^k)\}$ is bounded; moreover, if w^∞ is any accumulation point of this w-sequence, then using the complementary cone argument, as done in the proof of part (b), we deduce the existence of a solution $z^\infty \in \mathrm{SOL}(q^\infty, \boldsymbol{M})$ such that $w^\infty = q^\infty + \boldsymbol{M} z^\infty$. This is enough to show by part (a) that the sequence $\{w(q^k)\}$ has a unique accumulation point which is equal to $w(q^\infty)$. Therefore the continuity of the map $w(q)$ at every vector $q \in \mathcal{R}(\boldsymbol{M})$ follows. Q.E.D.

Our goal is to apply the above theorem to the matrix

$$\boldsymbol{M}_\lambda = \begin{bmatrix} \Lambda & -\Lambda \\ -\Lambda & \Lambda \end{bmatrix} = \begin{bmatrix} \Pi^T \\ -\Pi^T \end{bmatrix} \Xi \,[\, \Pi \quad -\Pi \,].$$

For this purpose, we derive a corollary of Theorem 1 pertaining to a symmetric positive definite matrix of the above form.

Corollary 1 Let $\boldsymbol{M} \equiv A^T E A$, where E is a symmetric positive semidefinite $m \times m$ matrix and A is an arbitrary $m \times n$ matrix.

(a) For every $\boldsymbol{q} \in \mathcal{R}(\boldsymbol{M})$, if z^1 and z^2 are any two solutions of the LCP $(\boldsymbol{q}, \boldsymbol{M})$, then $EAz^1 = EAz^2$. Let $\tilde{w}(\boldsymbol{q})$ denote the common vector EAz for any solution z of the LCP $(\boldsymbol{q}, \boldsymbol{M})$.

(b) There exists a constant $c' > 0$ such that

$$\| \tilde{w}(\boldsymbol{q}) \| \leq c' \| \boldsymbol{q} \|, \quad \forall \boldsymbol{q} \in \mathcal{R}(\boldsymbol{M}).$$

(c) The function $\tilde{w} : \mathcal{R}(\boldsymbol{M}) \to \Re^n$ is continuous.

Proof. We note that for any nonzero symmetric positive semidefinite $\boldsymbol{M}$, we have

$$\frac{1}{\lambda^+_{\min}(\boldsymbol{M})} \| \boldsymbol{M} z \|^2 \geq z^T \boldsymbol{M} z \geq \frac{1}{\lambda_{\max}(\boldsymbol{M})} \| \boldsymbol{M} z \|^2, \quad \forall z \in \Re^n,$$

where $\lambda^+_{\min}(\boldsymbol{M})$ is the smallest positive eigenvalue of $\boldsymbol{M}$ and $\lambda_{\max}(\boldsymbol{M})$ is the largest eigenvalue of $\boldsymbol{M}$. With $\boldsymbol{M} = A^T E A$, it follows that

$$\boldsymbol{M} z = 0 \Leftrightarrow EAz = 0.$$

Hence, for every $\boldsymbol{q} \in \mathcal{R}(\boldsymbol{M})$, EAz is a constant for all solutions of the LCP $(\boldsymbol{q}, \boldsymbol{M})$. Moreover, there exists a scalar $c > 0$ such that for every $\boldsymbol{q} \in \mathcal{R}(\boldsymbol{M})$,

$$\| \boldsymbol{M} z \| \leq (1 + c) \| \boldsymbol{q} \|,$$

for every solution z of the LCP $(\boldsymbol{q}, \boldsymbol{M})$. Since

$$\frac{1}{\lambda^+_{\min}(\boldsymbol{M})} \| \boldsymbol{M} z \|^2 \geq z^T \boldsymbol{M} z = (Az)^T EAz \geq \frac{1}{\lambda_{\max}(E)} \| EAz \|^2$$

for all $z \in \Re^n$, part (b) of the corollary follows readily. The proof of part (c) is very similar to that of the same part in Theorem 1. Q.E.D.

It can be shown, using the theory of piecewise affine functions, that both functions $w(q)$ and $\tilde{w}(q)$ are Lipschitz continuous on $\mathcal{R}(M)$. Since this Lipschitz continuity property is not needed in the subsequent analysis, we omit the details.

5.1 An existence result for a special VI

In what follows, we establish an existence result for a linearly constrained variational inequality (VI) of a special kind. This result will subsequently be applied to Models I and II of power market equilibria. The setup of the result is a VI (K, F), where K is the Cartesian product of two polyhedra $K_1 \subseteq \Re^{n_1}$ and $K_2 \subset \Re^{n_2}$ with K_2 being compact. The mapping F is of the form: for $(x,y) \in \Re^{n_1+n_2}$,

$$F(x,y) \equiv \begin{pmatrix} q \\ r \end{pmatrix} + \begin{bmatrix} M_{11} & M_{12} \\ M_{21} & M_{22} \end{bmatrix} \begin{pmatrix} x \\ y \end{pmatrix} + \begin{pmatrix} 0 \\ h(y) \end{pmatrix}, \tag{13}$$

where $h : \Re^{n_2} \to \Re^{n_2}$ is a continuous function and the matrix

$$\boldsymbol{M} \equiv \begin{bmatrix} M_{11} & M_{12} \\ M_{21} & M_{22} \end{bmatrix}$$

is positive semidefinite (not necessarily symmetric). In the following result, an AVI is a VI defined by an affine pair $(\hat{K}, \hat{F})$, i.e., $\hat{K}$ is a polyhedron and $\hat{F}$ is an affine map. (We refer the reader to the monograph [6] for a comprehensive treatment of the finite-dimensional variational inequalities and complementarity problems.)

Proposition 1 In addition to the above setting, assume that for all $\hat{y} \in K_2$, the AVI $(K, F^{\hat{y}})$ has a solution, where

$$F^{\hat{y}}(x,y) \equiv \begin{pmatrix} q \\ r + h(\hat{y}) \end{pmatrix} + \begin{bmatrix} M_{11} & M_{12} \\ M_{21} & M_{22} \end{bmatrix} \begin{pmatrix} x \\ y \end{pmatrix}, \quad (x,y) \in \Re^{n_1+n_2}.$$

The VI (K, F) has a solution.

Proof. We apply Kakutani's fixed-point theorem to the set-valued mapping $\Gamma : K_2 \to K_2$ defined as follows. For each $\hat{y} \in K_2$, $\Gamma(\hat{y})$ consists of all vectors $y \in K_2$ for which there exists a vector $x \in K_1$ such that the pair (x,y) solves the VI $(K, F^{\hat{y}})$. Clearly, $\Gamma(\hat{y})$ is a nonempty subset of K_2; $\Gamma(\hat{y})$ is convex because if y^1 and y^2 are any two elements in $\Gamma(\hat{y})$ and x^1 and x^2 are such that $(x^i, y^i) \in \mathrm{SOL}(K, F^{\hat{y}})$ for $i = 1, 2$, then $\tau(x^1, y^1) + (1-\tau)(x^2, y^2)$ remains a solution of the VI $(K, F^{\hat{y}})$ for all scalars $\tau \in (0,1)$, by the positive semidefiniteness of the matrix $\boldsymbol{M}$. We next verify that Γ is a closed map. For this purpose, let $\{\hat{y}^k\}$ be a sequence of vectors in K_2 converging to a vector $\hat{y}^\infty$ in K_2 and for each k let (x^k, y^k) be a solution of the VI $(K, F^{\hat{y}^k})$ such that the sequence $\{y^k\}$ converges to a vector y^∞. We need to show the existence of a vector x^∞ such that the pair (x^∞, y^∞) solves the VI $(K, F^{\hat{y}^\infty})$. Write

$$K_1 \equiv \{ x \in \Re^{n_1} : Ax \leq b \}$$

and

$$K_2 \equiv \{ y \in \Re^{n_2} : Cy \leq d \}.$$

For each k, there exist multipliers μ^k and η^k such that

$$\begin{pmatrix} q \\ r + h(\hat{y}^k) \end{pmatrix} + \begin{bmatrix} M_{11} & M_{12} \\ M_{21} & M_{22} \end{bmatrix} \begin{pmatrix} x^k \\ y^k \end{pmatrix} + \begin{pmatrix} A^T \mu^k \\ C^T \eta^k \end{pmatrix} = 0$$

$$0 \leq \mu^k \perp Ax^k - b \leq 0$$
$$0 \leq \eta^k \perp Cy^k - d \leq 0.$$

Again by a standard complementary cone argument, we can deduce the existence of μ^∞, η^∞, and x^∞, which are not necessarily the limits of the sequences $\{\mu^k\}$, $\{\eta^k\}$ and $\{x^k\}$, respectively, such that

$$\begin{pmatrix} q \\ r + h(\hat{y}^\infty) \end{pmatrix} + \begin{bmatrix} M_{11} & M_{12} \\ M_{21} & M_{22} \end{bmatrix} \begin{pmatrix} x^\infty \\ y^\infty \end{pmatrix} + \begin{pmatrix} A^T \mu^\infty \\ C^T \eta^\infty \end{pmatrix} = 0$$

$$0 \leq \mu^\infty \perp Ax^\infty - b \leq 0$$
$$0 \leq \eta^\infty \perp Cy^\infty - d \leq 0.$$

This establishes that Γ is a closed map. In particular, it follows that $\Gamma(\hat{y})$ is a closed subset of K_2 for all $\hat{y}$ in K_2. Thus Γ satisfies all the assumptions required for the applicability of Kakutani's fixed-point theorem. By this theorem, Γ has a fixed point, which can easily be seen to be a solution of the VI (K, F). Q.E.D.

6. Properties of Models I and II

Returning to the mixed NCPs (10) and (12), we consider the following LCP in the variables $\lambda^\mp$, parameterized by S and $\boldsymbol{g}$:

$$\begin{aligned} 0 \leq \lambda^- \perp q^- + \Lambda(\lambda^- - \lambda^+) + \Pi^T \left(\rho S - \sum_{f \in \mathcal{F}} g_f \right) \geq 0 \\ 0 \leq \lambda^+ \perp q^+ + \Lambda(\lambda^+ - \lambda^-) - \Pi^T \left(\rho S - \sum_{f \in \mathcal{F}} g_f \right) \geq 0. \end{aligned} \tag{14}$$

We want to derive a sufficient condition under which the above LCP will have a solution for all "feasible" sales and generations. Specifically, let

$$Y \equiv \prod_{f \in \mathcal{F}} \Big\{ \ (s_f, g_f) \in \Re_+^{2|\mathcal{N}|} : \\ \sum_{i \in \mathcal{N}} (s_{fi} - g_{fi}) = 0, \ g_{fi} \leq \text{CAP}_{fi}, \ \forall (f,i) \in \mathcal{F} \times \mathcal{N} \Big\},$$

be the set of such sales and generations. The set Y is a compact polyhedron in $\Re^{2|\mathcal{F}|\times|\mathcal{N}|}$. We have for all pairs $(\boldsymbol{s}, \boldsymbol{g}) \in Y$ and every $j \in \mathcal{N}$,

$$\rho_j \, S - \sum_{f\in\mathcal{F}} g_{fj} \; = \; - \sum_{f\in\mathcal{N}} \sum_{i\in\mathcal{N}} (\, \delta_{ij} - \rho_j \,) \, g_{fi}.$$

Thus

$$\sum_{j\in\mathcal{N}} \rho_j \, \mathrm{PDF}_{jk} \, S - \sum_{j\in\mathcal{N}} \sum_{f\in\mathcal{F}} \mathrm{PDF}_{jk} \, g_{fj}$$

$$= -\sum_{f\in\mathcal{F}} \sum_{i\in\mathcal{N}} \sum_{j\in\mathcal{N}} \mathrm{PDF}_{jk} \, (\, \delta_{ij} - \rho_j \,) \, g_{fi}$$

$$= -\sum_{f\in\mathcal{F}} \sum_{i\in\mathcal{N}} \left(\sum_{j\in\mathcal{N}} \mathrm{PDF}_{jk} \, \zeta_{ij} \right) \frac{P_i^0}{Q_i^0} \, g_{fi}.$$

Therefore, the LCP (14) can be written as:

$$0 \leq \begin{pmatrix} \lambda^- \\ \lambda^+ \end{pmatrix} \perp \begin{pmatrix} q^- \\ q^+ \end{pmatrix} - \begin{bmatrix} \Pi^T \\ -\Pi^T \end{bmatrix} \Xi \, D \sum_{f\in\mathcal{F}} g_f$$

$$+ \begin{bmatrix} \Pi^T \\ -\Pi^T \end{bmatrix} \Xi \, [\, \Pi \quad -\Pi \,] \begin{pmatrix} \lambda^- \\ \lambda^+ \end{pmatrix} \geq 0, \tag{15}$$

where D is the $|\mathcal{N}| \times |\mathcal{N}|$ diagonal matrix with diagonal entries P_i^0/Q_i^0.

Proposition 2 If there exists a vector $\lambda \in \Re^{|\mathcal{A}|}$ satisfying

$$-T^- \leq \Pi^T \Xi P^0 + \Pi^T \Xi \Pi \lambda \leq T^+, \tag{16}$$

then the LCP (14) has a solution $\lambda^{\mp}$ for every pair $(\boldsymbol{s}, \boldsymbol{g}) \in Y$.

Proof. The LCP (15) is of the form:

$$0 \leq \boldsymbol{z} \perp \boldsymbol{q} + A^T E \boldsymbol{r} + A^T E A \boldsymbol{z} \geq 0,$$

where E is a symmetric positive semidefinite matrix. It follows from LCP theory that if there exists a vector $\hat{\boldsymbol{z}}$ satisfying $\boldsymbol{q} + A^T E \hat{\boldsymbol{z}} \geq 0$, then the LCP $(\boldsymbol{q} + A^T E \boldsymbol{r}, \boldsymbol{M})$, where $\boldsymbol{M} \equiv A^T E A$, has a solution for all vectors $\boldsymbol{r}$. Q.E.D.

Throughout the following discussion, we assume that condition (16) holds. Thus the LCP (15) has a solution for all vectors $\boldsymbol{g}$. Moreover, specializing Corollary 1 to the matrix $\boldsymbol{M}_\lambda$, we deduce that for any vector $\boldsymbol{g}$, if $(\lambda^{-,i}, \lambda^{+,i})$ for $i = 1, 2$ are any two solutions of (15), we have

$$\Xi\,\Pi\,(\,\lambda^{+,1} - \lambda^{-,1}\,) \;=\; \Xi\,\Pi\,(\,\lambda^{+,2} - \lambda^{-,2}\,).$$

Furthermore, if $\Phi(\boldsymbol{g})$ denotes this common vector, then Φ is a Lipschitz continuous function from $\Re^{|\mathcal{F}|\times|\mathcal{N}|}$ into $\Re^{|\mathcal{N}|}$. In terms of this function, we have

$$\pi_i \;=\; \sigma - \omega\, S + \Phi_i(\boldsymbol{g}), \quad \forall\, i \in \mathcal{N}, \tag{17}$$

which shows that π_i is a function of the total sales S and the generations $\boldsymbol{g}$. Since each Φ is continuous and Y is compact, it follows that for each $i \in \mathcal{N}$, the scalar

$$\varsigma_i \;\equiv\; \min\{\, \sigma - \omega\, S + \Phi_i(\boldsymbol{g}) \;:\; (\,\boldsymbol{s}, \boldsymbol{g}\,) \in Y\,\}$$

is finite. Therefore, if the intercepts α_i satisfy

$$\alpha_i < \varsigma_i, \quad \forall\, i \in \mathcal{N},$$

then the denominators in

$$h_{\mathrm{I},fi}(\boldsymbol{s},\boldsymbol{g}) \;\equiv\; \left(\sum_{j\in\mathcal{N}} s_{fj}\right) \Bigg/ \left(\sum_{j\in\mathcal{N}} \left(\frac{Q_j^0}{P_j^0} + \frac{s_{-fj}}{\pi_j - \alpha_j}\right)\right)$$

and

$$h_{\mathrm{II},fi}(\boldsymbol{s},\boldsymbol{g}) \;\equiv\; s_{fi} \Bigg/ \left(\frac{Q_i^0}{P_i^0} + \frac{s_{-fi}}{\pi_i - \alpha_i}\right)$$

are positive for all $(\boldsymbol{s}, \boldsymbol{g}) \in Y$ and all $f \in \mathcal{F}$. Notice that as a result of (17), we can replace the dependence on $\lambda^{\pm}$ in the two functions h_{I} and h_{II} by the dependence on $\boldsymbol{g}$ instead. The computation of each scalar ς_i requires the solution of an mathematical program with equilibrium constraints [10] that has a linear objective function and a parametric, monotone LCP constraint.

6.1 Existence of solutions

Both NCPs (10) and (12) are equivalent to a VI of the type considered in Proposition 1. More specifically, define the following principal submatrix of $\boldsymbol{M}_{\mathrm{oli}}$ by removing the last row and column:

$$\tilde{\boldsymbol{M}}_{\mathrm{oli}} \;\equiv\; \begin{bmatrix} \boldsymbol{M}_\lambda & \boldsymbol{M}_{\lambda s} & \boldsymbol{M}_{\lambda g} \\ -(\,\boldsymbol{M}_{\lambda s}\,)^T & \boldsymbol{M}_s & 0 \\ -(\,\boldsymbol{M}_{\lambda g}\,)^T & 0 & 0 \end{bmatrix};$$

the matrix $\tilde{\boldsymbol{M}}_{\text{oli}}$ is of order $(2|\mathcal{A}| + 2|\mathcal{N}| \times |\mathcal{F}|)$. Define a reduced vector $\tilde{\boldsymbol{q}}_{\text{oli}}$ accordingly:

$$\tilde{\boldsymbol{q}}_{\text{oli}} \equiv (q^-, \; q^+, \; -\sigma \, \mathbf{1}_{|\mathcal{N}|\times|\mathcal{F}|}, \; c)^T \in \Re^{(2|\mathcal{A}|+2|\mathcal{N}|\times|\mathcal{F}|)}.$$

Let $n_1 \equiv 2|\mathcal{A}|$ and $n_2 \equiv 2|\mathcal{N}| \times |\mathcal{F}|$. Identify the vectors x and y with $\lambda^{\mp}$ and $(\boldsymbol{s}, \boldsymbol{g})$, respectively, the constant vector (q, r) with $\tilde{\boldsymbol{q}}_{\text{oli}}$, the matrix $\boldsymbol{M}$ with $\tilde{\boldsymbol{M}}_{\text{oli}}$ and the function $h(y)$ with either

$$\begin{pmatrix} h_{\text{I}}(\boldsymbol{s}, \boldsymbol{g}) \\ 0 \end{pmatrix} \quad \text{or} \quad \begin{pmatrix} h_{\text{II}}(\boldsymbol{s}, \boldsymbol{g}) \\ 0 \end{pmatrix}.$$

Furthermore, let K_1 be the nonnegative orthant of $\Re^{2|\mathcal{A}|}$ and K_2 be the set Y:

$$K_2 \equiv \prod_{f \in \mathcal{F}} \Big\{ \; (s_f, g_f) \in \Re_+^{2|\mathcal{N}|} : \\ \sum_{i \in \mathcal{N}} (s_{fi} - g_{fi}) = 0, \; g_{fi} \le \text{CAP}_{fi}, \; \forall (f, i) \in \mathcal{F} \times \mathcal{N} \Big\}.$$

Under the above identifications, models I and II can therefore be formulated as the VI $(K_1 \times K_2, F)$, where F is given by (13). We can readily apply Proposition 1 to establish the following existence result for the two models.

Theorem 2 Suppose that there exists a vector $\lambda \in \Re^{|\mathcal{A}|}$ satisfying (16). If

$$\alpha_i < \min \{ \sigma - \omega S + \Phi_i(\boldsymbol{g}) : (\boldsymbol{s}, \boldsymbol{g}) \in Y \}, \quad \forall i \in \mathcal{N}, \qquad (18)$$

then solutions exist to Models I and II.

Proof. Under the assumption on the intercepts α_i, the function $h(\boldsymbol{s}, \boldsymbol{g})$ is well defined on the set K_2. For every $\hat{y} \equiv (\hat{\boldsymbol{s}}, \hat{\boldsymbol{g}}) \in K_2$, the VI $(K, F^{\hat{y}})$ is equivalent to the mixed LCP in the variable $(\boldsymbol{x}, \varphi)$:

$$\begin{aligned} 0 &\le \boldsymbol{x} \perp \boldsymbol{q}_{\text{oli}} + \boldsymbol{M}_{\text{oli}} \boldsymbol{x} + \boldsymbol{N} \varphi + (0, 0, h(\hat{\boldsymbol{s}}, \hat{\boldsymbol{g}}))^T \ge 0 \\ 0 &= \boldsymbol{N}^T \boldsymbol{x}. \end{aligned}$$

This mixed LCP is clearly feasible and monotone. It therefore has a solution. The existence of solutions to Models I and II follows readily from Proposition 1. Q.E.D.

The next result identifies two important properties of the solutions to Model II. It shows in particular that Model I can be solved by solving Model II.

Theorem 3 Under the assumptions of Theorem 2, if $(\lambda^{\mp}, \boldsymbol{s}, \boldsymbol{g}, \varphi)$ is a solution to Model II, then $\forall\, i \neq j$ and $\forall\, f \in \mathcal{F}$

$$s_{fi} \Big/ \left(\frac{Q_i^0}{P_i^0} + \frac{s_{-fi}}{\pi_i - \alpha_i} \right) = s_{fj} \Big/ \left(\frac{Q_j^0}{P_j^0} + \frac{s_{-fj}}{\pi_j - \alpha_j} \right).$$

Therefore every solution to Model II is a solution to Model I.

Proof. It suffices to show that for all $i \neq j$ and all $f \in \mathcal{F}$,

$$s_{fi} \Big/ \left(\frac{Q_i^0}{P_i^0} + \frac{s_{-fi}}{\pi_i - \alpha_i} \right) \geq s_{fj} \Big/ \left(\frac{Q_j^0}{P_j^0} + \frac{s_{-fj}}{\pi_j - \alpha_j} \right).$$

This is because by reversing the role of i and j, we obtain the reverse inequality and equality therefore must hold. The above inequality is clearly valid if $s_{fj} = 0$. So we may assume that $s_{fj} > 0$. By complementarity, we have

$$\begin{aligned} \frac{s_{fj}}{\dfrac{Q_j^0}{P_j^0} + \dfrac{s_{-fj}}{\pi_j - \alpha_j}} &= \sigma - \omega\, S - \varphi_f - \sum_{k \in \mathcal{A}} \left[\sum_{j' \in \mathcal{N}} \rho_{j'}\, \mathrm{PDF}_{j'k} \right] (\lambda_k^+ - \lambda_k^-) \\ &\leq \frac{s_{fi}}{\dfrac{Q_i^0}{P_i^0} + \dfrac{s_{-fi}}{\pi_i - \alpha_i}}. \end{aligned}$$

This establishes the first assertion of the theorem. To prove the second assertion, we note that by what has just been proved, it follows that if $(\lambda^{\mp}, \boldsymbol{s}, \boldsymbol{g}, \varphi)$ is a solution to Model II, then we must have

$$s_{fi} \Big/ \left(\frac{Q_i^0}{P_i^0} + \frac{s_{-fi}}{\pi_i - \alpha_i} \right) = \sum_{j \in \mathcal{N}} s_{fj} \Big/ \sum_{j \in \mathcal{N}} \left[\frac{Q_j^0}{P_j^0} + \frac{s_{-fj}}{\pi_j - \alpha_j} \right],$$

This shows that $(\lambda^{\mp}, \boldsymbol{s}, \boldsymbol{g})$ is also a solution to Model I. Q.E.D.

6.2 Uniqueness in Model II

In this subsection, we show that if each price intercept α_i is suitably restricted, then the firms' sales in the market model II are unique. The cornerstone to this uniqueness property of the model solutions is the expression (9). Based on this expression, we show that the mapping

$$\boldsymbol{F}_{\mathrm{II}}(\boldsymbol{x},\varphi) \equiv \begin{pmatrix} \boldsymbol{q}_{\mathrm{oli}} \\ 0 \end{pmatrix} + \begin{bmatrix} \boldsymbol{M}_{\mathrm{oli}} & \boldsymbol{N} \\ -\boldsymbol{N} & 0 \end{bmatrix} \begin{pmatrix} \boldsymbol{x} \\ \varphi \end{pmatrix} + \begin{pmatrix} 0 \\ 0 \\ h_{\mathrm{II}}(\boldsymbol{s},\lambda^{\pm}) \\ 0 \\ 0 \end{pmatrix}$$

is monotone. Throughout the following analysis, we restrict the pair $(S,\lambda^{\pm})$ so that

$$\sigma - \omega\, S + \sum_{j\in\mathcal{N}} \sum_{k\in\mathcal{A}} (\,\delta_{ij} - \rho_j\,)\,\mathrm{PDF}_{jk}\,(\,\lambda_k^+ - \lambda_k^-\,) > \alpha_i, \quad \forall\, i \in \mathcal{N}.$$

To establish the desired monotonicity of $\boldsymbol{F}_{\mathrm{II}}$, we first compute the Jacobian matrix of the function $h_{\mathrm{II}}(\boldsymbol{s},\lambda^{\pm})$.

We begin by noting the following partial derivatives:

$$\frac{\partial \pi_i}{\partial s_{fi'}} = -\omega, \quad \forall\, f \in \mathcal{F},\, i,i' \in \mathcal{N}$$

and

$$\frac{\partial \pi_i}{\partial \lambda_k^{\pm}} = \pm\frac{P_i^0}{Q_i^0} \sum_{j\in\mathcal{N}} \zeta_{ij}\mathrm{PDF}_{jk} \quad \forall\, i \in \mathcal{N},\, k \in \mathcal{A}.$$

Next, recalling that

$$h_{\mathrm{II},fi}(\boldsymbol{s},\lambda^{\pm}) = \frac{s_{fi}}{\dfrac{Q_i^0}{P_i^0} + \dfrac{s_{-fi}}{\pi_i - \alpha_i}}, \quad \forall\, f \in \mathcal{F},\, i \in \mathcal{N},$$

we have for all $f \in \mathcal{F}$,

$$\frac{\partial h_{\mathrm{II},fi}}{\partial s_{fi'}} = \begin{cases} \dfrac{1}{\dfrac{Q_i^0}{P_i^0} + \dfrac{s_{-fi}}{\pi_i - \alpha_i}} - \dfrac{s_{fi}}{\left(\dfrac{Q_i^0}{P_i^0} + \dfrac{s_{-fi}}{\pi_i - \alpha_i}\right)^2} \dfrac{\omega\, s_{-fi}}{(\pi_i - \alpha_i)^2} & \text{if } i = i' \\[2em] -\dfrac{s_{fi}}{\left(\dfrac{Q_i^0}{P_i^0} + \dfrac{s_{-fi}}{\pi_i - \alpha_i}\right)^2} \dfrac{\omega\, s_{-fi}}{(\pi_i - \alpha_i)^2} & \text{if } i \neq i' \end{cases}$$

and for all $f \neq f'$,

$$\frac{\partial h_{\mathrm{II},fi}}{\partial s_{f'i'}} = \begin{cases} -\dfrac{s_{fi}}{\left(\dfrac{Q_i^0}{P_i^0} + \dfrac{s_{-fi}}{\pi_i - \alpha_i}\right)^2} \left(\dfrac{1}{\pi_i - \alpha_i} + \dfrac{\omega\, s_{-fi}}{(\pi_i - \alpha_i)^2}\right) & \text{if } i = i' \\ -\dfrac{s_{fi}}{\left(\dfrac{Q_i^0}{P_i^0} + \dfrac{s_{-fi}}{\pi_i - \alpha_i}\right)^2} \dfrac{\omega\, s_{-fi}}{(\pi_i - \alpha_i)^2} & \text{if } i \neq i'. \end{cases}$$

Moreover, for all $f \in \mathcal{F}$, $i \in \mathcal{N}$ and $k \in \mathcal{A}$,

$$\frac{\partial h_{\mathrm{II},fi}}{\partial \lambda_k^{\pm}} = \mp \frac{s_{fi}}{\left(\dfrac{Q_i^0}{P_i^0} + \dfrac{s_{-fi}}{\pi_i - \alpha_i}\right)^2} \frac{s_{-fi}}{(\pi_i - \alpha_i)^2} \frac{P_i^0}{Q_i^0} \sum_{j \in \mathcal{N}} \zeta_{ij} \mathrm{PDF}_{jk}$$

Therefore, the Jacobian matrix of $h_{\mathrm{II}}(\boldsymbol{s}, \lambda^{\pm})$ has the following partitioned form:

$$\begin{bmatrix} B_1^{\lambda^-} & B_1^{\lambda^+} & A_{11}^s & \cdots & A_{1|\mathcal{F}|}^s \\ \vdots & \vdots & \vdots & \ddots & \vdots \\ B_{|\mathcal{F}|}^{\lambda^-} & B_{|\mathcal{F}|}^{\lambda^+} & A_{|\mathcal{F}|1}^s & \cdots & A_{|\mathcal{F}||\mathcal{F}|}^s \end{bmatrix}$$

where each $A_{ff'}^s$ is an $|\mathcal{N}| \times |\mathcal{N}|$ matrix with entries

$$(A_{ff'}^s)_{ii'} = \frac{\partial h_{\mathrm{II}\,fi}}{\partial s_{f'i'}}, \quad \forall i, i' \in \mathcal{N}$$

and each $B_f^{\lambda^\pm}$ is an $|\mathcal{N}| \times |\mathcal{A}|$ matrix with entries

$$(B_f^{\lambda^\pm})_{ik} = \frac{\partial h_{\mathrm{II}\,fi}}{\partial \lambda_k^{\pm}}, \quad \forall i \in \mathcal{N},\, k \in \mathcal{A}.$$

Consequently, the Jacobian matrix of $\boldsymbol{F}_{\mathrm{II}}(\boldsymbol{x}, \varphi)$ can be written as the sum of two matrices L_1 and L_2, where

$$L_1 \equiv \begin{bmatrix} \Lambda & -\Lambda & 0 & \cdots & 0 & 0 & 0 \\ -\Lambda & \Lambda & 0 & \cdots & 0 & 0 & 0 \\ B_1^{\lambda^-} & B_1^{\lambda^+} & A_{11}^s & \cdots & A_{1|\mathcal{F}|}^s & 0 & 0 \\ \vdots & \vdots & \vdots & \ddots & \vdots & & \\ B_{|\mathcal{F}|}^{\lambda^-} & B_{|\mathcal{F}|}^{\lambda^+} & A_{|\mathcal{F}|1}^s & \cdots & A_{|\mathcal{F}||\mathcal{F}|}^s & 0 & 0 \\ 0 & 0 & 0 & 0 & 0 & 0 & 0 \\ 0 & 0 & 0 & 0 & 0 & 0 & 0 \end{bmatrix}$$

and

$$L_2 \equiv \begin{bmatrix} 0 & \boldsymbol{M}_{\lambda s} & \boldsymbol{M}_{\lambda g} & 0 & 0 \\ -(\boldsymbol{M}_{\lambda s})^T & \boldsymbol{M}_s & 0 & 0 & \boldsymbol{J} \\ -(\boldsymbol{M}_{\lambda g})^T & 0 & 0 & \boldsymbol{I} & -\boldsymbol{J} \\ 0 & 0 & -\boldsymbol{I} & 0 & 0 \\ 0 & -\boldsymbol{J} & \boldsymbol{J} & 0 & 0 \end{bmatrix}.$$

The matrix L_2 is skew-symmetric, thus positive semidefinite. To show that L_1 is also positive semidefinite, recall that $\Lambda = \Pi^T \Xi \Pi$. Furthermore, we have

$$B_f^{\lambda\pm} = \pm D_f \Xi \Pi, \quad \forall f \in \mathcal{F},$$

for some $|\mathcal{N}| \times |\mathcal{N}|$ diagonal matrix D_f with

$$(D_f)_{ii} \equiv \frac{s_{fi}}{\left(\dfrac{Q_i^0}{P_i^0} + \dfrac{s_{-fi}}{\pi_i - \alpha_i} \right)^2} \frac{s_{-fi}}{(\pi_i - \alpha_i)^2} \frac{P_i^0}{Q_i^0}, \quad \forall i \in \mathcal{N}.$$

Let

$$\boldsymbol{A} \equiv \begin{bmatrix} A_{11}^s & \cdots & A_{1|\mathcal{F}|}^s \\ \vdots & \ddots & \vdots \\ A_{|\mathcal{F}|1}^s & \cdots & A_{|\mathcal{F}||\mathcal{F}|}^s \end{bmatrix} \quad \text{and} \quad \boldsymbol{D} \equiv \begin{bmatrix} D_1 \\ \vdots \\ D_{|\mathcal{F}|} \end{bmatrix}.$$

Notice that each block $A_{ff'}^s$ is a function of $\boldsymbol{s}$ and π; so is each matrix D_f. Consider the matrix

$$\tilde{L}_1 \equiv \begin{bmatrix} \begin{bmatrix} \Pi^T \\ -\Pi^T \end{bmatrix} \Xi \, [\, \Pi \;\; \Pi \,] & 0 \\ \boldsymbol{D} \Xi \, [\, \Pi \;\; \Pi \,] & \boldsymbol{A} \end{bmatrix}.$$

Clearly, L_1 is positive semidefinite if and only if $\tilde{L}_1$ is so. The next lemma shows that the latter matrix is positive semidefinite.

Lemma 1 For every compact set $\Omega \subset \Re^{(|\mathcal{F}| \times |\mathcal{N}|) + |\mathcal{N}|}$, there exists $\bar{\alpha}$ such that if

$$\alpha_i < \bar{\alpha}, \quad \forall i \in \mathcal{N},$$

the matrix $\tilde{L}_1$ is positive semidefinite for all $(\boldsymbol{s}, \pi) \in \Omega$.

Proof. The symmetric part of the matrix $\tilde{L}_1$ is equal to

$$\left[\begin{array}{cc} \left[\begin{array}{c} \Pi^T \\ -\Pi^T \end{array}\right] \Xi \left[\begin{array}{cc} \Pi & -\Pi \end{array}\right] & \frac{1}{2}\left[\begin{array}{c} \Pi^T \\ -\Pi^T \end{array}\right] \Xi \boldsymbol{D}^T \\ \frac{1}{2}\boldsymbol{D}\Xi \left[\begin{array}{cc} \Pi & -\Pi \end{array}\right] & \boldsymbol{A} \end{array}\right],$$

which we can write as the sum of two matrices:

$$\left[\begin{array}{c} \Pi^T \\ -\Pi^T \\ \frac{1}{2}\boldsymbol{D} \end{array}\right] \Xi \left[\begin{array}{ccc} \Pi & -\Pi & \frac{1}{2}\boldsymbol{D}^T \end{array}\right] + \left[\begin{array}{ccc} 0 & 0 & 0 \\ 0 & 0 & 0 \\ 0 & 0 & \boldsymbol{A} - \frac{1}{4}\boldsymbol{D}\Xi\boldsymbol{D}^T \end{array}\right].$$

The first summand is clearly positive semidefinite. Provided that the pair $(\boldsymbol{s}, \pi)$ is bounded, the matrix $\boldsymbol{A} - \frac{1}{4}\boldsymbol{D}\Xi\boldsymbol{D}^T$ is positive definite for all α_i with $|\alpha_i|$ sufficiently large. Q.E.D.

Each pair $(\boldsymbol{s}, \boldsymbol{g})$ in the compact set Y induces a price vector π via the expression (9), where $\lambda^{\pm}$ is a solution of the LCP (14). The induced prices are bounded by the continuity of the function Φ and the boundedness of Y; cf. (17). Let Ω be a compact convex subset of $\Re^{(|\mathcal{F}|\times|\mathcal{N}|)+|\mathcal{N}|}$ containing all such pairs $(\boldsymbol{s}, \pi)$. Corresponding to this set Ω, we may choose $\bar{\alpha}$ such that the Jacobian matrix of $\boldsymbol{F}_{\mathrm{II}}(\boldsymbol{x}, \varphi)$ is positive semidefinite for all pairs $(\boldsymbol{x}, \varphi)$ belonging to a convex set that contains all solutions of Model II. Thus $\boldsymbol{F}_{\mathrm{II}}$ is monotone on this set. Based on this monotonicity property, we can establish the desired uniqueness of the sales and other variables in Model II.

Theorem 4 Under the assumptions of Theorem 2, there exists $\bar{\alpha}$ such that if

$$\alpha_i < \bar{\alpha}, \quad \forall i \in \mathcal{N},$$

the following variables are unique in the solutions of Model II:

(a) the sales s_{fi} for all $f \in \mathcal{F}$ and $i \in \mathcal{N}$;

(b) the prices π_i for all $i \in \mathcal{N}$;

(c) the total generations $\sum_{i\in\mathcal{N}} g_{fi}$ for all $f \in \mathcal{F}$; and

(d) the profits for each firm.

Proof. Let $(\boldsymbol{x}^1, \varphi^1)$ and $(\boldsymbol{x}^2, \varphi^2)$ be two solutions of Model II. Let π^1 and π^2 be the induced prices. By the monotonicity of $\boldsymbol{F}_{\mathrm{II}}$, it follows that

$$\left(\begin{array}{c} \boldsymbol{x}^1 - \boldsymbol{x}^2 \\ \varphi^1 - \varphi^2 \end{array}\right)^T (\boldsymbol{F}_{\mathrm{II}}(\boldsymbol{x}^1, \varphi^1) - \boldsymbol{F}_{\mathrm{II}}(\boldsymbol{x}^2, \varphi^2)) = 0.$$

Let $\tilde{\lambda}^i \equiv \lambda^{+,i} - \lambda^{-,i}$ and $S^i \equiv \sum_{f \in \mathcal{F}} \sum_{i \in \mathcal{N}} s_{fi}$, for $i = 1, 2$. We have

$$(\tilde{\lambda}^1 - \tilde{\lambda}^2)^T \Pi^T \Xi \Pi (\tilde{\lambda}^1 - \tilde{\lambda}^2) + \omega (S^1 - S^2)^2 +$$
$$(s^1 - s^2)^T (h_{\mathrm{I}}(s^1, \lambda^{\pm,1}) - h_{\mathrm{I}}(s^2, \lambda^{\pm,2})) = 0.$$

By the mean-value theorem, it follows that for some triple $(\tilde{s}, \tilde{\pi})$ on the line segment joining (s^1, π^1) and (s^2, π^2),

$$\begin{pmatrix} \lambda^{+,1} - \lambda^{+,2} \\ \lambda^{-,1} - \lambda^{-,2} \\ s^1 - s^2 \end{pmatrix}^T \begin{bmatrix} \begin{bmatrix} \Pi^T \\ -\Pi^T \end{bmatrix} \Xi \, [\, \Pi \;\; \Pi \,] & 0 \\ D \Xi \, [\, \Pi \;\; \Pi \,] & A \end{bmatrix} \begin{pmatrix} \lambda^{+,1} - \lambda^{+,2} \\ \lambda^{-,1} - \lambda^{-,2} \\ s^1 - s^2 \end{pmatrix} +$$
$$\omega (S^1 - S^2)^2 = 0,$$

where the matrices D and A are evaluated at $(\tilde{s}, \tilde{\pi})$. By the proof of Lemma 1, it follows that

$$s^1 = s^2 \quad \text{and} \quad \Xi \Pi \tilde{\lambda}^1 = \Xi \Pi \tilde{\lambda}^2.$$

This yields $\pi^1 = \pi^2$. Since

$$\sum_{i \in \mathcal{N}} g_{fi} = \sum_{i \in \mathcal{N}} s_{fi}, \quad \forall f \in \mathcal{F},$$

it follows that each firm's total generation is unique.

Finally, to show that the profit for each firm is unique, note that the profit of firm f is equal to

$$p_f(s, \pi, w) \sum_{i \in \mathcal{N}} s_{fi} - \sum_{i \in \mathcal{N}} (c_{fi} - w_i) g_{fi} = \sum_{i \in \mathcal{N}} (\pi_i - c_{fi}) g_{fi},$$

because $\pi_i = p_f(s, \pi, w) + w_i$ (by (6)) and the sum of s_{fi} over all i in $\mathcal{N}$ is equal to the sum of g_{fi} over all i in $\mathcal{N}$. Let $(\hat{x}, \hat{\varphi})$ be an arbitrary solution of Model II. Consider the linear program in the variable $g_f \equiv (g_{fi} : i \in \mathcal{N})$:

$$\begin{array}{lll} \text{maximize} & \displaystyle\sum_{i \in \mathcal{N}} (\hat{\pi}_i - c_{fi}) g_{fi} & \\ \text{subject to} & 0 \leq g_{fi} \leq \mathrm{CAP}_{fi}, \quad \forall i \in \mathcal{N} & (19) \\ \text{and} & \displaystyle\sum_{i \in \mathcal{N}} g_{fi} = \sum_{i \in \mathcal{N}} \hat{s}_{fi}. & \end{array}$$

Since $\hat{\pi}_i$ for $i \in \mathcal{N}$ and $\sum_{i \in \mathcal{N}} \hat{s}_{fi}$ are constants of Model II, it follows that the above linear program depends only on the firm f and does not depend on the pair $(\hat{x}, \hat{\varphi})$ of solution to Model II. The optimal objective value of the linear program gives the profit of firm f. Q.E.D.

Acknowledgments. The work of this research was partially supported by the National Science Foundation under grant ECS-0080577.

References

[1] R. Baldick, R. Grant, and E. Kahn, "Linear Supply Function Equilibrium: Generalizations, Application, and Limitations," PWP-078, University of California Energy Institute, Berkeley, CA (August 2000).

[2] C.A. Berry, B.F. Hobbs, W.A. Meroney, R.P. O'Neill, and W.R. Stewart, Jr., "Analyzing Strategic Bidding Behavior in Transmission Networks," *Utilities Policy* 8 (1999) 139–158.

[3] R.W. Cottle, J.S. Pang, and R.E. Stone, *The Linear Complementarity Problem*, Academic Press (Boston 1992).

[4] C. Day, B.F. Hobbs, and J.S. Pang, "Oligopolistic Competition in Power Networks: A Conjectured Supply Function Approach," *IEEE Transactions on Power Systems* 17 (2002) 97–107.

[5] R. Gilbert and E. Kahn, editors, International Comparisons of Electricity Regulation, Cambridge University Press (New York 1996).

[6] F. Facchinei and J.S. Pang, Finite-Dimensional Variational Inequalities and Complementarity Problems, Springer-Verlag (New York 2003).

[7] B.F. Hobbs, "Linear Complementarity Models of Nash-Cournot Competition in Bilateral and POOLCO Power Markets," *IEEE Transactions on Power Systems* 16 (2001) 194–202.

[8] E. Kahn, "Numerical Techniques for Analyzing Market Power in Electricity," *Electricity Journal* 11 (July 1998) 34–43.

[9] P.D. Klemperer and M.A. Meyer, "Supply Function Equilibria," *Econometrica* 57 (1989) 1243–1277.

[10] Z.Q. Luo, J.S. Pang, and D. Ralph, *Mathematical Programs with Equilibrium Constraints*, Cambridge University Press (Cambridge 1996).

[11] C. Metzler, B.F. Hobbs, and J.S. Pang, "Nash-Cournot Equilibria in Power Markets on a Linearized DC Network with Arbitrage: Formulations and Properties", *Network and Spatial Economics* 3 (2003) 123–150.

[12] R.E. Schuler, "Analytic and Experimentally Derived Estimates of Market Power in Deregulated Electricity Systems: Policy Implications for the Management and Institutional Evolution of the Industry," *Decision Support Systems* 30 (January 2001) 341–355.

[13] F.C. Schweppe, M.C. Caramanis, R.E. Tabors, and R.E. Bohn, *Spot Pricing of Electricity*, Kluwer Academic Publishers, (Norwell 1988).

[14] Y. Smeers and J.Y. Wei, "Spatially Oligopolistic Model with Opportunity Cost Pricing for Transmission Capacity Reservations–A Variational Inequality Approach," CORE Discussion Paper 9717, Université Catholique de Louvain (February 1997).

RISK CONTROL AND OPTIMIZATION FOR STRUCTURAL FACILITIES

Rüdiger Rackwitz
Technische Universität München, Munich, Germany
Arcisstr. 21, D-80290 München
rackwitz@mb.bv.tum.de

Abstract Optimization techniques are essential ingredients of reliability-oriented optimal designs of technical facilities. Although many technical aspects are not yet solved and the available spectrum of models and methods in structural reliability is still limited many practical problems can be solved. A special one-level optimization is proposed for general cost-benefit analysis and some technical aspects are discussed. However, focus is on some more critical issues, for example, "what is a reasonable replacement strategy for structural facilities?", "how safe is safe enough?" and "how to discount losses of material, opportunity and human lives?". An attempt has been made to give at least partial answers.

Keywords: Structural reliabiltiy, optimization, risk acceptability, discount rates

1. Introduction

The theory of structural reliability has been developed to fair maturity within the last 30 years. The inverse problem, i.e. how to determine certain parameters in the function describing the boundary between safe and failure states for given reliability, has been addressed only recently. It is a typical optimization problem. Designing, erecting and maintaining structural facilities may be viewed as a decision problem where maximum benefit and least cost are sought and the reliability requirements are fulfilled simultaneously. In what follows the basic formulations of the various aspects of the decision problem are outlined making use of some more recent results in the engineering literature. The structure of a suitable objective function is first discussed. A renewal model proposed as early as 1971 by Rosenblueth/Mendoza [42], further developed in [17], [40] and extended in [36], [18] is presented in some detail. Theory and methods of structural reliability are reviewed next where it is pointed out that the calculation of suitable reliability measures is essentially an optimization problem. Focus is on the concepts of modern first- and second reliability methods [20]. The problem of the value of human life is then discussed in the context of modern health-related economic theories. Some remarks are made about appropriate discount rates. Finally, details of a special version of modern reliability-oriented optimization techniques based on work in [26] are outlined followed by an illustrative example.

2. Optimal Structures

A structure is optimal if the following objective is maximized:

$$Z(\mathbf{p}) = B(\mathbf{p}) - C(\mathbf{p}) - D(\mathbf{p}) \tag{1}$$

Without loss of generality it is assumed that all quantities in eq. (1) can be measured in monetary units. $B(\mathbf{p})$ is the benefit derived from the existence of the structure, $C(\mathbf{p})$ is the cost of design and construction and $D(\mathbf{p})$ is the cost in case of failure. $\mathbf{p}$ is the vector of all safety relevant parameters. Statistical decision theory dictates that expected values are to be taken. In the following it is assumed that $B(\mathbf{p}), C(\mathbf{p})$ and $D(\mathbf{p})$ are differentiable in each component of $\mathbf{p}$. The cost may differ for the different parties involved, e.g. the owner, the builder, the user and society. A structural facility makes sense only if $Z(\mathbf{p})$ is positive for all parties involved within certain parameter ranges. The intersection of these ranges defines reasonable structures.

The structure which eventually will fail after a long time will have to be optimized at the decision point, i.e. at time $t = 0$. Therefore, all cost need to be discounted. We assume a continuous discounting function $\delta(t) = \exp[-\gamma t]$ which is accurate enough for all practical purposes and where γ is the interest rate.

It is useful to distinguish between two replacement strategies, one where **the facility is given up after failure** and one where **the facility is systematically replaced after failure**. Further we distinguish between structures which **fail upon completion or never** and structures which **fail at a random point in time** much later due to service loads, extreme external disturbances or deterioration. The first option implies that loads on the structure are time invariant. Reconstruction times are assumed to be negligibly short. At first sight there is no particular preference for either of the replacement strategies. For infrastructure facilities the second strategy is a natural strategy. Structures only used once, e.g. special auxiliary construction structures or boosters for space transport vehicles fall into the first category.

3. The Renewal Model

3.1 Failure upon completion due to time-invariant loads

The objective function for a **structure given up after failure at completion** due to time-invariant loads (essentially dead weight) is

$$Z(\mathbf{p}) = B^* R_f(\mathbf{p}) - C(\mathbf{p}) - HP_f(\mathbf{p}) = B^* - C(\mathbf{p}) - (B^* + H)P_f(\mathbf{p}) \quad (2)$$

$R_f(\mathbf{p})$ is the reliability and $P_f(\mathbf{p}) = 1 - R_f(\mathbf{p})$ the failure probability, respectively. H is the direct cost of failure including demolition and debris removal cost. For **failure at completion and systematic reconstruction** we have

$$\begin{aligned} Z(\mathbf{p}) &= B^* - C(\mathbf{p}) - (C(\mathbf{p}) + H) \sum_{i=1}^{\infty} i P_f(\mathbf{p})^i R_f(\mathbf{p}) \\ &= B^* - C(\mathbf{p}) - (C(\mathbf{p}) + H) \frac{P_f(\mathbf{p})}{1 - P_f(\mathbf{p}))} \end{aligned} \quad (3)$$

After failure one, of course, investigates its causes and redesigns the structure. However, we will assume that the first design was already optimal so that there is no reason to change the design rules leading to the same $P_f(\mathbf{p})$. If each structural realization is independent of each other formula (3) holds.

A certain ambiguity exists when assessing the benefit B^* taken here and in the following as independent of $\mathbf{p}$. If the intended time of use of the facility is t_s it is simply

$$B^* = B(t_s) = \int_0^{t_s} b(t)\delta(t)dt \quad (4)$$

For constant benefit per time unit $b(t) = b$ one determines

$$B^* = B(t_s) = \frac{b}{\gamma}\left[1 - \exp\left[-\gamma t_s\right]\right] \underset{t_s \to \infty}{=} \frac{b}{\gamma} \quad (5)$$

3.2 Random Failure in Time

Assume now random failure events in time. The time to the first event has distribution function $F_1(t, \mathbf{p})$ with probability density $f_1(t, \mathbf{p})$. If the **structure is given up after failure** it is obviously

$$B(t_s) = \int_0^{t_s} b(t)\delta(t)R_1(t, \mathbf{p})dt \quad (6)$$

$$D(t_s) = \int_0^{t_s} f_1(t, \mathbf{p})\delta(t) H dt \tag{7}$$

and therefore

$$Z(\mathbf{p}) = \int_0^{t_s} b(t)\delta(t)R_1(t, \mathbf{p})dt - C(\mathbf{p}) - \int_0^{t_s} \delta(t) f_1(t, \mathbf{p}) H dt \tag{8}$$

For $t_s \to \infty$ and $f_1^*(\gamma, \mathbf{p}) = \int_0^\infty e^{-\gamma t} f_1(t, \mathbf{p})dt$ the Laplace transform of $f_1(t, \mathbf{p})$ it is instead

$$Z(\mathbf{p}) = \frac{b}{\gamma}[1 - f_1^*(\gamma, \mathbf{p})] - C(\mathbf{p}) - H f_1^*(\gamma, \mathbf{p}) \tag{9}$$

For the more important case of **systematic reconstruction** we generalize our model slightly. Assume that the time to first failure has density $f_1(t)$ while all other times between failure are independent of each other and have density $f(t)$, i.e. failures and subsequent renewals follow a modified renewal process [11]. This makes sense because extreme loading events usually are not controllable, i.e. the time origin lies somewhere between the zeroth and first event. The independence assumption is more critical. It implies that the structures are realized with independent resistances at each renewal according to the same design rules and the loads on the structures are independent, at least asymptotically. For constant benefit per time unit $b(t) = b$ we now derive by making use of the convolution theorem for Laplace transforms

$$\begin{aligned} Z(\mathbf{p}) &= \int_0^\infty b e^{-\gamma t} dt - C(\mathbf{p}) - (C(\mathbf{p}) + H) \sum_{n=1}^{\infty} \int_0^\infty e^{-\gamma t} f_n(t, \mathbf{p}) dt \\ &= \frac{b}{\gamma} - C(\mathbf{p}) - (C(\mathbf{p}) + H) \frac{f_1^*(\gamma, \mathbf{p})}{1 - f^*(\gamma, \mathbf{p})} \\ &= \frac{b}{\gamma} - C(\mathbf{p}) - (C(\mathbf{p}) + H) h_1^*(\gamma, \mathbf{p}) \end{aligned} \tag{10}$$

where $h_1^*(\gamma, \mathbf{p})$ is the Laplace transform of the renewal intensity $h_1(t, \mathbf{p})$. For regular renewal processes one replaces $f_1^*(\gamma, \mathbf{p})$ by $f^*(\gamma, \mathbf{p})$. For the renewal intensity and its Laplace transform there is an important asymptotic result [11]:

$$\lim_{t \to \infty} h(t, \mathbf{p}) = \lim_{\gamma \to 0} \gamma h^*(\gamma, \mathbf{p}) = \frac{1}{m(\mathbf{p})} \tag{11}$$

where $m(\mathbf{p})$ is the mean of the renewal times.

If, in particular, the events follow a stationary Poisson process with intensity λ we have

$$f_1^*(\gamma) = f^*(\gamma) = \int_0^\infty \exp[-\gamma t]\,\lambda \exp[-\lambda t]\,dt = \frac{\lambda}{\gamma + \lambda} \tag{12}$$

and

$$h^*(\gamma) = \frac{\lambda}{\gamma} \tag{13}$$

This result is of great importance because structural failures should, in fact, be rare, independent events. Then, the Poisson intensity λ can be replaced by the so-called outcrossing rate ν^+ to be described below - even in the locally non-stationary case. Finally, if at an extreme loading event (e.g. flood, wind storm, earthquake, explosion) failure occurs with probability $P_f(\mathbf{p})$ and $f_1(t)$ and $f(t)$, respectively, denote the densities of the times between the loading events one obtains by similar considerations

$$g_1^*(\gamma, \mathbf{p}) = \sum_{n=1}^{\infty} f_1^*(\gamma) f_{n-1}^*(\gamma) P_f(\mathbf{p}) R_f(\mathbf{p}))^{n-1} = \frac{P_f(\mathbf{p}) f_1^*(\gamma)}{1 - R_f(\mathbf{p}) f^*(\gamma)} \tag{14}$$

For the case treated in eq. (13) we have for stationary Poissonian load occurrences:

$$h^*(\gamma, \mathbf{p}) = \frac{g_1^*(\gamma, \mathbf{p})}{1 - g^*(\gamma, \mathbf{p})} = \frac{P_f(\mathbf{p})\lambda}{\gamma} \tag{15}$$

Unfortunately, Laplace transforms are rarely analytic. Taking Laplace transforms numerically requires some effort but taking the inverse Laplace transform must simply be considered as an numerically ill-posed problem. Then, however, one always can resort to the asymptotic result which can be shown to be accurate enough for all practical purposes.

The foregoing results can be generalized to cover multiple mode failure, loss of serviceability, obsolescence of the facility and inspection and maintenance. Also, the case of non-constant benefit, a case of obsolescence, or non-constant damage has been addressed. Further developments are under way.

4. Computation of Failure Probabilities and Failure Rates

4.1 Time-invariant Reliabilities

The simplest problem of computing failure probabilities is given as a volume integral

$$P_f(\mathbf{p}) = P(F) = \int_{F(\mathbf{p})} dF_{\mathbf{X}}(\mathbf{x}, \mathbf{p}) = \int_{F(\mathbf{p})} f_{\mathbf{X}}(\mathbf{x}, \mathbf{p}) d\mathbf{x} \tag{16}$$

where the failure event is $F(\mathbf{p}) = \{h(\mathbf{x},\mathbf{p}) \le 0\}$ and the random vector $\mathbf{X} = (X_1, X_2, ..., X_n)^T$ has joint distribution function $F_{\mathbf{X}}(\mathbf{x})$. Since n usually is large and $P_f(\mathbf{p})$ is small serious numerical difficulties occur if standard methods of numerical integration are applied. However, if it is assumed that the density $f_{\mathbf{X}}(\mathbf{x},\mathbf{p})$ of $F_{\mathbf{X}}(\mathbf{x})$ exists everywhere and $h(\mathbf{x},\mathbf{p})$ is twice differentiable, then, the problem of computing failure probabilities can be converted into a problem of optimization and some simple algebra. For convenience, a probability preserving distribution transformation $\mathbf{U} = T^{-1}(\mathbf{X})$ is first applied [19]. Making use of Laplace integration methods [4] one can then show that with $h(\mathbf{x},\mathbf{p}) = h(T(\mathbf{u}),p) = g(\mathbf{u},\mathbf{p})$ [5], [20]

$$\begin{aligned} P_f(\mathbf{p}) &= \int_{h(\mathbf{x},\mathbf{p})<0} f_{\mathbf{X}}(\mathbf{x},\mathbf{p})d\mathbf{x} = \int_{g(\mathbf{u},\mathbf{p})<0} \varphi_{\mathbf{U}}(\mathbf{u},\mathbf{p})d\mathbf{u} \\ &\sim \Phi(-\beta)\prod_{i=1}^{n-1}(1-\beta\kappa_i)^{-1/2} \approx \Phi(-\beta) \end{aligned} \tag{17}$$

for $1 < \beta \to \infty$ with

$$\beta = \|\mathbf{u}^*\| = \min\{\mathbf{u}\} \text{ for } \{\mathbf{u} : g(\mathbf{u},\mathbf{p}) \le 0\}, \tag{18}$$

$\varphi_{\mathbf{U}}(\mathbf{u})$ the multinormal density, $\Phi(.)$ the one-dimensional normal integral, $g(\mathbf{0},\mathbf{p}) > 0$ and κ_i the main curvatures of the failure surface $\partial F = \{g(\mathbf{u},\mathbf{p}) = 0\}$. Of course, it is assumed that a unique "critical" point $\mathbf{u}^*$ exists but methods have been devised to also locate and consider appropriately multiple critical points. In line two the asymptotic result is given denoted by second order since the Hessian of $g(\mathbf{u},\mathbf{p}) = 0$ is involved. The last result represents a first-order result corresponding to a linearization of $g(\mathbf{u},\mathbf{p})$ in $\mathbf{u}^*$ already pointed out by [16]. Very frequently this is sufficiently accurate in practical applications.

4.2 Time-variant Reliabilities

Much more difficult is the computation of time-variant reliabilities. Here, the question is not that the system is in an adverse state at some arbitrary point in time but that it enters it for the first time given that it was initially at time $t = 0$ in a safe state. The problem is denoted by first passage problem in the engineering literature. But exact results for distributions of first passage times are almost inexistent. However, good approximations can be obtained by the so-called outcrossing approach [13]. The outcrossing rate is defined by

$$\nu^+(\tau) = \lim_{\Delta\to 0}\frac{1}{\Delta}P(N(\tau,\tau+\Delta) = 1) \tag{19}$$

or for the original vector process

$$\nu^+(\tau) = \lim_{\Delta \to 0} \frac{1}{\Delta} P_1(\{\mathbf{X}(\tau) \in \bar{F}\} \cap \{\mathbf{X}(\tau + \Delta) \in F\}) \tag{20}$$

One easily sees that the definition of the outcrossing rate coincides formally with the definition of the renewal intensity. The counting process $N(.)$ of outcrossings must be a regular process [12] so that the mean value of outcrossings in $[0, t]$ is given by

$$E[N(t)] = \int_0^t \nu^+(\tau) d\tau \tag{21}$$

One can derive an important upper bound. Failure occurs either if $\mathbf{X}(0) \in V$ or $N(t) > 0$. Therefore [28]

$$\begin{aligned}
P_f(t) &= 1 - P(\mathbf{X}(\tau) \in \overline{F}) \text{ for all } \tau \in [0, t] \\
&= P(\{\mathbf{X}(0) \in F\} \cup \{N(t) > 0\}) \\
&= P(\mathbf{X}(0) \in F) + P(N(t) > 0) - P(\{\mathbf{X}(0) \in F\} \cap \{N(t) > 0\}) \\
&\leq P(\mathbf{X}(0) \in F) + P(N(t) > 0) \\
&\leq P(\mathbf{X}(0) \in V) + E[N(t)]
\end{aligned} \tag{22}$$

If the original process is sufficiently mixing one can derive the asymptotic result [13]:

$$P_f(t) \sim 1 - \exp\left[-E[N(t)]\right] \tag{23}$$

justifying the remarks below eq. (13). A lower bound can also be given. It is less useful.

Consider a stationary vectorial rectangular wave renewal process each component having renewal rate λ_i and amplitude distribution function $F_i(x)$. The amplitudes X_i are independent. Regularity assures that only one component has a renewal in a small time interval with probability $\lambda_i \; \Delta$. Then [9]

$$\begin{aligned}
\nu^+(F)\Delta &= P(\bigcup_{i=1}^{n} \{\text{renewal in } [0, \Delta]\} \cap \{\mathbf{X}_i \in \overline{F}\} \cap \{\mathbf{X}_i^+ \in F\}) \\
&= \sum_{i=1}^{n} \Delta \; \lambda_i P(\{\mathbf{X}_i \in \overline{F}\} \cap \{\mathbf{X}_i^+ \in F\}) \\
&= \sum_{i=1}^{n} \Delta \; \lambda_i [P(\mathbf{X}_i^+ \in F) - P(\{\mathbf{X}_i \in F)\} \cap \{\mathbf{X}_i^+ \in F\})]
\end{aligned} \tag{24}$$

$\mathbf{X}_i$ denotes the process $\mathbf{X}$ before and $\mathbf{X}_i^+$ the process after a jump of the $i-$th component. If the components are standard normally distributed and the failure domain is a half-space $F = \left\{\alpha^T \mathbf{u} + \beta \leq 0\right\}$ one

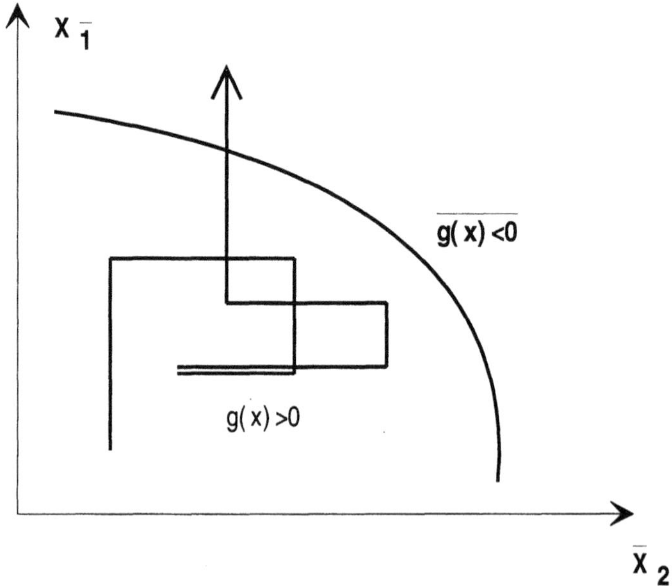

Figure 1. Outcrossings of a vectorial rectangular wave renewal process

determines

$$\nu^+(V) = \sum_{i=1}^{n} \lambda_i \left[P \left(\{ \alpha^T \mathbf{U}_i > -\beta \} \cap \{ \alpha^T \mathbf{U}_i^+ \leq -\beta \} \right) \right]$$
$$= \sum_{i=1}^{n} \lambda_i \left[\Phi_2(\beta, -\beta; \rho_i) \right] \leq \sum_{i=1}^{n} \lambda_i \Phi(-\beta) \quad (25)$$

where $\rho_i = 1 - \alpha_i^2$ is the correlation coefficient of the process before and after a jump and $\Phi_2(.,.;.)$ the bivariate normal integral. For general non-linear failure surfaces one can show that asymptotically [8]

$$\nu^+(F) = \sum_{i=1}^{n} \lambda_i \Phi(-\beta) \prod_{i=1}^{n-1} (1 - \beta\ \kappa_i)^{-1/2}; 1 < \beta \to \infty \quad (26)$$

with $\beta = \|\mathbf{u}^*\| = \min\{\|\mathbf{u}\|\}$ for $g(\mathbf{u}) \leq \mathbf{0}$ and κ_i the main curvatures in the solution point $\mathbf{u}^*$. This corresponds to the result in eq. (17). The same optimization problem as in the time-invariant case has to be solved. Rectangular wave renewal processes are used to model life loads, sea states, traffic loads, etc..

For stationary vector processes with differentiable sample paths it is useful to standardize the original process $\mathbf{X}(t)$ and its derivative (in mean square) process $\dot{\mathbf{X}}(t) = \frac{d}{dt}\mathbf{X}(t)$ such that $E[\mathbf{U}(t)] = E\left[\dot{\mathbf{U}}(t)\right] = 0, \mathbf{R}(0) = \mathbf{I}$ where $\mathbf{R}(\tau) = \mathbf{E}\left[\mathbf{U}(0)\mathbf{U}(\tau)^T\right]$ is the matrix of correlation functions and $\tau = |t_1 - t_2|$. A matrix of cross correlation functions between $\mathbf{U}(t)$ and $\dot{\mathbf{U}}(t)$, $\dot{\mathbf{R}}(\tau) = \mathbf{E}\left[\mathbf{U}(0)\dot{\mathbf{U}}(\tau)^T\right]$, as well as of the derivative process $\ddot{\mathbf{R}}(\tau) = \mathbf{E}\left[\dot{\mathbf{U}}(0)\dot{\mathbf{U}}(\tau)^T\right]$ also exists. The general outcrossing rate is defined by [38], [3]

$$\nu^+(t) = \lim_{\Delta\tau\to 0} \frac{P(\{\mathbf{U}(t) \in \mathbf{\Delta}(\partial F(t))\} \bigcap \left\{\dot{U}_N(t) > \partial\dot{F}(t)\right\} \text{ in } [\tau \leq t \leq \tau + \Delta\tau])}{\Delta\tau} \tag{27}$$

where $\dot{U}_N(t) = \mathbf{n}^T(\mathbf{u},t)\dot{\mathbf{U}}(t)$ the projection of $\dot{\mathbf{U}}(t)$ on the normal $\mathbf{n}(\mathbf{u},t) = -\alpha(\mathbf{u},t)$ of $\partial F(t)$ in $(\mathbf{u},t)$. $\mathbf{\Delta}(\partial F(t))$ is a thin layer around $\partial F(t)$ with thickness $(\dot{u}_N(t) - \partial\dot{F}(t))\Delta\tau$. Hence, it is:

$$\begin{aligned} &P(\{\mathbf{U}(t) \in \mathbf{\Delta}(\partial F(t))\} \bigcap \left\{\dot{U}_N(t) > \partial\dot{F}(t)\right\} \text{ in } [\tau \leq t \leq \tau + \Delta\tau]) \\ &= \int_{\mathbf{\Delta}(\partial F(t))} \int_{\dot{U}_N(t) > \partial\dot{F}(t)} \varphi_{n+1}(\mathbf{u},\dot{u}_N, t) d\mathbf{u} d\dot{u}_N \\ &= \Delta\tau \int_{\partial F(t)} \int_{\dot{U}_N(t) > \partial\dot{F}(t)} (\dot{u}_N - \partial\dot{F}(t))\varphi_{n+1}(\mathbf{u},\dot{u}_N, t) ds(\mathbf{u}) d\dot{u}_N \end{aligned} \tag{28}$$

In the stationary case one finds with $\partial F \equiv g(\mathbf{u}) = 0$

$$\begin{aligned} \nu^+(\partial F) &= \int_{\partial F} \int_0^\infty \dot{u}_N \varphi_{n+1}(\mathbf{u}, \dot{u}_N) d\dot{u}_N ds(\mathbf{u}) \\ &= \int_{\partial F} \int_0^\infty \dot{u}_N \varphi_1(\dot{u}_N | \mathbf{U} = \mathbf{u}) \varphi_n(\mathbf{u}) d\dot{u}_N ds(\mathbf{u}) \\ &= \int_{\partial F} E_0^\infty \left[\dot{U}_N | \mathbf{U} = \mathbf{u}\right] \varphi_n(\mathbf{u}) ds(\mathbf{u}) \\ &= \int_{\mathrm{R}^{n-1}} E_0^\infty \left[\dot{U}_N | \mathbf{U} = \mathbf{u}\right] \varphi_{n-1}(\tilde{\mathbf{u}}, p(\tilde{\mathbf{u}})) T(\tilde{\mathbf{u}}) d\tilde{\mathbf{u}} \end{aligned} \tag{29}$$

where $u_n = p(\tilde{\mathbf{u}}) = g^{-1}(u_1, u_2, ..., u_{n-1})$ a parameterization of the surface and $T(\tilde{\mathbf{u}})$ the corresponding transformation determinant.

Explicit results are available only for special forms of the failure surface. For example, if it is a hyperplane

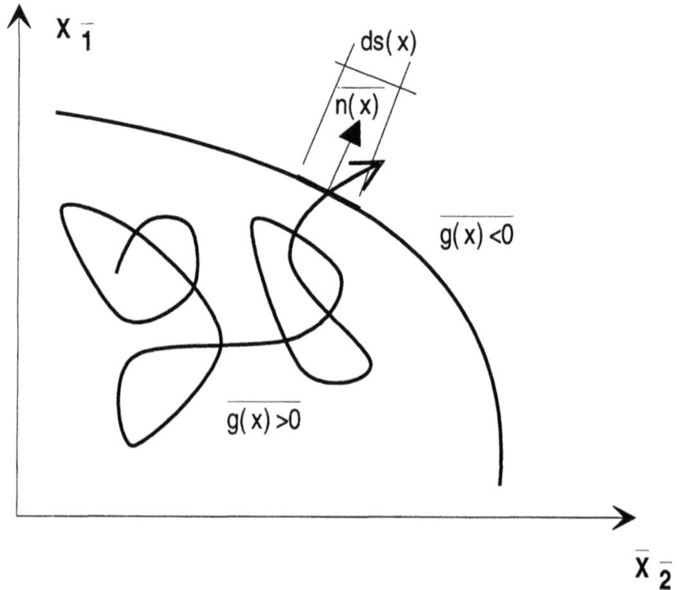

Figure 2. Outcrossing of a vectorial differentiable process

$$\partial V = \left\{ \sum_{i=1}^{n} \alpha_i u_i + \beta = 0 \right\} \tag{30}$$

the outcrossing rate of a stationary standardized Gaussian process is [51]:

$$\nu^+(\partial F) = E\left[\dot{U}_N\right] f(\partial F) = \frac{\kappa_N}{\sqrt{2\pi}} \varphi(\beta) \tag{31}$$

with $\kappa_N^2 = \alpha^T \ddot{\mathbf{R}}(\tau)\alpha$.. An asymptotic result for general non-linear surfaces has been derived in [7]:

$$\nu^+(\partial F) = \omega_0 \frac{\varphi(\beta)}{\sqrt{2\pi}} \prod_{i=1}^{n-1} (1 - \beta\kappa_i)^{-1/2} \tag{32}$$

with

$$\omega_0^2 = \mathbf{n}(\mathbf{u}^*)^T \left[\ddot{\mathbf{R}}(0) + \dot{\mathbf{R}}(0)^T \mathbf{G}(\mathbf{u}^*) \dot{\mathbf{R}}(0)\right] \mathbf{n}(\mathbf{u}^*)$$

provided that $g(\mathbf{0}) > 0$ and with $\dot{\mathbf{R}}(0) = \mathbf{E}\left[\mathbf{U}(0)\dot{\mathbf{U}}(0)^T\right]$ and

$$\mathbf{G}(\mathbf{u}^*) = \left\{ \nabla g(\mathbf{u}^*)^{-1} \frac{\partial^2 g(\mathbf{u}^*)}{\partial u_i \partial u_j}; i, j = 1, \ldots, n \right\}$$

Here again we have $\beta = \|\mathbf{u}^*\| = \min\{\|\mathbf{u}\|\}$ for $g(\mathbf{u}) \leq \mathbf{0}$ and κ_i are the main curvatures of ∂F in the solution point $\mathbf{u}^*$. Differentiable processes are used to model the turbulent natural wind, wind waves and earthquake excitations but also the output of dynamical systems.

Exact or approximate results have also been obtained for non-gaussian rectangular wave processes with or without correlated components [34], certain non-gaussian differentiable processes [14] and a variety of non-stationarities of the processes or the failure surfaces [35]. If one is not satisfied with the (asymptotic) approximations one can apply importance sampling methods in order to arrive at an arbitrarily exact result. Due to regularity of the crossings one can combine rectangular wave and differentiable processes. The processes can be intermittent [46], [22]. This allows the modelling of disturbances of short to very short duration (earthquakes, explosions). Such models have also been extended to deal with occurrence clustering [55], [45].

It is remarkable that the "critical" point $\mathbf{u}^*$, i.e. the magnitude of β, plays an important role in all cases as in the time-invariant case. It must be found by a suitable algorithm. Sequential quadratic programming algorithms tuned to the special problem of interest turned out to solve the optimization problem reliably and efficiently in practical applications [1].

However, it must be mentioned that in time-variant reliability more general models, e.g. renewal models with non-rectangular wave shapes, filtered Poisson process models, etc. can be easily formulated but hardly made practical from a computational point of view.

5. The Value of Human Life and Limb in the Public Interest

Two questions remain: a. Is it admissible to optimize benefits and cost if human lives are endangered and b. can we discount the "cost of human lives"? First of all, modern approaches to these questions do not speak of a monetary value of the human life but rather speak of the cost to save lives.. Secondly, any further argumentation must be within the framework of our moral and ethical principles as laid down in our constitutions and elsewhere. We quote as an example a few articles from the BASIC LAW of the Federal Republic of Germany:

- *Article 2: (1) Everyone has the right to the free development of his personality ...(2) Everyone has the right to life and to inviolability of his person*

- *Article 3: (1) All persons are equal before the law. (2) Men and women have equal rights. (3) No one may be prejudiced or favored because of his sex, his parentage, his race, his language, his homeland and origin, his faith or his religious or political opinions.*

Similar principles are found in all modern, democratic constitutions. But H. D. Thoreau (1817-1862 p.Chr.) realistically says about the value of human life: " *The cost of a thing is the amount of what I will call life which is required to be exchanged for it, immediately or in the long run.* ... [29].

Can these value fixings be transferred to engineering acceptability criteria? This is possible when starting from certain social indicators such as life expectancy, gross national product (GNP), state of health care, etc.. Life expectancy e is the area under the survivor curve $S(a)$ as a function of age a, i.e. $e = \int_0^\infty S(a)da$. A suitable measure for the quality of life is the GNP per capita, despite of some moral indignation at first sight. The GNP is created by labor and capital (stored labor). It provides the infrastructure of a country, its social structure, its cultural and educational offers, its ecological conditions among others but also the means for the individual enjoyment of life by consumption. Most importantly in our context, it creates the possibilities to "buy" additional life years through better medical care, improved safety in road traffic, more safety in or around building facilities or from hazardous technical activities, etc.. Safety of buildings via building codes is an investment into saving lives. The investments into structural safety must be efficient, however. Otherwise investments into other life saving activities are preferable. In all further considerations only about 60% of the GNP, i.e. g $\approx$ 0.6 GNP which is the part available for private use, are taken into account.

Denote by $c(\tau) > 0$ the consumption rate at age τ and by $u(c(\tau))$ the utility derived from consumption. Individuals tend to undervalue a prospect of future consumption as compared to that of present consumption. This is taken into account by some discounting. The life time utility for a person at age a until she/he attains age $t > a$ then is

$$\begin{aligned} U(a,t) &= \int_a^t u\left[c(\tau)\right] \exp\left[-\int_a^\tau \rho(\theta)d\theta\right] d\tau \\ &= \int_a^t u\left[c(\tau)\right] \exp\left[-\rho(\tau-a)\right] d\tau \end{aligned} \tag{33}$$

for $\rho(\theta) = \rho$. It is assumed that consumption is not delayed, i.e. incomes are not transformed into bequests. ρ should be conceptually distinguished from a financial interest rate and is referred to as rate of time preference of consumption. A rate $\rho > 0$ has been interpreted as the effect of human impatience, myopia, egoism, lack of telescopic faculty, etc.. Exponential population growth with rate n can be considered by replacing ρ by $\rho - n$ taking into account that families are by a factor $exp[nt]$ larger at a later time $t > 0$. The correction $\rho > n$ appears always necessary, simply because future generations are expected to be larger and wealthier. ρ is reported to be between 1 and 3% for health related investments, with tendency to lower values [53]. Empirical estimates reflecting pure consumption behavior vary considerably but are in part significantly larger [25].

The expected remaining present value life time utility at age a (conditional on having survived until a) then is (see [2] [43] [39] [15])

$$\begin{aligned} L(a) = E\left[U(a)\right] &= \int_a^{a_u} \frac{f(t)}{\ell(a)} U(a,t)dt \\ &= \int_a^{a_u} \frac{f(t)}{\ell(a)} \int_a^t u\left[c(\tau)\right] \exp\left[-(\rho - n)(\tau - a)\right] d\tau dt \\ &= \frac{1}{\ell(a)} \int_a^{a_u} u\left[c(t)\right] \exp\left[-(\rho - n)(t - a)\right] \ell(t)dt \\ &= u\left[c\right] e_d(a, \rho, n) \end{aligned} \tag{34}$$

where $f(t)dt = \left(\mu(\tau)\exp\left[-\int_0^t \mu(\tau)d\tau\right]\right) dt$ is the probability of dying between age t and $t + dt$ computed from life tables. The expression in the third line is obtained upon integration by parts. Also, a constant consumption rate c independent of t has been introduced which can be shown to be optimal under perfect market conditions [43]. The "discounted" life expectancy $e_d(a, \rho, n)$ at age a can be computed from

$$e_d(a, \rho, n) = \frac{\exp((\rho - n)a)}{\ell(a)} \int_a^{a_u} \exp\left[-\int_0^t (\mu(\tau) + (\rho - n))d\tau\right] dt \tag{35}$$

"Discounting" affects $e_d(a, \rho, n)$ primarily when $\mu(\tau)$ is small (i.e. at young age) while it has little effect for larger $\mu(\tau)$ at higher ages. It is important to recognize that "discounting" by ρ is initially with respect to $u\left[c(\tau)\right]$ but is formally included in the life expectancy term.

For $u\left[c\right]$ we select a power function

$$u\left[c\right] = \frac{c^q - 1}{q} \tag{36}$$

with $0 \leq q \leq 1$, implying constant relative risk aversion according to Arrow-Pratt. The form of eq. (36) reflects the reasonable assumption that marginal utility $\frac{du[c]}{dc} = c^{q-1}$ decays with consumption c. $u\left[c\right]$ is a concave function since $\frac{du[c]}{dc} > 0$ for $q \geq 0$ and $\frac{d^2u[c]}{dc^2} < 0$ for $q < 1$. The numerical value has been chosen to be about 0.2 (see [43] [15] and elsewhere as well as table 2 below). It may also be derived from from the work-leisure optimization principle as outlined in [29] where $q = \frac{w}{1-w}$ and w the average fraction of e devoted to (paid) work (see [37] for estimates derived from this principle). This magnitude has also been verified empirically (see, for example, [25]). For simplicity, we also take $c = g \gg 1$.

Shepard/Zeckhauser [43] now define the "value of a statistical life" at age a by converting eq. (34) into monetary units in dividing it by the marginal utility $\frac{du(c(t))}{dc(t)} = u'[c(t)]$:

$$\begin{aligned} VSL(a) &= \int_a^{a_u} \frac{u[c(t)]}{u'[c(t)]} \exp\left[-(\rho - n)(t-a)t\right] \frac{\ell(t)}{\ell(a)} dt \\ &= \frac{u[c]}{u'[c]} \frac{1}{\ell(a)} \int_a^{a_u} \exp\left[-(\rho - n)(t-a)\right] \ell(t) dt \\ &= \frac{g}{q} \frac{1}{\ell(a)} \int_a^{a_u} \exp\left[-(\rho - n)(t-a)\right] \ell(t) dt \\ &= \frac{g}{q} e_d(a, \rho, n) \end{aligned} \tag{37}$$

because $\frac{u[c(t)]}{u'[c(t)]} = \frac{g}{q}$. The "willingness-to-pay" has been defined as

$$WTP(a) = VSL(a)\, dm \tag{38}$$

In analogy to Pandey/Nathwani [31], and here we differ from the related economics literature, these quantities are averaged over the age distribution $h(a, n)$ in a stable population in order to take proper account of the composition of the population exposed to hazards in and from technical objects. One obtains the "societal value of a statistical life"

$$\overline{SVSL} = \frac{g}{q} \bar{E} \tag{39}$$

with

$$\bar{E} = \int_0^{a_u} e_d(a, \rho, n) h(a, n) da \tag{40}$$

and the "societal willingness-to-pay" as:

$$\overline{SWTP} = \overline{SVSL}\, dm \tag{41}$$

For $\rho = 0$ the averaged "discounted" life expectancy $\bar{E}$ is a quantity which is about 60% of e and considerably less than that for larger ρ. In this purely economic consideration it appears appropriate to define also the undiscounted average lost earnings in case of death, i.e. the so-called "human capital":

$$HC = \int_0^{a_u} g(e-a)h(a,n)da \tag{42}$$

Table 1 shows the $\overline{SVSL}$ for some selected countries as a function of ρ indicating the importance of a realistic assessment of ρ.

		France	Germany	Japan	Russia	USA
e		78	78	80	66	77
n		0.37%	0.27%	0.17%	-0.35	0.90%
g		14660	14460	15960	5440	22030
q		0.174	0.167	0.208	0.188	0.222
	0%	4.05	3.96	3.46	0.93	5.83
	1%	3.05	3.00	2.62	0.74	4.28
ρ	2%	2.38	2.36	2.06	0.61	3.27
	3%	1.92	1.92	1.67	0.51	2.59
	4%	1.59	1.59	1.39	0.54	2.11

Table 1: $\overline{SVSL}$ 10^6 in PPP US$ for some countries for various ρ (from recent complete life tables provided by national statistical offices)

It can reasonably be assumed that the life risk in and from technical facilities is uniformly distributed over the age and sex of those affected. Also, it is assumed that everybody uses such facilities and, therefore, is exposed to possible fatal accidents. The total cost of a safety related regulation per member of the group and year is $\overline{SWTP} = -dC_Y(p) = -\frac{1}{N}\sum_{i=1}^{r} dC_{Y,i}(p)$ where r is the total number of objects under discussion, each with incremental cost $dC_{Y,i}$ and N is the group size. For simplicity, the design parameter is temporarily assumed to be a scalar. This gives:

$$-dC_Y(p) + \overline{SVSL}\ dm = 0 \tag{43}$$

Let dm be proportional to the mean failure rate $dh(p)$, i.e. it is assumed that the process of failures and renewals is already in a stationary state that is for $t \to \infty$. Rearrangement yields

$$\frac{dC_Y(p)}{dh(p)} = -k\overline{SVSL} \tag{44}$$

where

$$dm = kdh(p), 0 < k \leq 1 \tag{45}$$

the proportionality constant k relating the changes in mortality to changes in the failure rate. Note that for any reasonable risk reducing intervention there is necessarily $dh(p) < 0$.

The criterion eq. (44) is derived for safety-related regulations for a larger group in a society or the entire society. Can it also be applied to individual technical projects? $\overline{SVSL}$ as well as HC were related to one anonymous person. For a specific project it makes sense to apply criterion (44) to the whole group exposed. Therefore, the "life saving cost" of a technical project with N_F potential fatalities is:

$$H_F = HC\ kN_F \tag{46}$$

The monetary losses in case of failure are decomposed into $H = H_M + H_F$ in formulations of the type eq. (10) with H_M the losses not related to human life and limb.

Criterion (44) changes accordingly into:

$$\frac{dC_Y(p)}{dh(p)} = -\overline{SVSL}kN_F \tag{47}$$

All quantities in eq. (47) are related to one year. For a particular technical project all design and construction cost, denoted by $dC(p)$, must be raised at the decision point $t = 0$. The yearly cost must be replaced by the erection cost $dC(p)$ at $t = 0$ on the left hand side of eq. (47) and discounting is necessary. The method of discounting is the same as for discharging an annuity. If the public is involved $dC_Y(p)$ may be interpreted as cost of societal financing of $dC(p)$. The interest rate to be used must then be a societal interest rate to be discussed below. Otherwise the interest rate is the market rate. g in $\overline{SVSL}$ also grows approximately exponentially with rate ζ, the rate of economic growth in a country. It can be taken into account by discounting. The acceptability criterion for individual technical projects then is (discount factor for discounted erection cost moved to the right hand side):

$$\begin{aligned} \frac{dC(p)}{dh(p)} &= -\frac{\exp[\gamma t] - 1}{\gamma \exp[\gamma t]} \overline{SVSL}kN_F \frac{\zeta \exp[\zeta t]}{\exp[\zeta t] - 1} \\ &\underset{t \to \infty}{\to} -\overline{SVSL}kN_F \frac{\zeta}{\gamma} \end{aligned} \tag{48}$$

It must be mentioned that a similar very convincing consideration about the necessary effort to reduce the risk for human life from technical objects has been given by Nathwani et al. [29] and in [31] producing estimates for the equivalent of the constant $\overline{SVSL}$ very close to those given in table 1. The estimates for $\overline{SVSL}$ are in good agreement with several other estimates in the literature (see, for example, [49], [43]; [52]; [24] and many others) which are between 1000000 and 10000000 PPP US$ with a clustering around 5000000 PPP US$.

6. Remarks about Interest Rates

A cost-benefit optimization must use interest rates. Considering the time horizon of some 20 to more than 100 years for most structural facilities but also for many risky industrial installations it is clear that average interest rates net of in/deflation must be chosen. If the option with systematic reconstruction is chosen one immediately sees from eq. (14) that the interest rate must be non-zero. For the same equation we see that there is a maximum interest rate $\gamma_{\max}$ for which $Z(\mathbf{p})$ becomes negative for any $\mathbf{p}$

$$\gamma_{\max} = \frac{m(\mathbf{p})b - (C(\mathbf{p}) + H)}{m(\mathbf{p})C(\mathbf{p})} \tag{49}$$

and, therefore, $0 < \gamma \leq \gamma_{\max}$. Also $m(\mathbf{p})b > C(\mathbf{p}) + H$ must be valid for any reasonable project which further implies that $b/\gamma > 1$. Very small interest rates, on the other hand, cause benefit and damage cost to dominate over the erection cost. Then, in the limit

$$Z(\mathbf{p}) = b - \frac{(C(\mathbf{p}) + H)}{m(\mathbf{p})} \tag{50}$$

where the interest rate vanishes. Erection cost are normally weakly increasing in the components of $\mathbf{p}$ but $m(\mathbf{p})$ grows significantly with $\mathbf{p}$. Consequently, the optimum is reached for $m(\mathbf{p}) \to \infty$, that is for perfect safety which is not attainable in practice. In other words the interest rate must be distinctly different from zero. Otherwise, the different parties involved in the project may use interest rates taken from the financial market at the decision point $t = 0$.

The cost for saving life years also enters into the objective function and with it the question of discounting those cost also arises. At first sight this is not in agreement with our moral value system. However, a number of studies summarized in [32] and [23] express a rather clear opinion based on ethical and economical arguments. The cost for saving life years must be discounted at the same rate as other investments, especially in view of the fact that our present value system should be maintained also for future generations. Otherwise serious inconsistencies cannot be avoided.

What should then the discount rate for public investments into life saving projects be? A first estimate could be based on the long term growth rate of the GNP. In most developed, industrial countries this was a little more than 2% over the last 50 years. The United Nations Human Development Report 2000 gives values between 1.2 and 1.9 % for industrialized countries during 1975-1998. If one extends the consideration to the last 120 years one finds an average growth rate ζ of about 1.8% (see table 1). Using data in [47], [27] and the UN Human Development Report 2000 [50] the following table has been compiled from a more detailed table.

	1850	1998								
Ctry.	GNP	GNP	g	e	$n\%$	q	$\rho\%$	$\zeta\%$	$\gamma\%$	$\overline{SVSL}$
UK	3109	23500	15140	77	0.23	0.19	0.5	1.3	1.3	$3.1 \cdot 10^6$
US	1886	34260	22030	78	0.90	0.22	1.3	1.8	2.3	$3.9 \cdot 10^6$
F	1840	24470	14660	78	0.37	0.17	0.7	1.9	1.9	$3.3 \cdot 10^6$
S	1394	23770	12620	79	0.02	0.18	0.3	1.9	1.6	$2.7 \cdot 10^6$
D	1400	25010	14460	77	0.27	0.17	0.6	1.9	1.9	$3.3 \cdot 10^6$
AUS	4027	25370	15750	80	0.99	0.21	0.7	1.2	1.9	$3.3 \cdot 10^6$
J	969	26460	15960	80	0.17	0.20	1.2	2.7	2.3	$2.8 \cdot 10^6$

Table 1: Social indices for some developed industrial countries (all monetary values are in US$, 1998)

It is noted that economic growth the first half of the last century was substantially below average while the second half was well above average. The above considerations can at least define the range of interest rates to be used in long term public investments into life saving operations. For the discount rates to be used in long term public investments the growth theory established by Solow [48] is applied, i.e.

$$n + \zeta(1 - \epsilon) < \rho < \gamma \leq \gamma_{\max} < n + \epsilon\zeta \tag{51}$$

where $\epsilon = 1 - q$ the so-called elasticity of marginal consumption (income). There is much debate about interest rates for long term public investments, especially if sustainability aspects are concerned. But there is an important mathematical result which may guide our choice. Weitzman [54] and others showed that the far-distant future should be discounted at the lowest possible rate > 0 if there are different possible scenarios each with a given probability of being true.

7. A One-Level Optimization for Structural Components

Let us now turn to the technical aspects of optimization. Cost-benefit optimization according to eq. (3) or (10) in principle requires two levels of optimization, one to minimize cost and the other to solve the reliability of optimization, one to minimize cost and the other to solve the reliability problem. However, it is possible to reduce it to one level by adding the Kuhn-Tucker condition of the reliability problem to the cost optimization task provided that the reliability task is formulated in the transformed standard space. For the task in eq. (3) we have

$$\begin{array}{ll}\textbf{Maximize:} & Z(\mathbf{p}) = B^* - C(\mathbf{p}) - (C(\mathbf{p}) + H_M + H_F) \cdot \frac{P_f(\mathbf{p})}{1-P_f(\mathbf{p}))} \\ \textbf{Subject to:} & \\ & g(\mathbf{u}, \mathbf{p}) = 0 \\ & u_i \|\nabla_{\mathbf{u}} g(\mathbf{u}, \mathbf{p})\| + \nabla_{\mathbf{u}} g(\mathbf{u}, \mathbf{p})_i \|\mathbf{u}\| = 0;\ i = 1, ..., n-1 \\ & h_k(\mathbf{p}) \leq 0, k = 1, \ldots, q \\ & \nabla_p C(\mathbf{p}) \geq k\overline{SVSL} N_F \frac{\varsigma}{\gamma} \nabla_p P_f(\mathbf{p}) \end{array} \tag{52}$$

where the first and second condition represent the Kuhn-Tucker condition for a valid "critical" point, the third condition some restrictions on the parameter vector $\mathbf{p}$ and the forth condition the human life criterion in eq. (48). Frequently, the term $\frac{P_f(\mathbf{p})}{1-P_f(\mathbf{p}))}$ in the objective can be replaced by $P_f(\mathbf{p})$. The failure probability is

$$P_f(\mathbf{p}) \approx \Phi(-\beta(\mathbf{p})) C_{SORM} \tag{53}$$

and we have to require that $\|\mathbf{u}\| \neq 0$ and $\|\nabla_{\mathbf{u}} g(\mathbf{u}, \mathbf{p})\| \neq 0$. It is assumed that the second-order correction C_{SORM} is nearly independent of $\mathbf{p}$. In fact, at the expense of some more numerical effort, one can use any update of the first-order result $\Phi(-\beta(\mathbf{p}))$, for example an update by importance sampling provided that the result of importance sampling is formulated as a correction factor to the first-order result. $\nabla_p C(\mathbf{p})$ usually must be determined numerically.

For time-variant problems as in eq. (10) one finds the outcrossing rate for a combination of rectangular wave and differentiable processes as:

$$\nu^+(\mathbf{p}) = \left(\sum_{i=1}^{n_J} \lambda_i \Phi(-\beta) + \omega_0 \frac{\varphi(\beta)}{\sqrt{2\pi}} \right) C_{SORM} \tag{54}$$

The optimization task is

$$
\begin{aligned}
&\textbf{Minimize:} \quad Z(\mathbf{p}) = \tfrac{b}{\gamma} - C(\mathbf{p}) - (C(\mathbf{p}) + H_M + H_F) \cdot \tfrac{\nu^+(\mathbf{p})}{\gamma} \\
&\textbf{Subject to:} \\
&\qquad g(\mathbf{u}, \mathbf{p}) = 0 \\
&\qquad u_i \|\nabla_{\mathbf{u}} g(\mathbf{u}, \mathbf{p})\| + \nabla_{\mathbf{u}} g(\mathbf{u}, \mathbf{p})_i \|\mathbf{u}\| = 0; \; i = 1, ..., n-1 \\
&\qquad h_k(\mathbf{p}) \leq 0, k = 1, \ldots, q \\
&\qquad \nabla_p C(\mathbf{p}) \geq k\overline{SVSL} N_F \tfrac{\zeta}{\gamma} \nabla_p \nu^+(\mathbf{p})
\end{aligned}
\tag{55a}
$$

For the case in eq. (15) one replaces $\frac{\nu^+(\mathbf{p})}{\gamma}$ by $\frac{\lambda P_f(\mathbf{p})}{\gamma}$ and $\nabla_p \nu^+(\mathbf{p})$ by $\nabla_p \lambda P_f(\mathbf{p})$.

The optimization tasks in eq. (52) or in (55a) are conveniently performed by suitable SQP-algorithms (for example, [44], [33]). For both formulations eq. (52) and (55a), respectively, gradient-based optimizers require the gradients of the objective as well as the gradients of all constraints. This means that second derivatives are required in order to calculate the gradient of second condition as well as of the human value criterion, in particular, the entries into the Hessian of $g(\mathbf{u}, \mathbf{p})$. This is also the most serious objection against this form of a one level approach. One can, however, proceed iteratively for well-behaved failure surfaces. Initially, one assumes a linear or linearized failure surface and sets $C_{SORM}^{(0)} = 1$. Then, all entries $\frac{\partial^2 g(\mathbf{u}.\mathbf{p})}{\partial u_i \partial u_j}$ are zero. After a first solution of problem (52) or (55a) one determines the Hessian once in the solution point $(\mathbf{u}^{*(1)}, \mathbf{p}^{(1)})$ and with it also calculates $C_{SORM}^{(1)}$. Problems (52) or (55a) are then solved a second time with fixed Hessian $\mathbf{G}(\mathbf{u}^{*(1)}, \mathbf{p}^{(1)})$ and so forth. This schemes is repeated until convergence is reached which usually is after a few steps. From a practical point of view it is frequently sufficient to use first-order reliability results and no iteration is necessary.

In closing this section it is important to note that the optimization tasks as formulated in eq. (52) and (55a) are among the easiest one can think of. In practice safety related design decisions additionally include changes in the lay-out, in the structural system or in the maintenance strategy. Optimization is over discrete sets of design alternatives. Clearly, this is more difficult and very little is known how to do it formally except in a heuristic, empirical manner in small dimensions.

8. Example

As an example we take a rather simple case of a system where failure is defined if the random resistance or capacity is exceeded by the random demand, i.e. the failure event is defined as $F = \{R - S(t) \leq 0\}$. The demand is modelled as a one-dimensional, stationary marked Poissonian renewal process of disturbances (earthquakes, wind storms, explosions, etc.) with stationary renewal rate λ and random, independent sizes of the disturbances $S_i, i = 1, 2, \ldots$..Random resistance is log-normally distributed with mean p and a coefficient of variation V_R. The disturbances are also independently log-normally distributed with mean equal to unity and coefficient of variation V_S. A disturbance causes failure with probability:

$$P_f(p) = \Phi\left(-\frac{\ln\left\{p\sqrt{\frac{1+V_S^2}{1+V_R^2}}\right\}}{\sqrt{\ln\left((1+V_R^2)(1+V_S^2)\right)}}\right) \tag{56}$$

Thus, the failure rate is $\lambda P_f(p)$ and the Laplace transform of the renewal density is:

$$h^*(\gamma, p) = \frac{\lambda P_f(p)}{\gamma} \tag{57}$$

An appropriate objective function given systematic reconstruction then is

$$\frac{Z(p)}{C_0} = \frac{b}{\gamma C_0} - \left(1 + \frac{C_1}{C_0}p^a\right) - \left(1 + \frac{C_1}{C_0}p^a + \frac{H_M}{C_0} + \frac{H_F}{C_0}\right)\frac{\lambda P_f(p)}{\gamma} \tag{58}$$

which is to be maximized. The criterion (62) has the form:

$$\frac{d}{dp}\left(1 + \frac{C_1}{C_0}p^a\right) \geq -k\overline{SVSL}N_F\frac{\zeta}{\gamma}\frac{d}{dp}\left(\lambda P_f(p)\right) \tag{59}$$

Some more or less realistic, typical parameter assumptions are: $C_0 = 10^6$, $C_1 = 10^4$, $a = 1.25$, $H_M = 3 \cdot C_0$, $V_R = 0.2$, $V_S = 0.3$, and $\lambda = 1$ $[1/year]$. The socio-economic demographic data are $e = 77$, $GDP = 25000$, $g = 15000$, $w = 0.15, N_F = 100, k = 0.1$ so that $H_F = HC$ $kN_F = 5.8 \cdot 10^6$ and $\overline{SVSL}kN_F = 3.3 \cdot 10^7$. The value of N_F is chosen relatively large for demonstration purposes. Monetary values are in US$. Optimization is performed for the public and for the owner separately.

For the public $b_S = 0.02C_0$ and $\gamma_S = 0.0185$ are chosen. Also, we take $\frac{\zeta}{\gamma_S} = 1$ for simplicity. In particular, benefit and discount rate are chosen such that the public does not make direct profit from an economic activity of its members. Optimization including the cost H_F gives $p_S^* = 4.35$, the corresponding failure rate is $1.2 \cdot 10^{-5}$. Criterion (48) is already fulfilled for $p_l = 3.48$ corresponding to a yearly failure rate of $1.6 \cdot 10^{-4}$ but $Z_S(p_l)/C_0$ being already negative. It is interesting to see that in this case the public can do better in adopting the optimal solution rather than just realizing the facility at its acceptability limit as pointed out already earlier.

The owner uses some typical values of $b_O = 0.07C_0$ and $\gamma_O = 0.05$ and does or does not include life saving cost. If he includes life saving cost the objective function is shifted to the right (dashed line). The calculations yield $p_O^* = 3.76$ and $p_O^* = 4.03$, respectively, and the corresponding failure rates are $7.1 \cdot 10^{-5}$ and $3.2 \cdot 10^{-5}$. The acceptability criterion limits the owner's region for reasonable designs. Inclusion of life saving cost has relatively little influence on the position of the optimum.

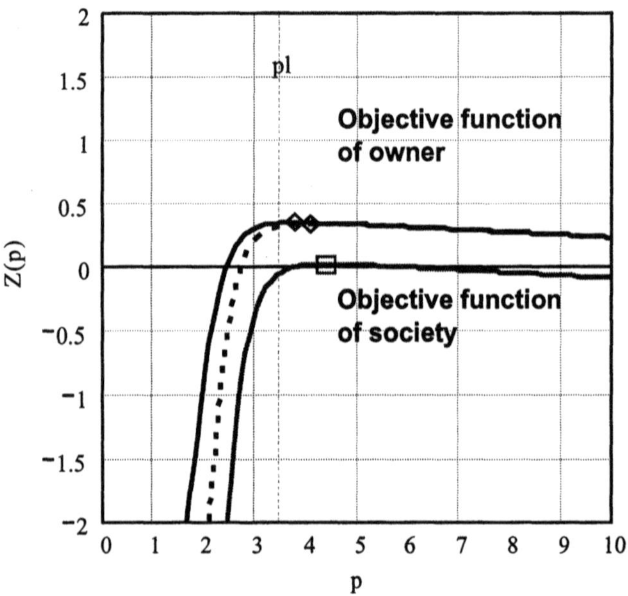

Figure 3. Objective function of owner and society

It is noted that the stochastic model and the variability of capacity and demand also play an important role for the magnitude and location of the optimum as well as the acceptability limit. The specific marginal cost (rate of change) of a safety measure and its effect on a reduction of the failure rate are equally important.

9. Conclusions

Optimization techniques are essential ingredients of reliability-oriented optimal designs of technical facilities. Although many technical aspects are not yet solved and the available spectrum of models and methods in structural reliability is still limited many practical problems can be solved. A special one-level optimization is proposed for general cost-benefit analysis. In this paper, however, focus is on some more critical issues, for example, "what is a reasonable replacement strategy for structural facilities?", "how safe is safe enough?" and "how to discount losses of material, opportunity and human lives?". An attempt has been made to give at least partial answers. Only if those issues have an answer overall optimization of technical facilities with respect to cost makes sense.

References

[1] β−point algorithms for large variable problems in time-invariant and time-variant reliability, Proc. 3rd IFIP WG 7.5 Working Conference, Berkeley, 1990, pp. 1-12, Springer, Berlin, 1990

[2] Arthur, W.B., The Economics of Risks to Life, American Economic Review, 71, pp. 54-64, 1981

[3] Belyaev, Y. K., On the Number of Exits across the Boundary of a Region by a Vector Stochastic Process, Theor. Prob. Appl., 1968, 13, pp. 320-324

[4] Bleistein, N., Handelsman, R.A., Asymptotic Expansions of Integrals, Holt, Rinehart and Winston, New York, 1975

[5] Breitung, K., Asymptotic Approximations for Multinormal Integrals, Journ. of the Eng. Mech. Div., 110, N3, 1984, pp. 357-366

[6] Breitung, K., Asymptotic Approximations for Probability Integrals, Prob. Eng. Mech., 1989, 4, 4, pp. 187-190

[7] Breitung, K., Asymptotic Crossing Rates for Stationary Gaussian Vector Processes, Stochastic Processes and their Applications, 29, 1988, pp. 195-207

[8] Breitung, K., Asymptotic Approximations for the Crossing Rates of Poisson Square Waves, Proc. of the Conf. on Extreme Value Theory and Applications, Gaithersburg/Maryland, NIST Special Publication 866, 3, 1993, pp. 75-80 1

[9] Breitung, K., Rackwitz, R., Nonlinear Combination of Load Processes, Journ. of Struct. Mech., 10, 2, 1982, pp. 145-166

[10] Cantril, H., The Pattern of Human Concerns, New Brunswik, N.J., Rutgers University Press, 1965

[11] Cox, D.R., Renewal Theory, Methuen, 1962

[12] Cox, D.R., Isham, V., Point Processes, Chapman & Hall, London, 1980

[13] Cramer, H., Leadbetter, M.R., Stationary and Related Stochastic Processes. Wiley, New York, 1967

[14] Grigoriu, M., Crossings of Non-Gaussian Translation Processes, Journal of the Engineering Mechanics Division, ASCE, 110, EM4, 1984, pp. 610-620

[15] Cropper, M.L., Sussman, F.G., Valuing Future Risks to Life, Journ. Environmental Economics and Management, 19, pp. 160-174, 1990

[16] Hasofer, A.M., Lind, N.C., An Exact and Invariant First Order Reliability Format, Journ. of Eng. Mech. Div., ASCE, 100, EM1, 1974, pp. 111-121

[17] Hasofer, A.M., Design for Infrequent Overloads, Earthquake Eng. and Struct. Dynamics, 2, 4, 1974, pp. 387-388

[18] Hasofer, A.M., Rackwitz, R., Time-dependent models for code optimization, Proc. ICASP'99, (ed. R.E. Melchers & M. G. Stewart), Balkema, Rotterdam, 2000, 1, pp. 151-158

[19] Hohenbichler, M., Rackwitz, R., Non-Normal Dependent Vectors in Structural Safety, Journ. of the Eng. Mech. Div., ASCE, 107, 6, 1981, pp. 1227-1249

[20] Hohenbichler, M., Gollwitzer, S., Kruse, W., Rackwitz, R., New Light on First- and Second-Order Reliability Methods, Structural Safety, 4, pp. 267-284, 1987

[21] Hohenbichler, M.; Rackwitz,.R.: Sensitivity and Importance Measures in Structural Reliability, Civil Engineering Systems, 3, 4, 1986, pp 203-209

[22] Iwankiewicz, R., Rackwitz, R., Non-stationary and stationary coincidence probabilities for intermittent pulse load processes, Probabilistic Engineering Mechanics, 2000, 15, pp. 155-167

[23] Lind, N.C., Target Reliabilities from Social Indicators, Proc. ICOSSAR93, Balkema, 1994, pp. 1897-1904

[24] Lutter, R., Morrall, J.F., Health-Health Analysis, A New Way to Evaluate Health and Safety Regulation, Journ. Risk and Uncertainty, 8, pp. 43-66, 1994

[25] Kapteyn, A., Teppa, F., Hypothetical Intertemporal Consumption Choices, Working paper, CentER, Tilburg University, Netherlands, 2002

[26] Kuschel, N., Rackwitz, R., Two Basic Problems in Reliability-Based Structural Optimization, Mathematical Methods of Operations Research, 46, 1997, 309-333

[27] Maddison, A., Monitoring the World Economy 1820-1992, OECD, Paris, 1995

[28] Madsen, H.O., Lind, N., Krenk, S., Methods of Structural Safety, Prentice-Hall, Englewood Cliffs, 1987

[29] Nathwani, J.S., Lind, N.C., Pandey, M.D., Affordable Safety by Choice: The Life Quality Method, Institute for Risk Research, University of Waterloo, Waterloo, Canada, 1997

[30] Paez, A., Torroja, E., La determinacion del coefficiente de seguridad en las distintas obras, Instituto Tecnico de la Construccion y del Cemento, Madrid, 1952

[31] Pandey, M.D., Nathwani, J.S., Canada Wide Standard for Particulate Matter and Ozone: Cost-Benefit Analysis using a Life-Quality Index, to be published in Journ. Risk Analysis, 2002

[32] Pate-Cornell, M.E., Discounting in Risk Analysis: Capital vs. Human Safety, Proc. Symp. Structural Technology and Risk, University of Waterloo Press, Waterloo, ON, 1984

[33] The Linearization Method for Constrained Optimization, Springer, Berlin, 1994

[34] Rackwitz, R., Reliability of Systems under Renewal Pulse Loading, Journ. of Eng. Mech., ASCE, 111, 9, 1985, pp. 1175-1184

[35] Rackwitz, R., On the Combination of Non-stationary Rectangular Wave Renewal Processes, Structural Safety, 13, 1+2, 1993, pp 21-28

[36] Rackwitz, R., Optimization - The Basis of Code Making and Reliability Verification, Structural Safety, 22, 1, 2000, pp.27-60

[37] Rackwitz, R., Optimization and Risk Acceptability based on the Life Quality Index, Structural Safety, 24, pp. 297-331, 2002

[38] Rice, S.O., Mathematical Analysis of Random Noise, Bell System Tech. Journ., 32, 1944, pp. 282 and 25, 1945, pp. 46

[39] Rosen, S., The Value of Changes in Life Expectancy, Journ. Risk and Uncertainty, 1, pp. 285-304, 1988

[40] Rosenblueth, E., Optimum Design for Infrequent Disturbances, Journ, Struct. Div., ASCE, 102, ST9, 1976, pp. 1807-1825

[41] Rosenblueth, E., Esteva, L., Reliability Basis for some Mexican Codes, in: ACI Spec. Publ., SP-31, Detroit, 1972

[42] Rosenblueth, E., Mendoza, E., Reliability Optimization in Isostatic Structures, Journ. Eng. Mech. Div., ASCE, 97, EM6, 1971, pp. 1625-1642

[43] Shepard, D.S., Zeckhauser, R.J., Survival versus Consumption, Management Science, 30, 4, pp. 423-439, 1984

[44] Schittkowski, K., Theory, Implementation, and Test of a Nonlinear Programming Algorithm. In: Eschenauer, H., Olhoff, N. (eds.), Optimization Methods in Structural Design, Proc. Euromech Colloquium 164, Universität Siegen, Oct. 12-14, 1982, Zürich 1983

[45] Schrupp, K., Rackwitz, R., Outcrossing Rates of Marked Poisson Cluster Processes in Structural Reliability, Appl. Math. Modelling, 12, 1988, Oct., 482-490

[46] Shinozuka, M., Stochastic Characterization of Loads and Load Combinations, Proc. 3rd ICOSSAR, Trondheim 32-25 June, 1981, Structural Safety and Reliability, T. Moan and M. Shinozuka (Eds.), Elsevier, Amsterdam, 1981

[47] Steckel, R.H., Floud, R., Health and Welfare during Industrialization, University of Chicago Press, Chicago, 1997

[48] Solow, R.M., Growth Theory, Clarendon Press, Oxford, 1970

[49] Tengs, T.O., Adams, M.E., Pliskin, J.S., Safran, D.G., Siegel, J.E., Weinstein, M.C., Graham, J.D., Five-Hundred Life-Saving Interventions and Their Cost-Effectiveness, Risk Analysis, 15, 3, pp. 369-390, 1995

[50] United Nations, HDR 2000, http://www.undp.org/hdr2000/english/HDR2000.html

[51] Veneziano, D., Grigoriu, M., Cornell, C.A., Vector-Process Models for System Reliability, Journ. of Eng. Mech. Div., ASCE, 103, EM 3, 1977, pp. 441-460

[52] Viscusi, W.K., The Valuation of Risks to Life and Health, Journ. Economic Literature, XXXI, pp. 1912-1946, 1993

[53] Viscusi, W.K., Discounting health effects on medical decision, in: Valuing Health Care, Costs, benefits and effectiveness of pharmaceuticals and other medical technologies, F.A. Sloan (ed), Cambridge University Press, pp. 125-147, 1996

[54] Weitzman, M.L., Why the Far-Distant Future Should Be Discounted at Its Lowest Possible Rate, Journal of Environmental Economics and Management, 36, pp. 201-208, 1998

[55] Wen, Y.K., A Clustering Model for Correlated Load Processes, Journ. of the Struct. Div., ASCE, 107, ST5, 1981, pp. 965-983

PROBABILITY OBJECTIVES IN STOCHASTIC PROGRAMS WITH RECOURSE

Rüdiger Schultz
Institute of Mathematics
University Duisburg-Essen
Lotharstr. 65
D-47048 Duisburg, Germany
schultz@math.uni-duisburg.de

Abstract Traditional models in multistage stochastic programming are directed to minimizing the expected value of random optimal costs arising in a multistage, non-anticipative decision process under uncertainty. Motivated by risk aversion, we consider minimization of the probability that the random optimal costs exceed some preselected threshold value. For the two-stage case, we analyse structural properties and propose algorithms both for models with integer decisions and for those without. Extension of the modeling to the multistage situation concludes the paper.

Keywords: Stochastic programming, mixed-integer optimization.

1. Introduction

Stochastic programs with recourse arise as deterministic equivalents to random optimization problems. In the present paper the main accent will be placed at the two-stage situation, and the most general random optimization problems to be considered are random mixed-integer linear programs. These are accompanied by a two-stage scheme of alternating decision and observation. After having decided on parts of the variables in a first stage, the random data infecting the problem are observed, and in turn the remaining (second-stage or recourse) variables are fixed. In our present analysis two basic assumptions underly this scheme. First, and naturally, the first-stage decision has to be taken on a "here-and-now" basis, i.e., it must not depend on (or anticipate) the outcome of the random data. Secondly, and providing some modeling restriction, the first-stage decision does not influence the probability distribution of

the random data.
In multistage stochastic programs the above two-stage scheme is extended into a finite horizon sequential decision process under uncertainty. Again we have to maintain nonanticipativity of decisions, and, so far, almost all results concern problems where the decisions do not influence the probability distribution of the random data. In the final section of the present paper we will return to multistage stochastic programs.
After having sketched the rules for how to make decisions, let us now discuss criteria for how to select a "best" decision. In this respect, the existing literature on stochastic programs with recourse (cf. the textbooks [5, 15, 20] and the references therein) almost unanimously suggests to start out from expectations of objective function values of the random optimization problem. For two-stage models (in a cost minimization framework) this implies that the deterministic first-stage decision is selected such that the expectation of the sum of the deterministic first-stage costs and the random second-stage costs (induced by the random data and an optimal second-stage decision) becomes minimal. Such a criterion has proven useful in many applications. In case the random optimization problem is a linear program without integer requirements, the resulting stochastic program with recourse enjoys convexity in the first-stage variables. This enabled application of powerful tools from convex analysis, both for structural investigations and algorithm design (cf. [4, 5, 15, 20, 32]).
In the present paper, we will discuss recourse stochastic programs where the optimization is based on minimizing the probability that the above sum of deterministic and random costs exceeds a given threshold value. Such models provide an opportunity to address risk aversion in the framework of recourse stochastic programming.
The proposal to replace the usual expectation-based objective function in recourse stochastic programming by a probability objective seemingly dates back to Bereanu [2] and, hitherto, has not been elaborated in much detail. Reformulating the stochastic program by adding another variable and including level sets of the objective into the constraints leads to a chance constrained stochastic program which is nonconvex in general. We will see that, along this line, some structural knowledge on chance constraints (cf. [5, 15, 16, 20, 29]) reappears in the structural analysis of our models. Algorithmically, we will view several well-established techniques from a fresh perspective. Among them there are cutting planes from convex subgradient optimization, Lagrangian relaxation of mixed-integer programs, and decomposition techniques for block-angular stochastic programs.
The paper is organized as follows. In Section 2 we formalize the mod-

eling outlined above, collect some prerequisites, and compare with the usual expectation-based modeling in recourse stochastic programming. Section 3 is devoted to structural results. In Section 4 we present some first algorithmic approaches. Separate attention is paid to models without integer decisions since they allow for an algorithmic shortcut. As already announced, the final section will discuss the extension of our modeling to multistage stochastic programs.

2. Modeling

Consider the following random mixed-integer linear program

$$\begin{aligned} & \min_{x,y,y'} c^T x + q^T y + q'^T y' \\ s.t. \quad & Tx + Wy + W'y' = h(\omega), \\ & x \in X,\ y \in \mathbb{Z}_+^{\bar{m}},\ y' \in \mathbb{R}_+^{m'} \end{aligned} \tag{1}$$

We assume that all ingredients above have conformal dimensions, that W, W' are rational matrices, and that $X \subseteq \mathbb{R}^m$ is a nonempty closed polyhedron. Integer requirements to components of x are formally possible but will not be imposed for ease of exposition. For the same reason, randomness is kept as simple as possible by claiming that only the right-hand side $h(\omega) \in \mathbb{R}^s$ is random, i.e., a random vector on some probability space $(\Omega, \mathcal{A}, \mathbb{P})$. Decision variables are divided into two groups: first-stage variables x to be fixed before and second-stage variables (y, y') to be fixed after observation of $h(\omega)$.
Let us denote

$$\Phi(t) := \min\{q^T y + q'^T y' \ :\ Wy + W'y' = t,\ y \in Z_+^{\bar{m}},\ y' \in \mathbb{R}_+^{m'}\}. \tag{2}$$

According to integer programming theory([19]), this function is real-valued on $\mathbb{R}^s$ provided that $W(\mathbb{Z}_+^{\bar{m}}) + W'(\mathbb{R}_+^{m'}) = \mathbb{R}^s$ and $\{u \in \mathbb{R}^s \ :\ W^T u \leq q,\ W'^T u \leq q'\} \neq \emptyset$ which, therefore, will be assumed throughout.
The classical expectation-based stochastic program with recourse now is the optimization problem

$$\min\Big\{ \int_\Omega (c^T x + \Phi(h(\omega) - Tx))\, \mathbb{P}(d\omega) \ :\ x \in X \Big\}. \tag{3}$$

The recourse stochastic program with probability objective reads

$$\min\Big\{ \mathbb{P}(\{\omega \in \Omega \ :\ c^T x + \Phi(h(\omega) - Tx) > \varphi_o\}) \ :\ x \in X \Big\} \tag{4}$$

where $\varphi_o \in \mathbb{R}$ denotes some preselected threshold (some ruin level in a cost framework, for instance). For convenience, we will call (3) the

expectation-based and (4) the probability-based recourse model. In doing so, we are well aware of the fact that, of course, (4) is expectation-based too, if probabilities are understood as expectations of indicator functions.
We will see in a moment, that both (3) and (4) are well-defined non-linear optimization problems. Their objective functions are denoted by $Q_{\mathbb{E}}(x)$ and $Q_{\mathbb{P}}(x)$, respectively. To detect their structure, the function Φ is crucial, which arises as a value function of a mixed-integer linear program. From parametric optimization ([1, 6]) the following is known

Proposition 2.1 *Assume that* $W(\mathbb{Z}_+^{\bar{m}}) + W'(\mathbb{R}_+^{m'}) = \mathbb{R}^s$ *and* $\{u \in \mathbb{R}^s : W^T u \leq q,\ {W'}^T u \leq q'\} \neq \emptyset$. *Then it holds*

(i) Φ *is real-valued and lower semicontinuous on* $\mathbb{R}^s$,

(ii) there exists a countable partition $\mathbb{R}^s = \cup_{i=1}^\infty \mathcal{T}_i$ *such that the restrictions of* Φ *to* $\mathcal{T}_i$ *are piecewise linear and Lipschitz continuous with a uniform constant* $L > 0$ *not depending on* i,

(iii) each of the sets $\mathcal{T}_i$ *has a representation* $\mathcal{T}_i = \{t_i + \mathcal{K}\} \setminus \cup_{j=1}^N \{t_{ij} + \mathcal{K}\}$ *where* $\mathcal{K}$ *denotes the polyhedral cone* $W'(\mathbb{R}_+^{m'})$ *and* t_i, t_{ij} *are suitable points from* $\mathbb{R}^s$, *moreover,* N *does not depend on* i,

(iv) there exist positive constants β, γ *such that* $|\Phi(t_1) - \Phi(t_2)| \leq \beta \|t_1 - t_2\| + \gamma$ *whenever* $t_1, t_2 \in \mathbb{R}^s$.

In case $\bar{m} = 0$, i.e., if there are no integer requirements in the second stage, Φ becomes the value function of a linear program. Under the assumptions of Proposition 2.1, Φ is real-valued on $\mathbb{R}^s$. By linear programming duality it is convex, piecewise linear, and adopts a representation

$$\Phi(t) = \max_{j=1,\dots,J} d_j^T t$$

where $d_1, \dots, d_J$ are the vertices of $\{u \in \mathbb{R}^s : {W'}^T u \leq q'\}$, which is a compact set in this case.
As an immediate conclusion we obtain, that, without integer requirements in the second stage, $1 - Q_{\mathbb{P}}(x)$ coincides with the probability of a closed polyhedron, providing a direct link to chance constrained stochastic programming ([5, 15, 20]).
Before we will turn our attention to $Q_{\mathbb{P}}(x)$, we review some properties of $Q_{\mathbb{E}}(x)$. For convenience we denote by μ the image measure $\mathbb{P} \circ h^{-1}$ on $\mathbb{R}^s$. Without integer requirements ($\bar{m} = 0$), convexity of Φ extends to $Q_{\mathbb{E}}$ under mild conditions. A standard result of stochastic linear programming reads

Proposition 2.2 *Assume $\bar{m} = 0$, $W'(\mathbb{R}_+^{m'}) = \mathbb{R}^s, \{u \in \mathbb{R}^s : W'^T u \leq q'\} \neq \emptyset$, and $\int_{\mathbb{R}^s} \|h\| \, \mu(dh) < \infty$. Then $Q_{\mathbb{E}} : \mathbb{R}^m \longrightarrow \mathbb{R}$ is a real-valued convex function.*

As already mentioned in the introduction, convexity has been exploited extensively in stochastic linear programming. For further reading we refer to the textbooks [5, 15, 20]. The remaining models, both expectation- and probability-based, to be discussed in the present paper enjoy convexity merely in exceptional situations. Straightforward examples (cf. e.g. [35]) confirm that convexity in (3) is lost already for very simple models as soon as integer requirements enter the second stage. In [33] the following is shown.

Proposition 2.3 *Assume that $W(\mathbb{Z}_+^{\bar{m}}) + W'(\mathbb{R}_+^{m'}) = \mathbb{R}^s$, $\{u \in \mathbb{R}^s : W^T u \leq q,\ W'^T u \leq q'\} \neq \emptyset$, and $\int_{\mathbb{R}^s} \|h\| \, \mu(dh) < \infty$. Then it holds*

(i) $Q_{\mathbb{E}} : \mathbb{R}^m \longrightarrow \mathbb{R}$ is a real-valued lower semicontinuous function,

(ii) if μ has a density, then $Q_{\mathbb{E}}$ is continuous on $\mathbb{R}^m$.

3. Structure

To analyse the structure of $Q_{\mathbb{P}}$ we introduce the notation

$$M(x) := \{h \in \mathbb{R}^s : c^T x + \Phi(h - Tx) > \varphi_o\}, \quad x \in \mathbb{R}^m.$$

By $\liminf_{x_n \to x} M(x_n)$ and $\limsup_{x_n \to x} M(x_n)$ we denote the (set theoretic) limes inferior and limes superior, i.e., the set of all points belonging to all but a finite number of the sets $M(x_n)$, $n \in \mathbb{N}$, and to infinitely many of the sets $M(x_n)$, respectively. Moreover, we denote

$$\begin{aligned} M_e(x) &:= \{h \in \mathbb{R}^s : c^T x + \Phi(h - Tx) = \varphi_o\}, \\ M_d(x) &:= \{h \in \mathbb{R}^s : \Phi \text{ is discontinuous at } h - Tx\}. \end{aligned}$$

Note that, by Proposition 2.1, both $M_e(x)$ and $M_d(x)$ are measurable sets for all $x \in \mathbb{R}^m$.

Lemma 3.1 *For all $x \in \mathbb{R}^m$ there holds*

$$M(x) \subseteq \liminf_{x_n \to x} M(x_n) \subseteq \limsup_{x_n \to x} M(x_n) \subseteq M(x) \cup M_e(x) \cup M_d(x).$$

Proof: Let $h \in M(x)$. The lower semicontinuity of Φ (Proposition 2.1) yields

$$\liminf_{x_n \to x} (c^T x_n + \Phi(h - Tx_n)) \geq c^T x + \Phi(h - Tx) > \varphi_o.$$

Therefore, there exists an $n_o \in I\!N$ such that $c^T x_n + \Phi(h - Tx_n) > \varphi_o$ for all $n \geq n_o$, implying $h \in M(x_n)$ for all $n \geq n_o$. Hence, $M(x) \subseteq \liminf_{x_n \to x} M(x_n)$.
Let $h \in \limsup_{x_n \to x} M(x_n) \setminus M(x)$. Then there exists an infinite subset $\tilde{I\!N}$ of $I\!N$ such that

$$c^T x_n + \Phi(h - Tx_n) > \varphi_o \; \forall n \in \tilde{I\!N} \quad \text{and} \quad c^T x + \Phi(h - Tx) \leq \varphi_o.$$

Now two cases are possible. First, Φ is continuous at $h - Tx$. Passing to the limit in the first inequality then yields that $c^T x + \Phi(h - Tx) \geq \varphi_o$, and $h \in M_e(x)$. Secondly, Φ is discontinuous at $h - Tx$. In other words, $h \in M_d(x)$. □

Proposition 3.2 *Assume that* $W(\mathbb{Z}_+^{\bar{m}}) + W'(I\!R_+^{m'}) = I\!R^s$ *and* $\{u \in I\!R^s : W^T u \leq q,\ W'^T u \leq q'\} \neq \emptyset$. *Then* $Q_{I\!P} : I\!R^m \longrightarrow I\!R$ *is a real-valued lower semicontinuous function.*
If in addition $\mu(M_e(x) \cup M_d(x)) = 0$, *then* $Q_{I\!P}$ *is continuous at* x.

Proof: The lower semicontinuity of Φ ensures that $M(x)$ is measurable for all $x \in I\!R^m$, and hence $Q_{I\!P}$ is real-valued on $I\!R^m$. By Lemma 3.1 and the (semi-) continuity of the probability measure on sequences of sets we have for all $x \in I\!R^m$

$$\begin{aligned} Q_{I\!P}(x) &= \mu(M(x)) \leq \mu(\liminf_{x_n \to x} M(x_n)) \leq \\ &\leq \liminf_{x_n \to x} \mu(M(x_n)) = \liminf_{x_n \to x} Q_{I\!P}(x_n), \end{aligned}$$

establishing the asserted lower semicontinuity. In case

$$\mu(M_e(x) \cup M_d(x)) = 0$$

this argument extends as follows

$$\begin{aligned} Q_{I\!P}(x) &= \mu(M(x)) = \mu(M(x) \cup M_e(x) \cup M_d(x)) \geq \\ &\geq \mu(\limsup_{x_n \to x} M(x_n)) \geq \limsup_{x_n \to x} \mu(M(x_n)) = \limsup_{x_n \to x} Q_{I\!P}(x_n), \end{aligned}$$

and $Q_{I\!P}$ is continuous at x. □

Proposition 2.1 now reveals that, for given $x \in I\!R^m$, both $M_e(x)$ and $M_d(x)$ are contained in a countable union of hyperplanes. The latter being of Lebesgue measure zero we obtain that $\mu(M_e(x) \cup M_d(x)) = 0$ is valid for all $x \in I\!R^m$ provided that μ has a density. This proves

Conclusion 3.3 *Assume that $W(\mathbb{Z}_+^{\bar{m}}) + W'(\mathbb{R}_+^{m'}) = \mathbb{R}^s$, $\{u \in \mathbb{R}^s : W^T u \leq q,\ W'^T u \leq q'\} \neq \emptyset$, and that μ has a density. Then $Q_{\mathbb{P}}$ is continuous on $\mathbb{R}^m$.*

This analysis can be extended towards Lipschitz continuity of $Q_{\mathbb{P}}$. In [36], Tiedemann has shown

Proposition 3.4 *Assume that q, q' are rational vectors, $W(\mathbb{Z}_+^{\bar{m}}) + W'(\mathbb{R}_+^{m'}) = \mathbb{R}^s$, $\{u \in \mathbb{R}^s : W^T u \leq q, W'^T u \leq q'\} \neq \emptyset$, and that for any nonsingular linear transformation $B \in L(\mathbb{R}^s, \mathbb{R}^s)$ all one-dimensional marginal distributions of $\mu \circ B$ have bounded densities which, outside some bounded interval, are monotonically decreasing with growing absolute value of the argument. Then $Q_{\mathbb{P}}$ is Lipschitz continuous on any bounded subset of $\mathbb{R}^m$.*

From numerical viewpoint, the optimization problems (3) and (4) pose the major difficulty that their objective functions are given by multidimensional integrals with implicit integrands. If $h(\omega)$ follows a continuous probability distribution the computation of $Q_{\mathbb{E}}$ and $Q_{\mathbb{P}}$ has to rely on approximations. Here, it is quite common to approximate the probability distribution of $h(\omega)$ by discrete distributions, turning the integrals in (3), (4) into sums this way. In the next section we will see that discrete distributions, despite the poor analytical properties they imply for $Q_{\mathbb{E}}$ and $Q_{\mathbb{P}}$, are quite attractive algorithmically, since they allow for integer programming techniques.
Approximating the underlying probability measures in (3) and (4) raises the question whether "small" perturbations in the measures result in only "small" perturbations of optimal values and optimal solutions. Subjective assumptions and incomplete knowledge on $\mu = \mathbb{P} \circ h^{-1}$ in many practical modeling situations provide further motivation for asking this question. Therefore, stability analysis has gained some interest in stochastic programming (for surveys see [9, 35]).
For the models (3) and (4) qualitative and quantitative continuity of $Q_{\mathbb{E}}$, $Q_{\mathbb{P}}$ jointly in the decision variable x and the probability measure μ becomes a key issue then. Once established, the continuity, together with well-known techniques from parametric optimization, lead to stability in the spirit sketched above. In the present paper, we will not pursue stability analysis, but show how to arrive at qualitative joint continuity of $Q_{\mathbb{P}}$. For continuity results on $Q_{\mathbb{E}}$ we refer to [14, 24, 30, 33, 34], for extensions towards stability to [35] and the references therein.
For the rest of this section, we consider $Q_{\mathbb{P}}$ as a function mapping from $\mathbb{R}^m \times \mathcal{P}(\mathbb{R}^s)$ to $\mathbb{R}$. By $\mathcal{P}(\mathbb{R}^s)$ we denote the set of all Borel probability measures on $\mathbb{R}^s$. While $\mathbb{R}^s$ is equipped with the usual topology, the set

$\mathcal{P}(I\!R^s)$ is endowed with weak convergence of probability measures. This has proven both sufficiently general to cover relevant applications and sufficiently specific to enable substantial statements. A sequence $\{\mu_n\}$ in $\mathcal{P}(I\!R^s)$ is said to converge weakly to $\mu \in \mathcal{P}(I\!R^s)$, written $\mu_n \xrightarrow{w} \mu$, if for any bounded continuous function $g : I\!R^s \to I\!R$ we have

$$\int_{I\!R^s} g(\xi)\mu_n(d\xi) \to \int_{I\!R^s} g(\xi)\mu(d\xi) \quad \text{as} \quad n \to \infty. \tag{5}$$

A basic reference for weak convergence of probability measures is Billingsley's book [3].

Proposition 3.5 *Assume that* $W(Z\!\!Z_+^{\bar{m}}) + W'(I\!R_+^{m'}) = I\!R^s$ *and* $\{u \in I\!R^s : W^T u \leq q,\ W'^T u \leq q'\} \neq \emptyset$. *Let* $\mu \in \mathcal{P}(I\!R^s)$ *be such that* $\mu(M_e(x) \cup M_d(x)) = 0$. *Then* $Q_{I\!P} : I\!R^m \times \mathcal{P}(I\!R^s) \longrightarrow I\!R$ *is continuous at* (x, μ).

Proof: Let $x_n \longrightarrow x$ and $\mu_n \xrightarrow{w} \mu$ be arbitrary sequences. By $\chi_n, \chi : I\!R^s \longrightarrow \{0,1\}$ we denote the indicator functions of the sets $M(x_n), M(x), n \in I\!N$. In addition, we introduce the exceptional set

$$E := \{h \in I\!R^s : \exists h_n \to h \text{ such that } \chi_n(h_n) \not\to \chi(h)\}.$$

Now we have $E \subseteq M_e(x) \cup M_d(x)$. To see this, assume that $h \in (M_e(x) \cup M_d(x))^c = (M_e(x))^c \cap (M_d(x))^c$ where the superscript c denotes the set-theoretic complement. Then Φ is continuous at $h - Tx$, and either $c^T x + \Phi(h - Tx) > \varphi_o$ or $c^T x + \Phi(h - Tx) < \varphi_o$. Thus, for any sequence $h_n \to h$ there exists an $n_o \in I\!N$ such that for all $n \geq n_o$ either $c^T x_n + \Phi(h_n - Tx_n) > \varphi_o$ or $c^T x_n + \Phi(h_n - Tx_n) < \varphi_o$. Hence, $\chi_n(h_n) \to \chi(h)$ as $h_n \to h$, implying $h \in E^c$.
In view of $E \subseteq M_e(x) \cup M_d(x)$ and $\mu(M_e(x) \cup M_d(x)) = 0$ we obtain that $\mu(E) = 0$. A theorem on weak convergence of image measures attributed to Rubin in [3], p. 34, now yields that the weak convergence $\mu_n \xrightarrow{w} \mu$ implies the weak convergence $\mu_n \circ \chi_n^{-1} \xrightarrow{w} \mu \circ \chi^{-1}$.
Note that $\mu_n \circ \chi_n^{-1}, \mu \circ \chi^{-1}, n \in I\!N$ are probability measures on $\{0,1\}$. Their weak convergence then particularly implies that

$$\mu_n \circ \chi_n^{-1}(\{1\}) \longrightarrow \mu \circ \chi^{-1}(\{1\}).$$

In other words, $\mu_n(M(x_n)) \longrightarrow \mu(M(x))$ or $Q_{I\!P}(x_n, \mu_n) \longrightarrow Q_{I\!P}(x, \mu)$.
$\square$

As done for the expectation-based model (3) in [33], continuity of optimal values and upper semicontinuity of optimal solution sets of the probability-based model (4) can be derived from Proposition 3.5.

Remark 3.6 *(probability-based model without integer decisions)* *Without integer second-stage variables the set $M_d(x)$ is always empty, and Propositions 3.2 and 3.5 readily specify. A direct approach to these models including stability analysis and algorithmic techniques has been carried out in [23]. Lower semicontinuity of $Q_{I\!P}$ in the absence of integer variables can already be derived from Proposition 3.1 in [29], a statement concerning chance constrained stochastic programs. Some early work on continuity properties of general probability functionals has been done by Raik ([21, 22], see also [16, 20]).*

4. Algorithms

In the present section we will review two algorithms for solving the probability-based recourse problem (4) provided the underlying measure μ is discrete, say with realizations h_j and probabilities $\pi_j, j = 1, \ldots, J$. The algorithms were first proposed in [23] and [36], respectively, where further details can be found.

4.1 Linear Recourse

We assume that there are no integer requirements to second-stage variables which is usually referred to as linear recourse in the literature. Suppose that μ is the above discrete measure and consider problem (4) with

$$\Phi(t) := \min\{q^T y \; : \; Wy \geq t, \; y \in I\!R^{m'}_+\}. \tag{6}$$

For ease of exposition let $X \subseteq I\!R^m$ be a nonempty compact polyhedron. Let $e \in I\!R^s$ denote the vector of all ones and consider the set

$$D \; := \; \{(u, u_o) \in I\!R^{s+1} \; : 0 \leq u \leq e, \; 0 \leq u_o \leq 1, \; W^T u - u_o q \leq 0\}$$

together with its extreme points $(d_k, d_{ko}), k = 1, \ldots, K$. Furthermore, consider the indicator function

$$\chi(x, h) := \begin{cases} 1 & , \quad h \in M(x) \\ 0 & , \quad \text{otherwise.} \end{cases} \tag{7}$$

The key idea of the subsequent algorithm is to represent χ by a binary variable and a number of optimality cuts which enables exploitation of cutting plane techniques from convex subgradient optimization. The latter have proven very useful in classical two-stage linear stochastic programming, see e.g. [4, 32].

Lemma 4.1 *There exists a sufficiently large constant $M_o > 0$ such that problem (4) can be equivalently restated as*

$$\min_{x,\theta}\{\sum_{j=1}^{J}\pi_j\theta_j : (h_j - Tx)^T d_k + (c^T x - \varphi_o)d_{ko} \le M_o\theta_j, \quad x \in X,\ \theta_j \in \{0,1\},\ k = 1,\dots,K,\ j = 1,\dots,J\}. \quad (8)$$

Proof: For any $x \in X$ and any $j \in \{1,\dots,J\}$ consider the feasibility problem

$$\min_{y \in \mathbb{R}_+^{m'},\ (t,t_o) \in \mathbb{R}_+^{s+1}} \{e^T t + t_o \ :\ Wy + t \ge h_j - Tx,\ q^T y - t_o \le \varphi_o - c^T x\} \quad (9)$$

and its linear programming dual

$$\max\{(h_j - Tx)^T u + (c^T x - \varphi_o)u_o : 0 \le u \le e,\ 0 \le u_o \le 1, W^T u - u_o q \le 0\}.$$

Clearly, both programs are always solvable. Their optimal value is equal to zero, if and only if $\chi(x,h_j) = 0$. In addition, D coincides with the feasible set of the dual. If M_o is selected as

$$M_o := \max_{x \in X,\ k \in \{1,\dots,K\}, j \in \{1,\dots,J\}} \{(h_j - Tx)^T d_k + (c^T x - \varphi_o)d_{ko}\},$$

then, for any $x \in X$, the vector (x,θ) with $\theta_j = 1, j = 1,\dots,J$ is feasible for (8).
If $\chi(x,h_j) = 1$ for some $x \in X$ and $j \in \{1,\dots,J\}$, then there has to exist some $k \in \{1,\dots,K\}$ such that

$$(h_j - Tx)^T d_k + (c^T x - \varphi_o)d_{ko} > 0.$$

Hence, given $x \in X$, $\theta_j = 0$ is feasible in (8) if and only if $\chi(x,h_j) = 0$. Therefore, (8) is equivalent to $\min\{\sum_{j=1}^{J}\pi_j\chi(x,h_j) \ :\ x \in X\}$. □

The algorithm progresses by sequentially solving a master problem and adding violated optimality cuts generated through the solution of subproblems (9). These cuts correspond to constraints in (8). Assuming that the cuts generated before iteration ν correspond to subsets $\mathcal{K}_\nu \subseteq \{1,\dots,K\}$ the current master problem reads

$$\min_{x,\theta}\{\sum_{j=1}^{J}\pi_j\theta_j \ :\ (h_j - Tx)^T d_k + (c^T x - \varphi_o)d_{ko} \le M_o\theta_j, \quad x \in X,\ \theta_j \in \{0,1\},\ k \in \mathcal{K}_\nu,\ j = 1,\dots,J\}. \quad (10)$$

The full algorithm proceeds as follows.

Algorithm 4.2
Step 1 *(Initialization): Set* $\nu = 0$ *and* $\mathcal{K}_o = \emptyset$.
Step 2 *(Solving the master problem): Solve the current master problem (10) and let* (x^ν, θ^ν) *be an optimal solution.*
Step 3 *(Solving subproblems): Solve the feasibility problem (9) for* $x = x^\nu$ *and all* $j \in \{1, \ldots, J\}$ *such that* $\theta_j^\nu = 0$. *Consider the following situations:*

1 *If all these problems have optimal value equal to zero, then the current* x^ν *is optimal for (8).*

2 *If some of these problems have optimal value strictly greater than zero, then, via the dual solutions, a subset* $(d_k, d_{ko}), k \in \tilde{\mathcal{K}} \subseteq \{1, \ldots, K\}$ *of extreme points of* D *is identified. The corresponding cuts are added to the master.*
Set $\mathcal{K}_{\nu+1} := \mathcal{K}_\nu \cup \tilde{\mathcal{K}}$ *and* $\nu := \nu + 1$; *go to Step 2.*

The algorithm terminates since D has a finite number of extreme points. For further details on correctness of the algorithm and first computational experiments we refer to [23].

4.2 Linear Mixed-Integer Recourse

In the present subsection we allow for integer requirements to second-stage variables. Again we assume that $X \subseteq \mathbb{R}^m$ is a nonempty compact polyhedron and that μ is the discrete measure introduced at the beginning of the present section. We consider problem (4) with

$$\Phi(t) := \min\{q^T y \ : \ Wy \geq t, \ y \in Y\}. \tag{11}$$

For notational convenience we have integrated the former vector (y, y') into one vector y now varying in $Y := \mathbb{Z}_+^{\bar{m}} \times \mathbb{R}_+^{m'}$. Accordingly, the former (q, q') and (W, W') are integrated into q and W. To be consistent with Subsection 4.1 we have inequality constraints in (11).

Lemma 4.3 *There exists a sufficiently large constant* $M_1 > 0$ *such that problem (4) can be equivalently restated as*

$$\min_{x,y,\theta} \{\sum_{j=1}^{J} \pi_j \theta_j \ : \ Wy_j \geq h_j - Tx, \ q^T y_j + c^T x - \varphi_o \leq M_1 \theta_j,$$
$$x \in X, \ \ y_j \in Y, \ \ \theta_j \in \{0, 1\}, \ \ j = 1, \ldots, J\}. \quad (12)$$

Proof: We choose M_1 by

$$M_1 \ := \ \sup\{c^T x + \Phi(h_j - Tx) \ : \ x \in X, \ j = 1, \ldots, J\}.$$

To see that this supremum is finite, recall the compactness of X and the general assumptions on Φ in the paragraph following formula (2). Part (iv) of Proposition 2.1 then confirms that $\Phi(h_j - Tx)$ remains bounded if x and j vary over X and $\{1, \ldots, J\}$, respectively.
The selection of M_1 guarantees that for any $x \in X$ and $y_j \in Y$ such that $Wy_j \geq h_j - Tx$ the selection $\theta_j = 1$ is feasible.
Given x, the selection $\theta_j = 0$ is feasible if and only if there exists a $y_j \in Y$ fulfilling $Wy_j \geq h_j - Tx$ and $c^T x + q^T y_j \leq \varphi_o$. The latter holds if and only if $c^T x + \Phi(h_j - Tx) \leq \varphi_o$ which is equvalent to $\chi(x, h_j) = 0$. This proves that (12) is equivalent to $\min\{\sum_{j=1}^{J} \pi_j \chi(x, h_j) \; : \; x \in X\}$. □

Compared with problem (8), problem (12) again arises by representing the indicator function χ from (7) by a binary variable. Lacking duality, however, prevents the usage of optimality cuts such that minimization with respect to y has to be carried out explicitly in (12). Hence, (8) is a variant of (12) where the linear programming nature of the second stage enables an algorithmic shortcut.

Problem (12) is a mixed-integer linear program that quickly becomes large-scale in practical applications. General purpose mixed-integer linear programming algorithms and software fail in such situations. As an alternative, we present a decomposition method based on Lagrangian relaxation of nonanticipativity. This decomposition method for block-angular stochastic integer programs has been elaborated for the first time in [7] for the expectation-based model (3).

Introduce in (12) copies $x_j, j = 1, \ldots, J$, according to the number of scenarios, and add the nonanticipativity constraints $x_1 = \ldots = x_J$ (or an equivalent system), for which we use the notation $\sum_{j=1}^{J} H_j x_j = 0$ with proper $(l, n)-$matrices $H_j, j = 1, \ldots, J$. Problem (12) then becomes

$$\min_{x,y,\theta} \{\sum_{j=1}^{J} \pi_j \theta_j \; : \; Tx_j + Wy_j \; \geq \; h_j, \; c^T x_j + q^T y_j - M_1 \theta_j \; \leq \; \varphi_o,$$

$$x_j \in X, \; y_j \in Y, \; \theta_j \in \{0, 1\}, \; j = 1, \ldots, J, \; \sum_{j=1}^{J} H_j x_j = 0\}. \quad (13)$$

This formulation suggests Lagrangian relaxation of the interlinking constraints $\sum_{j=1}^{J} H_j x_j = 0$. For $\lambda \in \mathbb{R}^l$ we consider the functions

$$L_j(x_j, y_j, \theta_j, \lambda) \; := \; \pi_j \theta_j \; + \; \lambda^T H_j x_j, \;\; j = 1, \ldots, J,$$

and form the Lagrangian

$$L(x,y,\theta,\lambda) := \sum_{j=1}^{J} L_j(x_j,y_j,\theta_j,\lambda).$$

The Lagrangian dual of (13) then is the optimization problem

$$\max\{D(\lambda) : \lambda \in \mathbb{R}^l\} \tag{14}$$

where

$$\begin{aligned} D(\lambda) &= \min\{\sum_{j=1}^{J} L_j(x_j,y_j,\theta_j,\lambda) : Tx_j + Wy_j \geq h_j, \\ &\qquad c^T x_j + q^T y_j - M_1\theta_j \leq \varphi_o, \\ &\qquad x_j \in X, \;\; y_j \in Y, \;\; \theta_j \in \{0,1\}, \;\; j = 1,\ldots,J\}. \end{aligned}$$

For separability reasons we have

$$D(\lambda) = \sum_{j=1}^{J} D_j(\lambda) \tag{15}$$

where

$$\begin{aligned} D_j(\lambda) &= \min\{L_j(x_j,y_j,\theta_j,\lambda) : Tx_j + Wy_j \geq h_j, \\ &\qquad c^T x_j + q^T y_j - M_1\theta_j \leq \varphi_o, \\ &\qquad x_j \in X, \;\; y_j \in Y, \;\; \theta_j \in \{0,1\}\}. \end{aligned} \tag{16}$$

$D(\lambda)$ being the pointwise minimum of affine functions in λ, it is piecewise affine and concave. Hence, (14) is a non-smooth concave maximization (or convex minimization) problem. Such problems can be tackled with advanced bundle methods, for instance with Kiwiel's proximal bundle method NOA 3.0, [17, 18]. At each iteration, these methods require the objective value and one subgradient of D. The structure of D, cf. (15), enables substantial decomposition, since the single-scenario problems (16) can be tackled separately. Their moderate size often allows application of general purpose mixed-integer linear programming codes. Altogether, the optimal value z_{LD} of (14) provides a lower bound to the optimal value z of problem (12). From integer programming ([19]) it is well-known, that in general one has to live with a positive duality gap. On the other hand, it holds that $z_{L!D} \geq z_{LP}$ where z_{LP} denotes the optimal value to the LP relaxation of (12). The lower bound obtained by the above procedure, hence, is never worse the bound obtained by

eliminating the integer requirements.
In Lagrangian relaxation, the results of the dual optimization often provide starting points for heuristics to find promising feasible points. Our relaxed constraints being very simple ($x_1 = \ldots = x_N$), ideas for such heuristics come up straightforwardly. For example, examine the x_j–components, $j = 1, \ldots, J$, of solutions to (16) for optimal or nearly optimal λ, and decide for the most frequent value arising or average and round if necessary.
If the heuristic yields a feasible solution to (12), then the objective value of the latter provides an upper bound $\bar{z}$ for z. Together with the lower bound z_{LD} this gives the quality certificate (gap) $\bar{z} - z_{LD}$.
The full algorithm improves this certificate by embedding the procedure described so far into a branch-and-bound scheme in the spirit of global optimization. Let $\mathcal{P}$ denote the list of current problems and $z_{LD} = z_{LD}(P)$ the Lagrangian lower bound for $P \in \mathcal{P}$. The algorithm then proceeds as follows.

Algorithm 4.4
Step 1 *(Initialization): Set* $\bar{z} = +\infty$ *and let* $\mathcal{P}$ *consist of problem (13).*
Step 2 *(Termination): If* $\mathcal{P} = \emptyset$ *then the solution* $\hat{x}$ *that yielded* $\bar{z} = Q_{I\!P}(\hat{x})$, *cf. (4), is optimal.*
Step 3 *(Node selection): Select and delete a problem* P *from* $\mathcal{P}$ *and solve its Lagrangian dual. If the optimal value* $z_{LD}(P)$ *hereof equals* $+\infty$ *(infeasibility of a subproblem) then go to Step 2.*
Step 4 *(Bounding): If* $z_{LD}(P) \geq \bar{z}$ *go to Step 2 (this step can be carried out as soon as the value of the Lagrangian dual rises above* $\bar{z}$*). Consider the following situations:*

1. *The scenario solutions* x_j, $j = 1, \ldots, J$, *are identical: If* $Q_{I\!P}(x_j) < \bar{z}$ *then let* $\bar{z} = Q_{I\!P}(x_j)$ *and delete from* $\mathcal{P}$ *all problems* P' *with* $z_{LD}(P') \geq \bar{z}$. *Go to Step 2.*

2. *The scenario solutions* x_j, $j = 1, \ldots, J$ *differ: Compute the average* $\bar{x} = \sum_{j=1}^{J} \pi_j x_j$ *and round it by some heuristic to obtain* $\bar{x}^R$. *If* $Q_{I\!P}(\bar{x}^R) < \bar{z}$ *then let* $\bar{z} = Q_{I\!P}(\bar{x}^R)$ *and delete from* $\mathcal{P}$ *all problems* P' *with* $z_{LD}(P') \geq \bar{z}$. *Go to Step 5.*

Step 5 *(Branching): Select a component* $x_{(k)}$ *of* x *and add two new problems to* $\mathcal{P}$ *obtained from* P *by adding the constraints* $x_{(k)} \leq \lfloor \bar{x}_{(k)} \rfloor$ *and* $x_{(k)} \geq \lfloor \bar{x}_{(k)} \rfloor + 1$, *respectively (if* $x_{(k)}$ *is an integer component), or* $x_{(k)} \leq \bar{x}_{(k)} - \varepsilon$ *and* $x_{(k)} \geq \bar{x}_{(k)} + \varepsilon$, *respectively, where* $\varepsilon > 0$ *is a tolerance parameter to have disjoint subdomains. Go to Step 3.*

The algorithm works both with and without integer requirements in the first stage. It is obviously finite in case X is bounded and all

$x-$components have to be integers. If x is mixed-integer (or continuous, as in the former presentation) some stopping criterion to avoid endless branching on the continuous components has to be employed. Some first computational experiments with Algorithm 4.4 are reported in [36].

5. Multistage Extension

The two-stage stochastic programs introduced in Section 2 are based on the assumptions that uncertainty is unveiled at once and that decisions subdivide into those before and those after unveiling uncertainty. Often, a more complex view is appropriate at this place. Multistage stochastic programs address the situation where uncertainty is unveiled stepwise with intermediate decisions.
The modeling starts with a finite horizon sequential decision process under uncertainty where the decision $x_t \in I\!R^{m_t}$ at stage $t \in \{1, \ldots, T\}$ is based on information available up to time t only. Information is modeled as a discrete time stochastic process $\{\xi_t\}_{t=1}^T$ on some probability space $(\Omega, \mathcal{A}, I\!P)$ with ξ_t taking values in $I\!R^{s_t}$. The random vector $\xi^t := (\xi_1, \ldots, \xi_t)$ then reflects the information available up to time t. Nonanticipativity, i.e., the requirement that x_t must not depend on future information, is formalized by saying that x_t is measurable with respect to the $\sigma-$algebra $\mathcal{A}_t \subseteq \mathcal{A}$ which is generated by $\xi^t, t = 1, \ldots, T$. Clearly, $\mathcal{A}_t \subseteq \mathcal{A}_{t+1}$ for all $t = 1, \ldots, T-1$. As in the two-stage case, the first-stage decision x_1 usually is deterministic. Therefore, $\mathcal{A}_1 = \{\emptyset, \Omega\}$. Moreover, we assume that $\mathcal{A}_T = \mathcal{A}$.

The constraints of our multistage extensions can be subdivided into three groups. The first group comprises conditions on x_t arising from the individual time stages:

$$x_t(\omega) \in X_t \quad , \quad B_t(\xi_t(\omega))x_t(\omega) \geq d_t(\xi_t(\omega)) \qquad (17)$$
$$I\!P-\text{almost surely}, \quad t = 1, \ldots, T.$$

Here, $X_t \subseteq I\!R^{m_t}$ is a set whose convex hull is a polyhedron. In this way, integer requirements to components of x_t are allowed for. For simplicity we assume that X_t is compact. The next group of constraints models linkage between different time stages:

$$\sum_{\tau=1}^{t} A_{t\tau}(\xi_t(\omega))x_\tau(\omega) \geq g_t(\xi_t(\omega)) \quad I\!P-\text{almost surely}, \quad t = 2, \ldots, T. \qquad (18)$$

Finally, there is the nonanticipativity of x_t, i. e.,

$$x_t \text{ is measurable with respect to } \mathcal{A}_t, \quad t = 1, \ldots, T. \qquad (19)$$

In addition to the constraints we have a linear objective function

$$\sum_{t=1}^{T} c_t(\xi_t(\omega))x_t(\omega).$$

The matrices $A_{t\tau}(.), B_t(.)$ as well as the right-hand sides $d_t(.), g_t(.)$ and the cost coefficients $c_t(.)$ all have conformal dimensions and depend affinely linearly on the relevant components of ξ.
The decisions x_t are understood as members of the function spaces $L_\infty(\Omega, \mathcal{A}, I\!P; I\!R^{m_t}), t = 1, \ldots, T$. The constraints (17), (18) then impose pointwise conditions on the x_t, whereas (19) imposes functional constraints, in fact, membership in a linear subspace of $\times_{t=1}^{T} L_\infty(\Omega, \mathcal{A}, I\!P; I\!R^{m_t})$, see e.g. [31] and the references therein.

Now we are in the position to formulate the multistage extensions to the expectation- and probability-based stochastic programs (3) and (4), respectively.
The multistage extension of (3) is the minimization of expected minimal costs subject to nonanticipativity of decisions:

$$\min_{x \text{ fulfilling } (19)} \int_\Omega \min_{x(\omega)} \{ \sum_{t=1}^{T} c_t(\xi_t(\omega))x_t(\omega) : (17),\ (18)\} I\!P(d\omega) \tag{20}$$

To have the integral in the objective well-defined, the additional assumption $\xi_t \in L_1(\Omega, \mathcal{A}, I\!P; I\!R^{s_t})$, $t = 1, \ldots, T$, is imposed in model (20), see [31] for further details.

The multistage extension of (4) is the minimization of the probability that minimal costs do not exceed a preselected threshold $\varphi_o \in I\!R$. Again this minimization takes place over nonanticipative decisions only:

$$\min_{x \text{ fulfilling } (19)} I\!P\Big(\Big\{\omega \in \Omega : \min_{x(\omega)} \{ \sum_{t=1}^{T} c_t(\xi_t(\omega))x_t(\omega) : (17),\ (18)\} > \varphi_o\Big\}\Big) \tag{21}$$

The minimization in the integrand of (20) being separable with respect to $\omega \in \Omega$, it is possible to interchange integration and minimization. Then the problem can be restated as follows:

$$\min \Big\{ \int_\Omega \sum_{t=1}^{T} c_t(\xi_t(\omega))x_t(\omega)\, I\!P(d\omega)\ :\ x \text{ fulfilling } (17),\ (18),\ (19)\Big\}. \tag{22}$$

Extending the argument from Lemma 4.3 we introduce an additional variable $\theta \in L_\infty(\Omega, \mathcal{A}, I\!P; \{0, 1\})$ as well as a sufficiently big constant

$M > 0$. Then problem (21) can be equivalently rewritten as:

$$\min \Big\{ \int_\Omega \theta(\omega) \, I\!P(d\omega) : \sum_{t=1}^{T} c_t(\xi_t(\omega)) x_t(\omega) \; - \; \varphi_o \; \leq \; M \cdot \theta(\omega),$$
$$\theta(\omega) \in \{0,1\} \; I\!P\text{−a.s.}, \; x \text{ fulfilling (17), (18), (19)} \Big\}. \qquad (23)$$

Problem (22) is the well-known multistage stochastic (mixed-integer) linear program. Without integer requirements, the problem has been studied intensively, both from structural and from algorithmic viewpoints. The reader may wish to sample from [4, 5, 8, 10, 13, 15, 20, 25, 26, 27, 32] to obtain insights into these developments. With integer requirements, problem (22) is less well-understood. Existing results are reviewed in [31].

To the best of our knowledge, the multistage extension (21) has not been addressed in the literature so far. Some basic properties of (21), (22) regarding existence and structure of optimal solutions can be derived by following arguments that were employed for the expectation-based model (22) in [31]. Their mathematical foundations are laid out in [11, 12, 28]. The arguments can be outlined as follows:
Problem (22) concerns the minimization of an abstract expectation over a function space, subject to measurability with respect to a filtered sequence of σ−algebras. Theorems 1 and 2 in [12] (whose assumptions can be verified for (22) using statements from [11, 28]) provide sufficient conditions for the solvability of such minimization problems and for the solutions to be obtainable recursively by dynamic programming. The stage-wise recursion rests on minimizing in the t−th stage the regular conditional expectation (with respect to $\mathcal{A}_t$) of the optimal value from stage $t+1$. When arriving at the first-stage, a deterministic optimization problem in x_1 remains (recall that $\mathcal{A}_1 = \{\emptyset, \Omega\}$). Its objective function $Q^m_{I\!P}(x_1)$ can be regarded the multistage counterpart to the function $Q_{I\!P}(x)$ that we have studied in Section 3.

Given that (22) is a well-defined and solvable optimization problem, Sections 3 and 4 provide several points of departure for future research. For instance, unveiling the structure of $Q^m_{I\!P}(x_1)$ may be possible by analysing the interplay of conditional expectations and mixed-integer value functions. Regarding solution techniques, the extension of Algorithm 4.4 to the multistage situation may be fruitful. Indeed, it is well-known that the nonanticipativity in (19) is a linear constraint. With a discrete distribution of ξ this leads to a system of linear equations. Lagrangian relaxation of these constraints produces single-scenario subproblems, and the scheme of Algorithm 4.4 readily extends. However, compared with the two-stage situation, the relaxed constraints are more complicated such that primal heuristics are not that obvious, and the dimension l of the Lagrangian dual (14) may require approximative instead of exact solution of (14). Further algorithmic ideas for (22) may arise from Lagrangian relaxation of either (17) or (18). In [31] this is discussed for the expectation-based model (22).

Acknowledgement. I am grateful to Morten Riis (University of Aarhus), Werner Römisch (Humboldt-University Berlin), and Stephan Tiedemann (Gerhard-Mercator University Duisburg) for stimulating discussions and fruitful cooperation.

References

[1] Bank, B.; Mandel, R.: Parametric Integer Optimization, Akademie-Verlag, Berlin 1988.

[2] Bereanu, B.: Minimum risk criterion in stochastic optimization, Economic Computation and Economic Cybernetics Studies and Research 2 (1981), 31-39.

[3] Billingsley, P.: Convergence of Probability Measures, Wiley, New York, 1968.

[4] Birge, J.R.: Stochastic programming computation and applications. INFORMS Journal on Computing 9 (1997), 111-133.

[5] Birge, J.R.; Louveaux, F.: Introduction to Stochastic Programming, Springer, New York, 1997.

[6] Blair, C.E.; Jeroslow, R.G.: The value function of a mixed integer program: I, Discrete Mathematics 19 (1977), 121-138.

[7] Carøe, C.C.; Schultz, R.: Dual decomposition in stochastic integer programming, Operations Research Letters 24 (1999), 37-45.

[8] Dempster, M.A.H.: On stochastic programming II: Dynamic problems under risk, Stochastics 25 (1988), 15-42.

[9] Dupačová, J.: Stochastic programming with incomplete information: a survey of results on postoptimization and sensitivity analysis, Optimization 18 (1987), 507-532.

[10] Dupačová, J.: Multistage stochastic programs: The state-of-the-art and selected bibliography, Kybernetika 31 (1995), 151-174.

[11] Dynkin, E.B., Evstigneev, I.V.: Regular conditional expectation of correspondences, Theory of Probability and Applications 21 (1976), 325-338.

[12] Evstigneev, I.: Measurable selection and dynamic programming, Mathematics of Operations Research 1 (1976), 267-272.

[13] Higle, J.L., Sen, S.: Duality in multistage stochastic programs, In: Prague Stochastics '98 (M. Hušková, P. Lachout, J.Á. Višek, Eds.), JČMF, Prague 1998, 233-236.

[14] Kall, P.: On approximations and stability in stochastic programming, Parametric Optimization and Related Topics (J. Guddat, H.Th. Jongen, B. Kummer, F. Nožička, Eds.), Akademie Verlag, Berlin 1987, 387-407.

[15] Kall, P.; Wallace, S.W.: Stochastic Programming, Wiley, Chichester, 1994.

[16] Kibzun, A.I.; Kan, Y.S.: Stochastic Programming Problems with Probability and Quantile Functions, Wiley, Chichester, 1996.

[17] Kiwiel, K. C.: Proximity control in bundle methods for convex nondifferentiable optimization, Mathematical Programming 46 (1990), 105-122.

[18] Kiwiel, K. C.: User's Guide for NOA 2.0/3.0: A Fortran Package for Convex Nondifferentiable Optimization, Systems Research Institute, Polish Academy of Sciences, Warsaw, 1994.

[19] Nemhauser, G.L.; Wolsey, L.A.: Integer and Combinatorial Optimization, Wiley, New York 1988.

[20] Prékopa, A.: Stochastic Programming, Kluwer, Dordrecht, 1995.

[21] Raik, E.: Qualitative research into the stochastic nonlinear programming problems, Eesti NSV Teaduste Akademia Toimetised / Füüsika, Matemaatica (News of the Estonian Academy of Sciences / Physics, Mathematics) 20 (1971), 8-14. In Russian.

[22] Raik, E.: On the stochastic programming problem with the probability and quantile functionals, Eesti NSV Teaduste Akademia Toimetised / Füüsika, Matemaatica (News of the Estonian Academy of Sciences / Physics, Mathematics) 21 (1971), 142-148. In Russian.

[23] Riis, M.; Schultz, R.: Applying the minimum risk criterion in stochastic recourse programs, Computational Optimization and Applications 24 (2003), 267-287.

[24] Robinson, S.M.; Wets, R.J-B: Stability in two-stage stochastic programming, SIAM Journal on Control and Optimization 25 (1987), 1409-1416.

[25] Rockafellar, R.T.: Duality and optimality in multistage stochastic programming, Annals of Operations Research 85 (1999), 1-19.

[26] Rockafellar, R.T., Wets, R.J-B: Nonanticipativity and L^1-martingales in stochastic optimization problems, Mathematical Programming Study 6 (1976), 170-187.

[27] Rockafellar, R.T., Wets, R.J-B: The optimal recourse problem in discrete time: L^1-multipliers for inequality constraints, SIAM Journal on Control and Optimization 16 (1978), 16-36.

[28] Rockafellar, R.T., Wets, R.J-B: Variational Analysis, Springer-Verlag, Berlin, 1997.

[29] Römisch, W.; Schultz, R.: Stability analysis for stochastic programs, Annals of Operations Research 30 (1991), 241-266.

[30] Römisch, W.; Wakolbinger, A.: Obtaining convergence rates for approximations in stochastic programming, Parametric Optimization and Related Topics (J. Guddat, H.Th. Jongen, B. Kummer, F. Nožička, Eds.), Akademie Verlag, Berlin 1987, 327-343.

[31] Römisch, W.; Schultz, R.: Multistage stochastic integer programs: an introduction, Online Optimization of Large Scale Systems (M. Grötschel, S.O. Krumke, J. Rambau, Eds.), Springer-Verlag Berlin, 2001, 581-600.

[32] Ruszczyński, A.: Decomposition methods in stochastic programming. Mathematical Programming 79 (1997), 333-353.

[33] Schultz, R.: On structure and stability in stochastic programs with random technology matrix and complete integer recourse, Mathematical Programming 70 (1995), 73-89.

[34] Schultz, R.: Rates of convergence in stochastic programs with complete integer recourse, SIAM Journal on Optimization 6 (1996), 1138-1152.

[35] Schultz, R.: Some aspects of stability in stochastic programming, Annals of Operations Research 100 (2000), 55-84.

[36] Tiedemann, S.: Probability Functionals and Risk Aversion in Stochastic Integer Programming, Diploma Thesis, Department of Mathematics, Gerhard-Mercator University Duisburg, 2001.

PARAMETRIC SENSITIVITY ANALYSIS: A CASE STUDY IN OPTIMAL CONTROL OF FLIGHT DYNAMICS

Christof Büskens
Lehrstuhl für Ingenieurmathematik, Universität Bayreuth
Universitätsstr. 30, D-95440 Bayreuth, Germany
christof.bueskens@uni-bayreuth.de

Kurt Chudej
Lehrstuhl für Ingenieurmathematik, Universität Bayreuth
Universitätsstr. 30, D-95440 Bayreuth, Germany
kurt.chudej@uni-bayreuth.de

Abstract Realistic optimal control problems from flight mechanics are currently solved by sophisticated direct or indirect methods in a fast and reliable way. Often one is not only interested in the optimal solution of *one* control problem, but is also strongly interested in the sensitivity of the optimal solution due to perturbations in certain parameters (constants or model functions) of the process. In the past this problem was solved by time-consuming parameter studies: A large number of almost similar optimal control problems were solved numerically. Sensitivity derivatives were approximated by finite differences. Recently a new approach, called *parametric sensitivity analysis*, was adapted to the direct solution of optimal control processes [3]. It uses the information gathered in the optimal solution of the unperturbed (nominal) optimal control problem to compute sensitivity differentials of all problem functions with respect to these parameters. This new approach is described in detail for an example from trajectory optimization.

Keywords: parametric sensitivity analysis, optimal control, direct methods, trajectory optimization.

Introduction

Realistically modelled optimal control problems can be solved efficiently and reliably by sophisticated direct and indirect methods (see e.g. the survey articles [1], [10]). Trajectory optimization problems for aircrafts and space vehicles usually pose hard challenges for the direct and indirect solution algorithms. A couple of direct algorithms have proved their ability to solve accurately and reliably trajectory optimization problems in the last decade, such as e.g. SOCS (Betts [2]), GESOP (Jänsch, Well, Schnepper [9]), DIRCOL (von Stryk [12]) and NUDOCCCS (Büskens [3]). Trajectory optimization problems use in general complicated models of the surrounding atmospheric effects, the performance and consumption of the engines and the ability to maneuver. Usually optimal solutions are computed at first for a nominal data set of the model. Later huge parameter studies are done for perturbed model data. This means that the whole optimization process is started again for the huge number of perturbed models.

We present a *new* approach of Büskens [3]: Exploiting already computed information during the solution of the nominal optimal control problem to derive sensitivity information. This substitutes the additional solution of perturbed optimal control problems.

We explain the new approach of parametric sensitivity analysis in detail for an example from flight mechanics. The trajectory optimization problem is concerned with minimizing the amount of fuel used per travelled range over ground with periodic boundary conditions. It is interesting that by periodic trajectories and controls savings in fuel consumption can be achieved in comparison to the steady-state solution.

In order to normalize the changing effects of the atmosphere due to the weather, one uses data of a reference atmosphere in the computational model. Unfortunately there exist a couple of reference atmospheres. Additionally one is also interested in realistic changes of the air density onto the computed optimal solution.

We therefore provide not only the nominal solution but also the sensitivity with respect to the air density as an example. Note that no parameter studies are needed. Information gathered during the computation of the nominal solution is used.

Applications to further perturbation parameters in the model are straight forward.

1. A Trajectory Optimiziation Problem

Aircraft usually use steady-state cruise to cover long distances. It is interesting, that these steady-state trajectories are non-optimal with respect to minimizing fuel [11].

The following optimal control problem from [8], enlarged by a perturbation parameter p, describes the problem of minimizing fuel per travelled range over ground for a realistically modelled aircraft flying in a vertical plane.

State variables are velocity v, flight path angle γ, altitude h and weight W. The range x is used as the independent variable. The lift coefficient C_L and the throttle setting δ are the control variables.

For a given value of the perturbation parameter p (nominal value is here $p_0 = 1$) find control functions $C_L(x;p)$ and $\delta(x;p)$ and the final range $x_f(p)$ such that the cost functional

$$I = [W_0 - W(x_f)]/x_f$$

is minimized and the following equations of motion, control constraints and boundary conditions are fulfilled.

$$\begin{aligned}
\frac{dv}{dx} &= \frac{g}{v\cos\gamma}\left[\frac{T(h,M)\,\delta - D(h,M,C_L)}{W_0} - \sin\gamma\right]\\
\frac{d\gamma}{dx} &= \frac{g}{v^2}\left[\frac{L(h,M,C_L)}{W_0\cos\gamma} - 1\right]\\
\frac{dh}{dx} &= \tan\gamma\\
\frac{dW}{dx} &= -T(h,M)\,\delta\,\frac{c(h,M)}{v\cos\gamma}
\end{aligned} \tag{1}$$

$$\begin{aligned}
0 \;\le\; &C_L \;\le\; C_{L,\max}\\
\delta_{\min} \;\le\; &\delta \;\le\; 1
\end{aligned} \tag{2}$$

$$v(0) = v(x_f),\;\; \gamma(0) = \gamma(x_f),\;\; h(0) = h(x_f),\;\; W(0) = W_0 \tag{3}$$

Model functions are the Mach number M, speed of sound a, thrust T, consumption c, lift L, drag D, air density ρ. S denotes the constant reference area. g denotes the gravitational constant.

$$\begin{aligned}
M(v,h) &= v/a(h)\\
a(h) &= \alpha_4\sqrt{\sum_{i=0}^{3}\alpha_i h^i}\\
T(h,M) &= c_1(h) + c_2(h)M + c_3(h)M^2 + c_4(h)M^3\\
c(h,M) &= d_1(h) + d_2(h)M + d_3(h)M^2 + d_4(h)M^3\\
L(v,h,C_L) &= \rho(h)\;S\;v^2\;C_L/2\\
D(v,h,C_L) &= \rho(h)\;S\;v^2\;[C_{D0}(M) + \Delta C_D(M,C_L)]/2\\
C_{D0}(M) &= a_1\arctan[a_2\;(M - a_3)] + a_4\\
\Delta C_D(M,C_L) &= b_1(M)C_L^2 + b_2(M)C_L^4 + b_3(M)C_L^6 + b_4(M)C_L^8\\
\rho(h) &= p\;\rho_0\;\exp\Big[\beta_7 + \beta_6 h + \beta_5\exp\Big(\sum_{i=1}^{4}\beta_i h^i\Big)\Big]
\end{aligned}$$

The coefficients of the polynomials $b_i(M)$, $c_i(h)$, $d_i(h)$, and the constants $\rho_0, \beta_i, a_i, S, \alpha_i, W_0, \delta_{\min}, C_{L,\max}$ can be found in the [8].

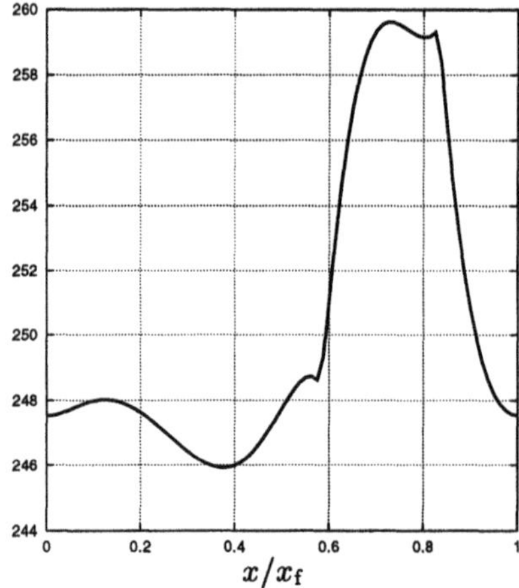

Figure 1. Nominal optimal state $v(x; p_0)$

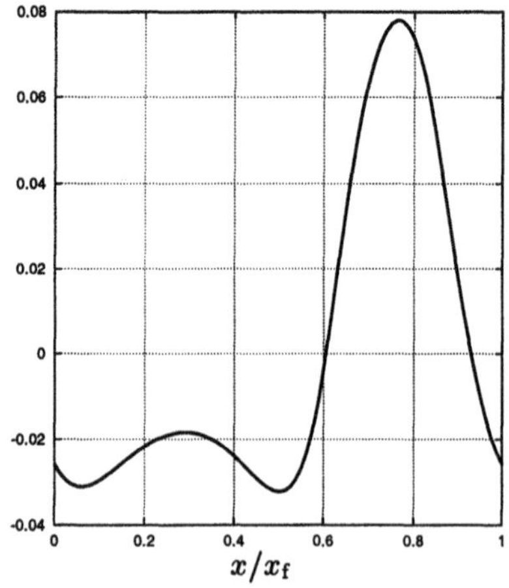

Figure 2. Nominal optimal state $\gamma(x; p_0)$

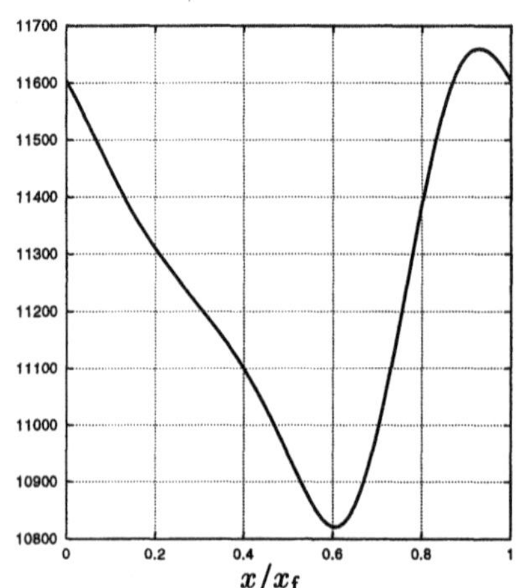

Figure 3. Nominal optimal state $h(x; p_0)$

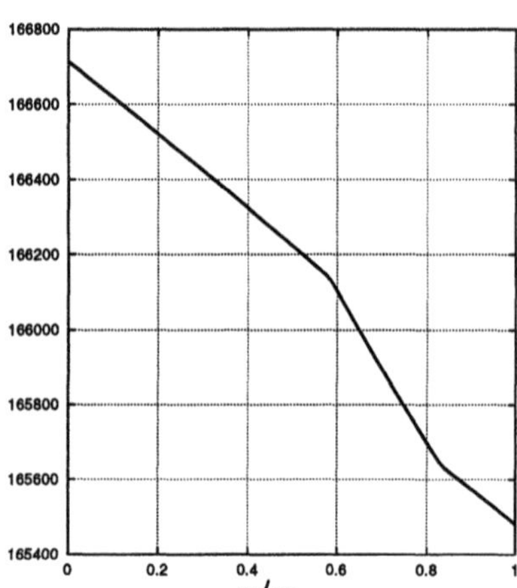

Figure 4. Nominal optimal state $W(x; p_0)$

As an additional perturbation parameter we use p. A solution by an indirect multiple shooting algorithm is presented in [8] for the nominal value of $p_0 = 1$. In these times, before sophisticated direct methods were developed, elaborated homotopies were required for the indirect

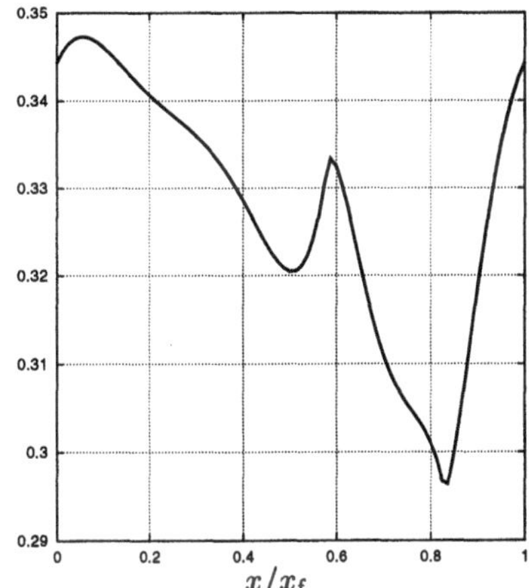

Figure 5. Nominal optimal control $C_L(x;p_0)$

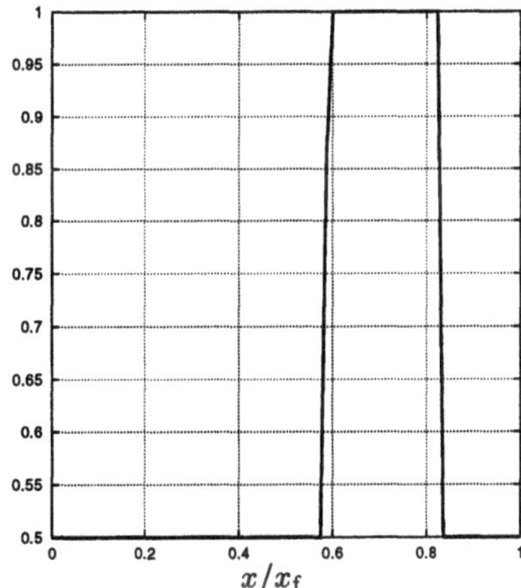

Figure 6. Nominal optimal control $\delta(x;p_0)$

solution. Today these initial estimates for the indirect method can be provided easily by direct methods. The figures 1–4 show the optimal states and the figures 5–6 show the optimal controls for the nominal value of $p = p_0 = 1$ computed by the direct method NUDOCCCS (Büskens [3]).

Moreover the direct method NUDOCCCS can compute in a post-processing step also the sensitivities of the optimal states $\frac{\partial v}{\partial p}(x;p), \ldots,$ $\frac{\partial W}{\partial p}(x;p)$ and the optimal controls $\frac{\partial C_L}{\partial p}(x;p)$, $\frac{\partial \delta}{\partial p}(x;p)$ with respect to perturbation parameters p.

2. Parametric Sensitivity Analysis

The general mathematical approach for a parametric sensitivity analysis of perturbed optimal control problems is based on NLP methods:

The following autonomous perturbed control problem of Mayer–form will be referred to as problem $\mathbf{OCP(p)}$:

For a given perturbation parameter $p \in P$ find control functions $u(x;p)$ and the final time $x_f(p)$ such that the cost functional

$$J = g(y(x_f), p) \tag{4}$$

is minimized subject to the following constraints

$$\begin{array}{lll} y'(x) = f(y(x), u(x), p) & , & x \in [0, x_f], \\ \psi(y(0), y(x_f), p) = 0 & , & \\ C(y(x), u(x), p) \leq 0 & , & x \in [0, x_f]. \end{array} \tag{5}$$

Herein $y(x) \in \mathbb{R}^n$ denotes the state of a system and $u(x) \in \mathbb{R}^m$ the control with respect to an independent variable x, which is often the time.

In the previously introduced trajectory optimization problem the independent variable x denotes the range, the state is given by $y := (v, \gamma, h, W)^\top$ and the control by $u := (C_L, \delta)^\top$.

The functions $g : \mathbb{R}^n \times P \to \mathbb{R}$, $f : \mathbb{R}^{n+m} \times P \to \mathbb{R}^n$, $\psi : \mathbb{R}^{2n} \times P \to \mathbb{R}^r$, and $C : \mathbb{R}^{n+m} \times P \to \mathbb{R}^k$ are assumed to be sufficiently smooth on appropriate open sets. The final time x_f is either fixed or free. Note that the formulation of mixed control-state constraints $C(y(x), u(x), p) \leq 0$ in (5) includes *pure control* constraints $C(u(x), p) \leq 0$ as well as *pure state* constraints $C(y(x), p) \leq 0$. It is well known, that problems of form **OCP(p)** can be solved efficiently by approximating the control functions $u^i \approx u(x_i)$ for given mesh points $x_i \in [0, x_f]$, $i = 1, \ldots, N$ and solving the state variables by standard integration methods. This leads to approximations $y(x_i; z, p) \approx y(x_i)$, $z := (u^1, \ldots, u^N)$ of the state at the mesh points x_i. For a more detailed discussion please refer to [3]–[6].

Therefore the optimal control problem **OCP(p)** is replaced by the finite dimensional perturbed nonlinear optimization problem **NLP(p)**

$$\begin{array}{c} \text{For a given } p \in P \\ \min\limits_{z} \quad g(y(x_N; z, p), p) \quad \text{s.t.} \\ \psi(y(x_N; z, p), p) = 0, \\ C(y(x_i; z, p), u^i, p) \leq 0, \quad i = 1, \ldots, N. \end{array} \tag{6}$$

Several reliable optimization codes have been developed for solving NLP problems (6), like e.g. SQP methods. This idea is implemented e.g. in the direct methods SOCS, GESOP, DIRCOL and NUDOCCCS.

An additional, and to our knowledge unique, feature of NUDOCCCS (Büskens [3]) is the ability to compute accurately the sensitivity differentials $\frac{\partial y}{\partial p}(x; p_0)$, $\frac{\partial u}{\partial p}(x; p_0)$ of the approximations

$$\begin{aligned} y(x; p_0 + \Delta p) &\approx y(x; p_0) + \tfrac{\partial y}{\partial p}(x; p_0) \cdot \Delta p, \\ u(x; p_0 + \Delta p) &\approx u(x; p_0) + \tfrac{\partial u}{\partial p}(x; p_0) \cdot \Delta p. \end{aligned} \tag{7}$$

This is done by the following idea:

Let z_0 denote the unperturbed solution of $\mathbf{NLP}(\mathbf{p_0})$ for a nominal parameter $p = p_0$ and let h^a denote the collection of active constraints in (6).

$$L(z, \mu, p) := g(y(x_N; z, p), p) + \mu^\top h^a(z, p)$$

is the Lagrangian function with the associated Lagrange multiplier μ. Then the following results hold [7]:

Solution Differentiability for NLP–problems: *Suppose that the optimal solution* (z_0, μ_0) *for the nominal problem* $\mathbf{NLP}(\mathbf{p_0})$ *satisfies a maximal rank condition for* $h_z^a(z_0, p_0)$*, second order sufficient optimality conditions and strict complementarity of the multiplier* μ*. Then the unperturbed solution* (z_0, μ_0) *can be embedded into a* C^1*-family of perturbed solutions* $(z(p), \mu(p))$ *for* $\mathbf{NLP}(\mathbf{p})$ *with* $z(p_0) = z_0$, $\mu(p_0) = \mu_0$.

The sensitivity differentials of the optimal solutions are given by the formula

$$\begin{pmatrix} \frac{dz}{dp}(p_0) \\ \frac{d\mu}{dp}(p_0) \end{pmatrix} = -\begin{pmatrix} L_{zz} & (h_z^a)^\top \\ h_z^a & 0 \end{pmatrix}^{-1} \begin{pmatrix} L_{zp} \\ h_p^a \end{pmatrix} \tag{8}$$

evaluated at the optimal solution. This formula provides good approximations for the sensitivity of the perturbed optimal controls at the mesh points, i.e. for the quantities $\frac{\partial u}{\partial p}(x_i; p_0)$, $i = 1, \ldots, N$. Then the state sensitivities $\frac{\partial y}{\partial p}(x_i; p_0)$ are obtained by differentiating the control–state relation in (6) $y(x_i) = y(x_i, z, p)$ with respect to the parameter p:

$$\frac{\partial y}{\partial p}(x_i; p_0) \approx \frac{\partial y}{\partial z}(x_i; z_0, p_0)\frac{dz}{dp}(p_0) + \frac{\partial y}{\partial p}(x_i; z_0, p_0). \tag{9}$$

The sensitivity differentials of the adjoint variables or objective functional can be calculated respectively.

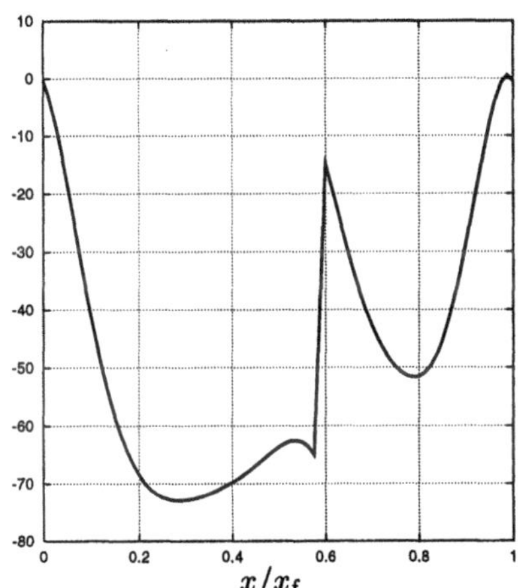

Figure 7. Sensitivity $\partial v/\partial p(x; p_0)$

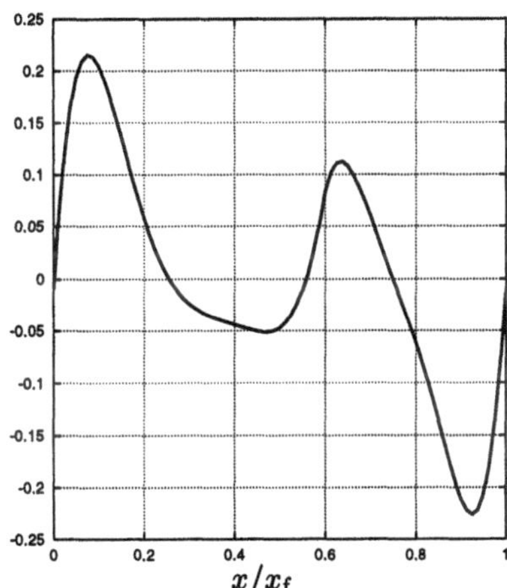

Figure 8. Sensitivity $\partial \gamma/\partial p(x; p_0)$

We return back to the example. In the first step the optimal nominal solution is calculated by the code NUDOCCCS of Büskens [3], see figures 1–6. In the second step the sensitivity differentials of the model functions (states, controls, adjoint variables, cost functional and further interesting model functions) are calculated from equations (8, 9), see figures 7–12.

These figures provide valuable information for the engineers. Additional perturbation parameters can be added to the model. Basically only an additional matrix vector multiplication is needed in order to compute the sensitivity differentials of the states and controls for each component of the perturbation parameter.

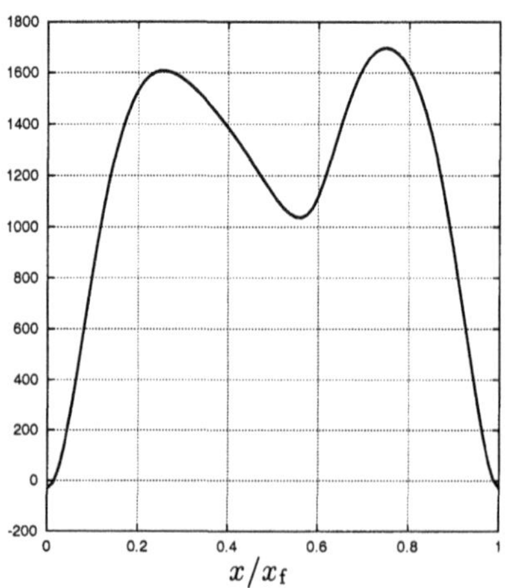

Figure 9. Sensitivity $\partial h/\partial p(x; p_0)$

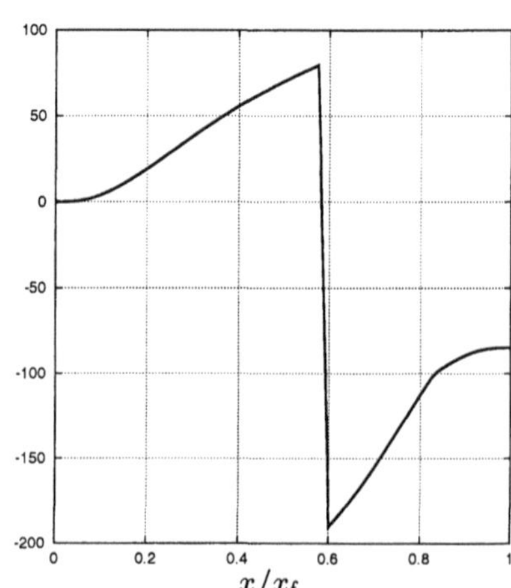

Figure 10. Sensitivity $\partial W/\partial p(x; p_0)$

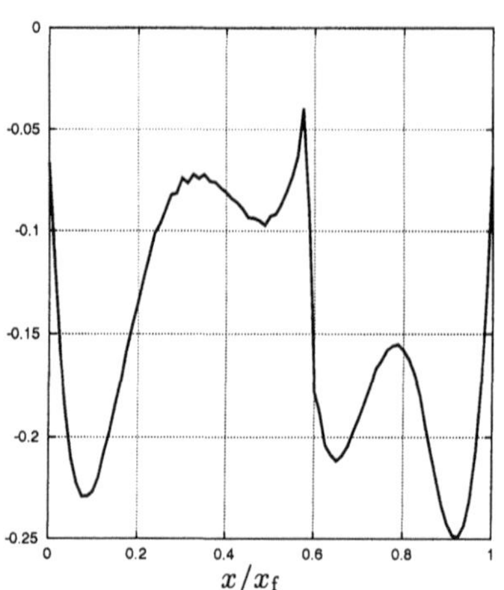

Figure 11. Sensitivity $\partial C_L/\partial p(x; p_0)$

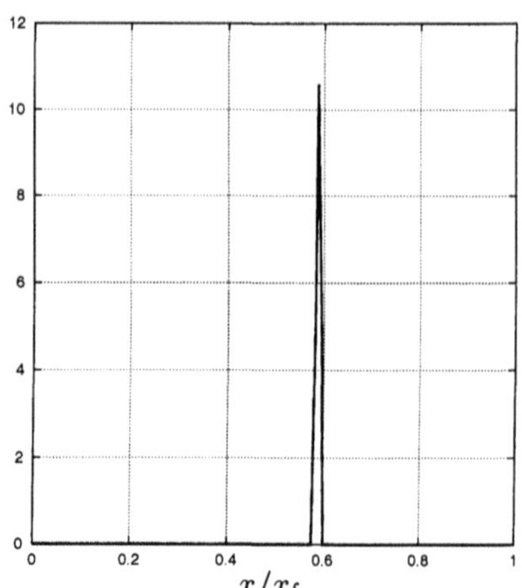

Figure 12. Sensitivity $\partial \delta/\partial p(x; p_0)$

References

[1] Betts, J. T. (1998) Survey of Numerical Methods for Trajectory Optimization. Journal of Guidance, Control, and Dynamics, Vol. 21, pp. 193–207.

[2] Betts, J. T. (2001) Practical Methods for Optimal Control Using Nonlinear Programming. SIAM, Philadelphia.

[3] Büskens, C. (1998) Optimierungsmethoden und Sensitivitätsanalyse für optimale Steuerprozesse mit Steuer- und Zustands-Beschränkungen. Dissertation, Universität Münster.

[4] Büskens, C., Maurer, H. (2000) SQP-Methods for Solving Optimal Control Problems with Control and State Constraints: Adjoint Variables, Sensitivity Analysis and Real-Time Control. Journal of Computational and Applied Mathematics, Vol. 120, pp. 85–108.

[5] Büskens, C., Maurer, H. (2001) Sensitivity Analysis and Real–Time Optimization of Parametric Nonlinear Programming Problems. – In: Grötschel, M., Krumke, S.O., Rambau, J. (Eds.): Online Optimization of Large Scale Systems: State of the Art. Springer Verlag, Berlin, pp. 3–16.

[6] Büskens, C., Maurer, H. (2001) Sensitivity Analysis and Real–Time Control of Parametric Optimal Control Problems Using Nonlinear Programming Methods. – In: Grötschel, M., Krumke, S.O., Rambau, J. (Eds.): Online Optimization of Large Scale Systems: State of the Art. Springer Verlag, Berlin, pp. 57–68.

[7] Fiacco, A.V. (1983) Introduction to Sensitivity and Stability Analysis in Nonlinear Programming. Academic Press, New York.

[8] Grimm, W., Well, K.H., Oberle, H.J. (1986) Periodic Control for Minimum-Fuel Aircraft Trajectories. Journal of Guidance, Vol. 9, 169–174.

[9] Jänsch, C., Well, K.H., Schnepper, K. (1994) GESOP – Eine Software Umgebung zur Simulation und Optimierung. In: Proc. des SFB 255 Workshops Optimalsteuerungsprobleme von Hyperschall-Flugsystemen, Ernst-Moritz-Arndt Universität Greifswald, pp. 15–23.

[10] Pesch, H.J. (1994) A Practical Guide to the Solution of Real-Life Optimal Control Problems. Control and Cybernetics, 23, pp. 7–60.

[11] Speyer, J.L. (1976) Nonoptimality of the Steady-State Cruise for Aircraft. AIAA Journal, Vol. 14, pp. 1604–1610.

[12] von Stryk, O. (1995) Numerische Lösung optimaler Steuerungsproblems: Diskretisierung, Parameteroptimierung und Berechnung der adjungierten Variablen. VDI-Verlag, Reihe 8, Nr. 441.

SOLVING QUADRATIC MULTI-COMMODITY PROBLEMS THROUGH AN INTERIOR-POINT ALGORITHM

Jordi Castro
Department of Statistics and Operations Research
Universitat Politècnica de Catalunya
Pau Gargallo 5, 08028 Barcelona, Spain *
jcastro@eio.upc.es

Abstract Standard interior-point algorithms usually show a poor performance when applied to multicommodity network flow problems. A recent specialized interior-point algorithm for linear multicommodity network flows overcame this drawback, and was able to efficiently solve large and difficult instances. In this work we perform a computational evaluation of an extension of that specialized algorithm for multicommodity problems with convex and separable quadratic objective functions. As in the linear case, the specialized method for convex separable quadratic problems is based on the solution of the positive definite system that appears at each interior-point iteration through a scheme that combines direct (Cholesky) and iterative (preconditioned conjugate gradient) solvers. The preconditioner considered for linear problems, which was instrumental in the performance of the method, has shown to be even more efficient for quadratic problems. The specialized interior-point algorithm is compared with the general barrier solver of CPLEX 6.5, and with the specialized codes PPRN and ACCPM, using a set of convex separable quadratic multicommodity instances of up to 500000 variables and 180000 constraints. The specialized interior-point method was, in average, about 10 times and two orders of magnitude faster than the CPLEX 6.5 barrier solver and the other two codes, respectively.

Keywords: Interior-point methods, network optimization, multicommodity flows, quadratic programming, large-scale optimization.

*Work supported by grant CICYT TAP99-1075-C02-02.

1. Introduction

Multicommodity flows are widely used as a modeling tool in many fields as, e.g., in telecommunications and transportation problems. The multicommodity network flow problem is a generalization of the minimum cost network flow one where k different items—the commodities—have to be routed from a set of supply nodes to a set of demand nodes using the same underlying network. This kind of models are usually very large and difficult linear programming problems, and there is a wide literature about specialized approaches for their efficient solution. However most of them only deal with the linear objective function case. In this work we consider a specialized interior-point algorithm for multicommodity network flow problems with convex and separable quadratic objective functions. The algorithm has been able to solve large and difficult quadratic multicommodity problems in a fraction of the time required by alternative solvers.

In the last years there has been a significant amount of research in the field of multicommodity flows, mainly for linear problems. The new solution strategies can be classified into four main categories: simplex-based methods [6, 15], decomposition methods [10, 12], approximation methods [13], and interior-point methods [4, 12]. Some of these algorithms were compared in [7] for linear problems.

The available literature for nonlinear multicommodity flows is not so extensive. For instance, of the above approaches, only the codes of [6] and [12] (named PPRN —nonlinear primal partitioning—and ACCPM—analytic center cutting plane method—, respectively) were extended to nonlinear (possibly non-quadratic) objective functions. In this work we compared the specialized interior-point algorithm with those two codes using a set of large-scale quadratic multicommodity problems. The specialized interior-point algorithm turned out to be the most efficient strategy for all the instances. A description and empirical evaluation of additional nonlinear multicommodity algorithms can be found in the survey [14].

The specialized-interior point method presented here is an extension for convex and separable quadratic objective functions of the algorithm introduced in [4] for linear multicommodity flows. The solution strategy suggested for linear problems (i.e., solving the positive definite system at each interior-point iteration through a scheme that combines direct and iterative solvers) can also be applied to convex and separable quadratic multicommodity problems. Moreover, as it will be shown in the computational results, this solution strategy turned out to be even more efficient for quadratic than for linear problems.

Up to now most applications of multicommodity flow models dealt with linear objective functions. Quadratic multicommodity problems are not usually recognized as a modeling tool, mainly due to the lack of an efficient solver for them. The specialized interior-point method can help to fill this void. The efficient solution of large and difficult quadratic multicommodity problems would open new modeling perspectives (e.g., they could be used in network design algorithms [9]).

The structure of the document is as follows. In Section 2 we formulate the quadratic multicommodity flow problem. In Section 3 we sketch the specialized interior-point algorithm for multicommodity flow problems, and show that it can also be applied to the quadratic case. Finally in Section 4 we perform an empirical evaluation of the algorithm using a set of large-scale quadratic multicommodity flow instances, and three alternative solvers (i.e., CPLEX 6.5, PPRN and ACCPM).

2. The quadratic multicommodity flow problem

Given a network of m nodes, n arcs and k commodities, the quadratic multicommodity network flow problem can be formulated as

$$\begin{array}{ll} \min & \displaystyle\sum_{i=0}^{k}((c^i)^T x^i + (x^i)^T Q^i x^i) \\ \text{subject to} & \begin{bmatrix} N & 0 & \dots & 0 & 0 \\ 0 & N & \dots & 0 & 0 \\ \vdots & \vdots & \ddots & \vdots & \vdots \\ 0 & 0 & \dots & N & 0 \\ \mathbb{1} & \mathbb{1} & \dots & \mathbb{1} & \mathbb{1} \end{bmatrix} \begin{bmatrix} x^1 \\ x^2 \\ \vdots \\ x^k \\ x^0 \end{bmatrix} = \begin{bmatrix} b^1 \\ b^2 \\ \vdots \\ b^k \\ u \end{bmatrix} \\ & 0 \le x^i \le u^i \quad i = 1 \dots k \\ & 0 \le x^0 \le u. \end{array} \tag{1}$$

Vectors $x^i \in \mathbb{R}^n, i = 1 \dots k$, are the flows for each commodity, while $x^0 \in \mathbb{R}^n$ are the slacks of the mutual capacity constraints. $N \in \mathbb{R}^{m \times n}$ is the node-arc incidence matrix of the underlying network, and $\mathbb{1}$ denotes the $n \times n$ identity matrix. $c^i \in \mathbb{R}^n$ are the arc linear costs for each commodity and for the slacks. $u^i \in \mathbb{R}^n$ and $u \in \mathbb{R}^n$ are respectively the individual capacities for each commodity and the mutual capacity for all the commodities. $b^i \in \mathbb{R}^m$ are the supply/demand vectors at the nodes of the network for each commodity. Finally $Q^i \in \mathbb{R}^{n \times n}$ are the arc quadratic costs for each commodity and for the slacks. We will restrict to the case where Q^i is a positive semidefinite diagonal matrix, thus having a convex and separable quadratic objective function. Note that (1) is a quadratic problem with $\tilde{m} = km+n$ constraints and $\tilde{n} = (k+1)n$ variables.

Most of the applications of multicommodity flows in the literature only involve linear costs. However, quadratic costs can be useful in the following situations:

- Adding a quadratic penalty term to the occupation of a line in a transmission/transportation network. In this case we would set $Q^i = \mathbb{1}, i = 1 \dots k$. This would penalize saturation of lines, guaranteeing a reserve capacity to redistribute the current pattern of flows when line failures occur.
- Replacing a convex and separable nonlinear function by its quadratic approximation.
- Finding the closest pattern of flows x to the currently used $\tilde{x}$, when changes in capacities/demands are performed. In this case the quadratic term would be $(x - \tilde{x})^T (x - \tilde{x})$
- Solution of the subproblems in an augmented Lagrangian relaxation scheme for the network design problem [9, 11]

3. The specialized interior-point algorithm

The multicommodity problem (1) is a quadratic program that can be written in standard form as

$$\min \left\{ c^T x + \frac{1}{2} x^T Q x \; : \; Ax = b, \; x + s = u, \; x, s \geq 0 \right\} , \tag{2}$$

where $x, s, u \in \mathbb{R}^{\tilde{n}}$, $Q \in \mathbb{R}^{\tilde{n} \times \tilde{n}}$ and $b \in \mathbb{R}^{\tilde{m}}$. The dual of (2) is

$$\max \left\{ b^T y - \frac{1}{2} x^T Q x - w^T u \; : \; A^T y - Qx + z - w = c, \; z, w \geq 0 \right\} , \tag{3}$$

where $y \in \mathbb{R}^{\tilde{m}}$ and $z, w \in \mathbb{R}^{\tilde{n}}$. For problem (1), matrix Q is made of $k+1$ diagonal blocks; blocks $Q^i, i = 1 \dots k$, are related to the flows for each commodity, while Q^0 are the quadratic costs of the mutual capacity slacks.

The solution of (2) and (3) by an interior-point algorithm is obtained through the following system of nonlinear equations (see [18] for details)

$$\begin{aligned} r_{xz} &\equiv \mu e - XZe = 0 \\ r_{sw} &\equiv \mu e - SWe = 0 \\ r_b &\equiv b - Ax = 0 \\ r_c &\equiv c - (A^T y - Qx + z - w) = 0 \\ & \quad (x, s, z, w) \geq 0 , \end{aligned} \tag{4}$$

where e is a vector of 1's of appropriate dimension, and matrices X, Z, S, W are diagonal matrices made from vectors x, z, s, w. The set of unique solutions of (4) for each μ value is known as the *central path*, and when $\mu \to 0$ these solutions superlinearly converge to those of (2) and (3) [18]. System (4) is usually solved by a damped version of Newton's method, reducing the μ parameter at each iteration. This procedure is known as the *path-following* algorithm [18]. Figure 1 shows the main steps of the path-following algorithm for quadratic problems.

Figure 1. Path-following algorithm for quadratic problems.

Algorithm *Path-following*(A, Q, b, c, u):
1 Initialize $x > 0, s > 0, y, z > 0, w > 0$;
2 **while** (x, s, y, z, w) is not solution **do**
3 $\quad \Theta = (X^{-1}Z + S^{-1}W + Q)^{-1}$;
4 $\quad r = S^{-1}r_{sw} + r_c - X^{-1}r_{xz}$;
5 $\quad (A\Theta A^T)\Delta y = r_b + A\Theta r$;
6 $\quad \Delta x = \Theta(A^T\Delta y - r)$;
7 $\quad \Delta w = S^{-1}(r_{sw} + W\Delta x)$;
8 $\quad \Delta z = r_c + \Delta w + Q\Delta x - A^T\Delta y$;
9 $\quad$ Compute $\alpha_P \in (0,1], \alpha_D \in (0,1]$;
10 $\quad x \leftarrow x + \alpha_P \Delta x$;
11 $\quad (y, z, w) \leftarrow (y, z, w) + \alpha_D(\Delta y, \Delta z, \Delta w)$;
12 **end_while**
End_algorithm

The specialized interior-point algorithm introduced in [4] for linear multicommodity problems exploited the constraints matrix structure of the problem for solving $(A\Theta A^T)\Delta y = \bar{b}$ (line 5 of Figure 1), which is by far the most computationally expensive step. Considering the structure of A in (1) and accordingly partitioning the diagonal matrix Θ defined in line 3 of Figure 1, we obtain

$$A\Theta A^T = \left[\begin{array}{c|c} B & C \\ \hline C^T & D \end{array}\right] = \left[\begin{array}{ccc|c} N\Theta^1 N^T & \dots & 0 & N\Theta^1 \\ \vdots & \ddots & \vdots & \vdots \\ 0 & \dots & N\Theta^k N^T & N\Theta^k \\ \hline \Theta^1 N^T & \dots & \Theta^k N^T & \sum_{i=0}^k \Theta^i \end{array}\right], \quad (5)$$

where $\Theta^i = ((X^i)^{-1}Z^i + (S^i)^{-1}W^i + Q^i)^{-1}, i = 0, 1 \dots k$. Note that the only difference between the linear and quadratic case is term Q^i of Θ^i.

Moreover, as we are assuming that Q^i is a diagonal matrix, Θ^i can be easily computed.

Using (5), and appropriately partitioning Δy and $\bar{b}$, we can write $(A\Theta A^T)\Delta y = \bar{b}$ as

$$\left[\begin{array}{c|c} B & C \\ \hline C^T & D \end{array}\right] \left[\begin{array}{c} \Delta y_1 \\ \hline \Delta y_2 \end{array}\right] = \left[\begin{array}{c} \bar{b}_1 \\ \hline \bar{b}_2 \end{array}\right]. \tag{6}$$

By block multiplication, we can reduce (6) to

$$(D - C^T B^{-1} C)\Delta y_2 = (\bar{b}_2 - C^T B^{-1} \bar{b}_1) \tag{7}$$

$$B\Delta y_1 = (\bar{b}_1 - C\Delta y_2). \tag{8}$$

System (8) is solved by performing a Cholesky factorization of each diagonal block $N\Theta^i N, i = 1 \dots k$, of B. System with matrix $H = D - C^T B^{-1} C$, the Schur complement of (5), is solved by a preconditioned conjugate gradient (PCG) method. A good preconditioner is instrumental for the performance of the method. In [4] it was proved that if

- D is positive semidefinite, an
- $D + C^T BC$ is positive semidefinite,

then the inverse of the Schur complement can be computed as

$$H^{-1} = \left(\sum_{i=0}^{\infty} (D^{-1}(C^T BC))^i \right) D^{-1}. \tag{9}$$

The preconditioner is thus obtained by truncating the infinite power series (9) at some term h (in practice $h = 0$ or $h = 1$; all the computational results in this work have been obtained with $h = 0$). Since

$$D_q = D_l + \sum_{i=0}^{k} (Q^i)^{-1},$$

D_q and D_l denoting the D matrix for a quadratic and linear problem respectively, it is clear that for quadratic multicommodity problems the above two conditions are also guaranteed, and then the same preconditioner can also be applied. Moreover, since we are assuming diagonal Q^i matrices, for $h = 0$ the preconditioner is equal to $H^{-1} = D^{-1}$, which is also diagonal, as for linear multicommodity problems. This is instrumental in the overall performance of the algorithm. More details about this solution strategy can be found in [4].

The effectiveness of the preconditioner is governed by the spectral radius of $D^{-1}(C^T BC))$, which is always in $[0, 1)$. The farthest from 1, the better the preconditioner. According to the computational results obtained, this value seems to be less for quadratic problems than for the equivalent linear problems without the quadratic term, since fewer conjugate gradient iterations are performed for solving (7). Moreover, the number of interior-point iterations also decreases in some instances. This can be observed in Figures 2, 3 and 4. Figures 2 and 3 show the overall number of PCG and IP iterations for the linear and quadratic versions of the Mnetgen problems in Table 1 of Section 4. Both versions only differ in the Q matrix. Clearly, for the quadratic problems fewer IP and PCG iterations are performed. The number of PCG iterations per IP iteration has also been observed to decrease for quadratic problems. For instance, Figure 4 shows the number of PCG iterations per IP iteration for the linear and quadratic versions of problem PDS20 in Table 2 of Section 4. We chose this instance because it can be considered a good representative of the general behavior observed and, in addition, the number of IP iterations is similar for the linear and quadratic problems. A better understanding of the relationship between the spectral radius of $D^{-1}(C^T BC))$ for the linear and quadratic problems is part of the further work to be done.

Figure 2. Overall number of PCG iterations for the quadratic and linear Mnetgen instances.

Figure 3. Overall number of IP iterations for the quadratic and linear Mnetgen instances.

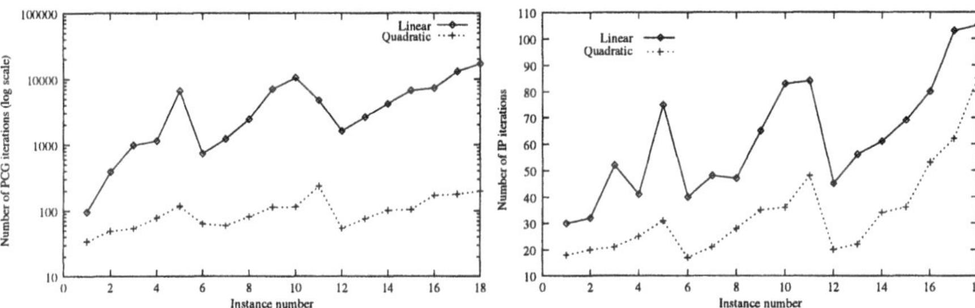

4. Computational results

The specialized algorithm of the previous section was tested using two sets of quadratic multicommodity instances. As far as we know, there is no standard set of quadratic multicommodity problems. Thus

Figure 4. Number of PCG iterations per interior-point iteration, for the quadratic and linear PDS20 instance.

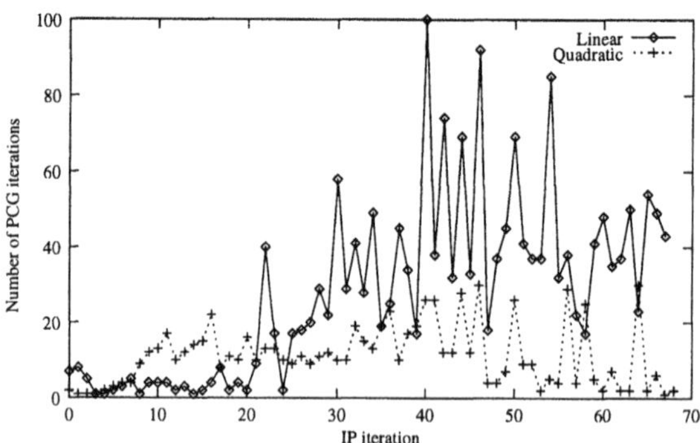

we developed a meta-generator that adds the quadratic term

$$\sum_{i=1}^{k}\sum_{j=1}^{n} q_j^i (x_j^i)^2$$

to the objective function of a linear multicommodity problem. Coefficients q_j^i are randomly obtained from an uniform distribution $U[0, C]$, where

$$C = \sqrt{\left|\frac{\sum_{i=1}^{k}\sum_{j=1}^{n} c_j^i}{kn}\right|},$$

in an attempt to guarantee that linear and quadratic terms are of the same order.

We applied our meta-generator to two sets of linear multicommodity instances obtained with the well-known Mnetgen [1] and PDS [3] generators. Tables 1 and 2 show the dimensions of the instances. Columns "m", "n", and "k" give the number of nodes, arcs and commodities of the network. Columns '$\tilde{n}$" and "$\tilde{m}$" give the number of variables and constraints of the quadratic problem. The Mnetgen and PDS generators can be downloaded from

`http://www.di.unipi.it/di/groups/optimize/Data/MMCF.html.`

We solved both sets with an implementation of the specialized interior-point algorithm, referred to as IPM [4], and with CPLEX 6.5 [8], a state-of-the-art interior-point code for quadratic problems. The IPM code, as well as a parallel version [5], can be downloaded for research purposes from

Table 1. Dimensions of the quadratic Mnetgen instances.

Instance	m	n	k	$\tilde{n}$	$\tilde{m}$
M_{64-4}	64	524	4	2620	780
M_{64-8}	64	532	8	4788	1044
M_{64-16}	64	497	16	8449	1521
M_{64-32}	64	509	32	16797	2557
M_{64-64}	64	511	64	33215	4607
M_{128-4}	128	997	4	4985	1509
M_{128-8}	128	1089	8	9801	2113
M_{128-16}	128	1114	16	18938	3162
M_{128-32}	128	1141	32	37653	5237
M_{128-64}	128	1171	64	76115	9363
$M_{128-128}$	128	1204	128	155316	17588
M_{256-4}	256	2023	4	10115	3047
M_{256-8}	256	2165	8	19485	4213
M_{256-16}	256	2308	16	39236	6404
M_{256-32}	256	2314	32	76362	10506
M_{256-64}	256	2320	64	150800	18704
$M_{256-128}$	256	2358	128	304182	35126
$M_{256-256}$	256	2204	256	566428	67740

Table 2. Dimensions of the quadratic PDS instances.

Instance	m	n	k	$\tilde{n}$	$\tilde{m}$
PDS1	126	372	11	4464	1758
PDS10	1399	4792	11	57504	20181
PDS20	2857	10858	11	130296	42285
PDS30	4223	16148	11	193776	62601
PDS40	5652	22059	11	264708	84231
PDS50	7031	27668	11	332016	105009
PDS60	8423	33388	11	400656	126041
PDS70	9750	38396	11	460752	145646
PDS80	10989	42472	11	509664	163351
PDS90	12186	46161	11	553932	180207

Table 3. Results for the quadratic Mnetgen problems

	CPLEX 6.5		IPM		
Instance	CPU	n.it.	CPU	n.it	$\frac{f^*_{CPLEX}-f^*_{IPM}}{1+f^*_{CPLEX}}$
M_{64-4}	0.7	12	0.3	18	−2.0e−6
M_{64-8}	3.1	12	0.8	20	5.6e−7
M_{64-16}	10.7	15	1.6	21	−1.6e−6
M_{64-32}	20.8	16	4.3	25	1.9e−6
M_{64-64}	46.8	14	10.7	31	−1.1e-6
M_{128-4}	2.8	11	0.8	17	1.2e−7
M_{128-8}	12.6	11	2.1	21	2.5e−6
M_{128-16}	80.5	13	5.9	28	6.9e−6
M_{128-32}	153.6	14	15.7	35	2.8e−6
M_{128-64}	305.5	14	35.3	36	−1.4e−6
$M_{128-128}$	741.9	15	98.8	48	−5.7e−7
M_{256-4}	13.1	13	2.7	20	−7.9e−6
M_{256-8}	73.8	14	6.7	22	2.4e−5
M_{256-16}	634.1	15	22.5	34	2.3e−5
M_{256-32}	1105.2	16	49.9	36	2.4e−6
M_{256-64}	2102.2	16	140.0	53	4.9e−7
$M_{256-128}$	4507.3	17	327.6	62	5.0e−6
$M_{256-256}$	11761.3	24	835.3	85	7.0e−6

Table 4. Results for the quadratic PDS problems

	CPLEX 6.5		IPM		
Instance	CPU	n.it.	CPU	n.it	$\frac{f^*_{CPLEX}-f^*_{IPM}}{1+f^*_{CPLEX}}$
PDS1	1.6	23	1.3	29	−2.7e−7
PDS10	234.8	43	78.6	62	−6.6e−7
PDS20	1425.6	55	271.0	69	1.9e−6
PDS30	5309.8	76	938.3	96	−6.0e−6
PDS40	10712.3	79	1965.2	105	−4.1e−6
PDS50	14049.7	80	3163.3	114	−4.1e−7
PDS60	17133.4	71	3644.2	95	3.6e−6
PDS70	25158.3	74	5548.7	101	−1.9e−7
PDS80	26232.1	74	7029.9	100	−1.3e−6
PDS90	32412.9	77	9786.7	109	−1.2e−6

`http://www-eio.upc.es/~jcastro.`

For each instance, Tables 3 and 4 give the CPU time in seconds required by IPM and CPLEX 6.5 (columns "CPU"), the number of interior-point iterations performed by IPM and CPLEX 6.5 (columns "n.it."), and the relative error $\frac{f^*_{CPLEX}-f^*_{\mathrm{IPM}}}{1+f^*_{CPLEX}}$ of the solution obtained with IPM (assuming CPLEX 6.5 provides the exact optimum). Executions were carried out on a Sun Ultra2 2200 workstation with 200MHz, 1Gb of main memory, and ≈45 Linpack Mflops.

Figures 5–8 summarize the information of Tables 3 and 4. Figures 5 and 6 show respectively the ratio between the CPU times of CPLEX 6.5 and IPM, and the number of interior-point iterations performed by CPLEX 6.5 and IPM, with respect to the dimension of the problem (i.e., number of variables), for the Mnetgen instances. The same information is shown in Figures 7 and 8 for the PDS problems.

Figure 5. Ratio of the execution times of CPLEX 6.5 and IPM for the quadratic Mnetgen problems.

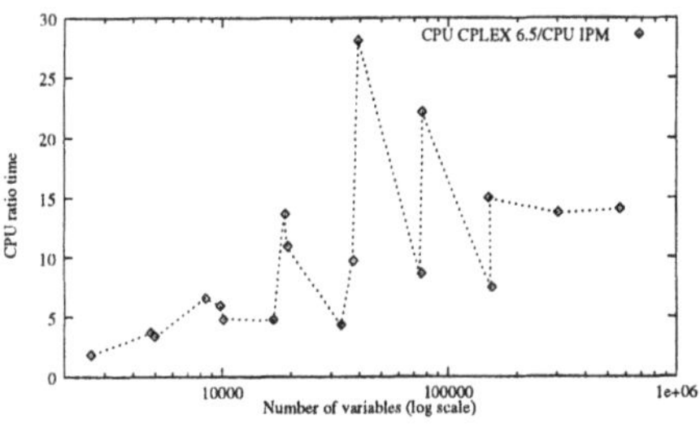

Figure 6. Number of IP iterations performed by CPLEX 6.5 and IPM for the quadratic Mnetgen problems.

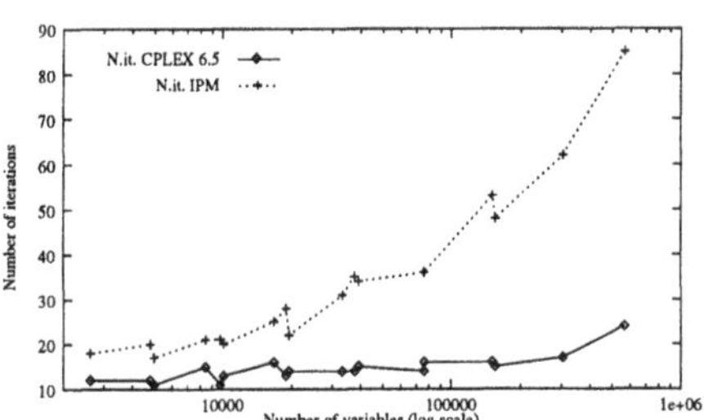

Figure 7. Ratio of the execution times of CPLEX 6.5 and IPM for the quadratic PDS problems.

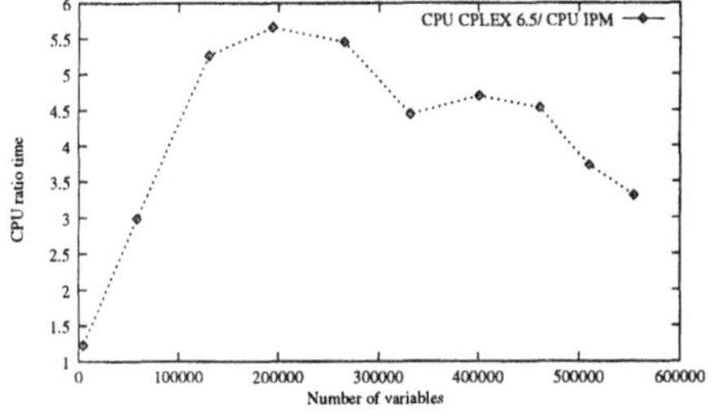

Figure 8. Number of IP iterations performed by CPLEX 6.5 and IPM for the quadratic PDS problems.

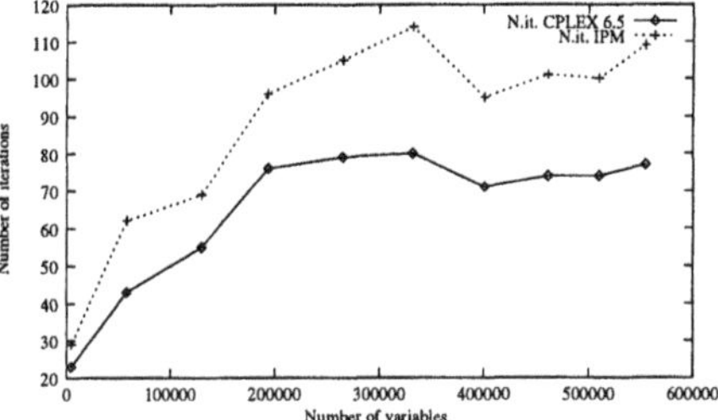

From Figures 5 and 7, IPM was in the all cases more efficient than CPLEX 6.5 (the ratio time was always greater than 1.0). For some Mnetgen and PDS instances IPM was about 20 and 5 times faster, respectively. It is important to note that IPM makes use of standard Cholesky routines [16], whereas CPLEX 6.5 includes a highly tuned and optimized factorization code [2]. Therefore, in principle, the performance of IPM could even be improved. Looking at Figures 6 and 8 it can be seen that IPM performed many more interior-point iterations than CPLEX 6.5. This is because, unlike CPLEX 6.5, the current version of IPM does not implement Mehrotra's predictor-corrector heuristic. In [4] it was shown that Mehrotra's heuristic was not appropriate for linear multicommodity problems. However, for quadratic problems, and because of the good behavior of the preconditioner, it could be an efficient option. Adding Mehrotra's strategy to IPM is part of the additional tasks to be performed.

Finally, we compared IPM and CPLEX 6.5 with PPRN [6] and with an implementation of the ACCPM [12] that we developed using the standard ACCPM library distribution [17]. For this purpose we chose some of the smallest Mnetgen and PDS instances, whose dimensions are shown in Tables 5 and 6 (columns m, n, k, $\tilde{m}$ and $\tilde{n}$, with the same meaning as before). These Tables also give the execution time in seconds (columns "CPU") for each solver. Clearly, CPLEX 6.5 and IPM outperformed both PPRN and ACCPM. Moreover, PPRN and ACCPM seemed not to be competitive approaches for quadratic multicommodity flows. (On the other hand, unlike CPLEX 6.5 and IPM, they can deal with nonlinear objective functions.)

Table 5. Dimensions and results for the small quadratic Mnetgen problems.

						CPU			
Instance	m	n	k	$\tilde{n}$	$\tilde{m}$	CPLEX	IPM	PPRN	ACCPM
M_{64-4}	64	524	4	2620	780	0.7	0.3	6.0	158.0
M_{64-8}	64	532	8	4788	1044	3.1	0.8	38.0	2116.9
M_{64-16}	64	497	16	8449	1521	10.7	1.6	184.6	5683.4
M_{64-32}	64	509	32	16797	2557	20.8	4.3	failed	15753.4
M_{64-64}	64	511	64	33215	4607	46.8	10.7	12710.1	34027.3

5. Conclusions and future tasks

From the computational experience reported, it can be stated that the specialized interior-point algorithm is a promising approach for separable quadratic multicommodity problems. Among the future tasks to be

Table 6. Dimensions and results for the small quadratic PDS problems.

						CPU			
Instance	m	n	k	$\tilde{n}$	$\tilde{m}$	CPLEX	IPM	PPRN	ACCPM
PDS1	126	372	11	4464	1758	1.6	1.3	75.5	failed
PDS2	252	746	11	8952	3518	5.2	7.9	293.3	failed
PDS3	390	1218	11	14616	5508	10.2	10.5	903.4	failed
PDS4	541	1790	11	21480	7741	22.4	33.9	1702.8	failed
PDS5	686	2325	11	27900	9871	53.2	44.7	2631.3	failed

performed we find a deep study of the behavior of the spectral radius of $D^{-1}(C^T BC)$, the addition of Mehrotra's predictor-corrector method, and using the algorithm in a network design framework.

References

[1] A. Ali and J.L. Kennington. (1977). Mnetgen program documentation, Technical Report 77003, Dept. of Ind. Eng. and Operations Research, Southern Methodist University, Dallas.

[2] R.E. Bixby, M. Fenelon, Z. Gu, E. Rothberg and R. Wunderling. (2000). MIP: Theory and practice—Closing the gap, in: *System Modelling and Optimization. Methods, Theory and Applications*, eds. M.J.D. Powell and S. Scholtes, Kluwer, 19–49.

[3] W.J. Carolan, J.E. Hill, J.L. Kennington, S. Niemi and S.J. Wichmann. (1990). An empirical evaluation of the KORBX algorithms for military airlift applications, *Operations Research*, 38, 240–248.

[4] J. Castro. (2000). A specialized interior-point algorithm for multicommodity network flows, *SIAM J. on Optimization*, 10(3), 852–877.

[5] J. Castro. (2000). Computational experience with a parallel implementation of an interior-point algorithm for multicommodity flows, in: *System Modelling and Optimization. Methods, Theory and Applications*, eds. M.J.D. Powell and S. Scholtes, Kluwer, 75–95.

[6] J. Castro and N. Nabona. (1996). An implementation of linear and nonlinear multicommodity network flows, *European J. of Operational Research*, 92, 37–53.

[7] P. Chardaire and A. Lisser. (1999). Simplex and interior point specialized algorithms for solving non-oriented multicommodity flow problems, Operations Research (to appear).

[8] ILOG CPLEX. (1999). *ILOG CPLEX 6.5 Reference Manual Library*, ILOG.

[9] A. Frangioni. (2000). Personal communication.

[10] A. Frangioni and G. Gallo. (1999). A bundle type dual-ascent approach to linear multicommodity min cost flow problems, *INFORMS J. on Comp.*, 11(4), 370–393.

[11] B. Gendron, T.G. Crainic, A. Frangioni. (1999). Multicommodity capacitated network design, in *Telecommunications Network Planning*, B. Sansó and P. Soriano (Eds.), Kluwer Academics Publishers, 1–19.

[12] J.-L. Goffin, J. Gondzio, R. Sarkissian and J.-P. Vial. (1996). Solving nonlinear multicommodity flow problems by the analytic center cutting plane method, *Math. Programming*, 76, 131–154.

[13] Andrew V. Goldberg, Jeffrey D. Oldham, Serge Plotkin and Cliff Stein. (1998). An implementation of a combinatorial approximation algorithm for minimum-cost multicommodity flow, in: *Lecture Notes in Computer Sciences. Proceedings of the 6th International Integer Programming and Combinatorial Optimization Conference*, eds. R.E. Bixby, E.A. Boyd and R.Z. Ríos-Mercado, Springer.

[14] A. Ourou, P. Mahey and J.-Ph. Vial. (2000). A survey of algorithms for convex multicommodity flow problems. *Management Science*, 46(1), 126–147.

[15] R.D. McBride. (1998). Progress made in solving the multicommodity flow problem, *SIAM J. on Opt.*, 8, 947–955.

[16] E. Ng and B.W. Peyton. (1993). Block sparse Cholesky algorithms on advanced uniprocessor computers, *SIAM J. Sci. Comput.*, 14, 1034–1056.

[17] O. Péton and J.-Ph. Vial. (2000). A tutorial on ACCPM, Technical Report, HEC/Logilab, University of Geneva.

[18] S.J. Wright. (1997). *Primal-Dual Interior-Point Methods*, SIAM, Philadelphia, PA.

STABILITY AND LOCAL GROWTH NEAR BOUNDED-STRONG OPTIMAL CONTROLS

Ursula Felgenhauer
Institute of Mathematics
Technical University Cottbus, Germany
felgenh@math.tu-cottbus.de

Abstract Nonlinear constrained optimal control problems as a rule suffer from the so-called two-norm discrepancy, which in particular says that under stable optimality conditions the objective functionals satisfy a quadratic local growth estimate in terms of the L_2 norms but in L_∞ neighborhoods of the solution only. Furthermore, in the case of weak local optima with continuous control functions, stability w.r.t. parameter changes usually can be expected to hold in L_∞ sense rather than in L_p.

Whenever we consider problems with discontinuous optimal control behavior, these results are too restrictive to discuss general variations of the solution including changes in the break points or switches in the active sets. In the paper we show how the use of certain integrated optimality criteria obtained via a duality approach allows for estimates also in the case of discontinuous controls. We consider L_2 and L_1 quadratic growth estimates and discuss consequences for the behavior of minimizing sequences.

Keywords: Constrained control problems, sufficient optimality conditions, stability.

1. Local optimality criteria in integrated form

Consider first a general nonlinear constrained optimal control problem (*primal* problem formulation):

$$\textbf{(P)} \qquad min \quad J(x,u) = k(x(0),x(T)) + \int_0^T r(t,x(t),u(t))\,dt$$

$$s.t. \qquad \dot{x} = f(t,x(t),u(t)) \qquad a.e. \text{ in } [0,T], \qquad (1)$$

$$\beta(x(0),x(T)) = 0, \qquad (2)$$

$$g(t,x(t),u(t)) \leq 0 \qquad a.e. \text{ in } [0,T]. \qquad (3)$$

The pair $(x,u) \in W^1_\infty(0,1;\mathbb{R}^n) \times L_\infty(0,1;\mathbb{R}^k)$ is called *admissible* for **(P)** if the state equation (1) (including the boundary conditions (2)) together with the inequality constraints (3) (where $g : [0,T] \times \mathbb{R}^n \times \mathbb{R}^k \to \mathbb{R}^m$, $\beta : [0,T] \times \mathbb{R}^n \times \mathbb{R}^k \to \mathbb{R}^s$) is fulfilled. All data functions are assumed to be sufficiently smooth.
Denote by H the HAMILTONian, and by $\hat{H}$ the *augmented* HAMILTONian related to the problem **(P)**:

$$H(t,x,u,p) = r(t,x,u) + p^T f(t,x,u) ,$$

$$\hat{H}(t,x,u,p,\mu) = H(t,x,u,p) + \mu^T g(t,x,u) , \quad \mu \geq 0.$$

Further, let W stand for the set

$$W = \{ (t,x,u) : t \in [0,1],\ g(t,x,u) \leq 0 \} .$$

Consider the dual variable S given by a function $S : [0,T] \times \mathbb{R}^n \to \mathbb{R}$ and the auxiliary functional θ for $\xi_{1,2} \in \mathbb{R}^n$:

$$\theta(\xi_1,\xi_2,\ S) = k(\xi_1,\xi_2) + S(0,\xi_1) - S(T,\xi_2) .$$

We assume that S is at least LIPSCHITZ continuous w.r.t. (t,x) whenever $(t,x,u) \in W$. Define

$$T(S) = \inf\{\theta(\xi_1,\xi_2,\ S) : \beta(\xi_1,\xi_2) = 0\}$$

Then the following problem is dual to the original control problem:

$$\textbf{(D)} \qquad \begin{array}{ll} max & T(S) \\ s.t. & H(t,x,u,S_x(t,x)) + S_t(t,x) \geq 0 \\ a.e. & \text{on } W = \{ (t,x,u) : t \in [0,1],\ g(t,x,u) \leq 0 \} \end{array}$$

If (x,u) is an admissible pair for problem **(P)** and S is feasible for **(D)** then the duality relation ([9], also [21] or [4]) holds, i.e.

$$J(x,u) \geq T(S) .$$

The relation turns into an equality if and only if for some admissible (x_0,u_0) and feasible dual S

$$\Psi(x_0,u_0,\ S) = \int_0^T [H(t,x(t),u(t),S_x(t,x(t))) + S_t(t,x(t))]\, dt = 0, \qquad (4)$$

$$\psi(x_0(0),x_0(T),\ S) = \theta(x_0(0),x_0(T),\ S) - T(S) = 0 .$$

In this case, the pair (x_0,u_0) is a solution of **(P)**.

The analysis of the behavior of Ψ and ψ can be further used to characterize local minima of **(P)** in detail including estimates for local growth terms if available (cf. e.g. [7]). It can be also distinguished between *weak* and *strong* local optima in dependence of the reference sets for which optimality holds. To this aim, let us introduce the sets

$$W_\epsilon = W \cap ([0,1] \times B_\epsilon(x_0, u_0)), \quad \hat{W}_\epsilon = W \cap ([0,1] \times B_\epsilon(x_0) \times L_\infty)$$
$$\text{or} \qquad \tilde{W}_\epsilon = W \cap ([0,1] \times B_\epsilon(x_0) \times B_M(0))$$

where $\tilde{W}_\epsilon$ with a given constant $M > 0$ is used to check for so-called *bounded-strong* local optima (see e.g. [18] or [20]).

Abstract optimality criteria have been given in [17], [21] or (in slightly generalized formulation) in [7], [6]. The results are summed up in the following Theorem:

Theorem 1 *Let (x_0, u_0) be admissible for* **(P)**. *Suppose that a function $S : [0,T] \times I\!R^n \to I\!R$ exists which is Lipschitz continuous w.r.t. x and piecewise continuously differentiable w.r.t. t such that for a suitably chosen positive constant ϵ the following relations hold with $\gamma = 1$, $D(t) = W_\epsilon(t)$ and $D_\pi = B_\epsilon(x_0(0), x_0(T))$:*

(R1) $\quad \Psi(x, u; S) > 0$
$\quad \forall\ (x,u) \neq (x_0, u_0)$ *with* $(x(t), u(t)) \in D(t)$ *a.e. in* $[0,T]$;

(R2) $\quad \Psi(x_0, u_0; S) = 0\,;$

(R3) $\quad \psi(\xi_1, \xi_2, S) > 0 \qquad \forall\, \xi \in D_\pi,\ \xi \neq (x_0(0), x_0(T)).$

Then (x_0, u_0) is a strict weak local minimizer of **(P)**.
If **(R1)** - **(R3)** *hold true for a certain constant $\epsilon > 0$ with $\gamma = 0$, $D(t) = \hat{W}_\epsilon(t)$ and $D_\pi = B_\epsilon(x_0(0), x_0(T))$, then (x_0, u_0) is a (strict) strong local minimizer.*
If the conditions are satisfied with $D(t) = \tilde{W}_\epsilon(t)$ for some $M > 0$, the point (x_0, u_0) is a (strict) bounded-strong local optimum.

The above Theorem differs from former results ([21], [17]) mainly by the consequent usage of the HAMILTON-JACOBI inequality (cf. **(D)**) in integrated form. This fact corresponds to some relaxation in the characterization of local minima by duality means and was first used in [4] for the theoretical convergence analysis of certain discretization methods.

Notice that in some cases it is possible to find estimates for $\Psi(x, u; S) - \Psi(x_0, u_0; S)$ in terms of $\|x - x_0\|_2^2 + \|u - u_0\|_2^2$, i.e. a local quadratic growth condition w.r.t. L_2 topology ([17], [7]) although the reference sets are neighborhoods w.r.t. L_∞ (at least with respect to x). This fact illustrates once more the effects of the so-called *two-norm discrepancy* appearing in optimal control problems (cf. [10] for details).

The optimality criteria of Theorem 1 in their original parametric formulation have been used in [17] to derive sufficient optimality conditions for general control problems. The approach leads to conditions which in a different way had been obtained via the investigation of so-called *stable* weak optima ([12] or [15], see also [14]).
In particular, it was shown that the following criteria ensure the strict *weak local* optimality of a solution pair (x_0, u_0):

(PMP) Pontryagin's maximum principle:

$$\begin{aligned} \dot{x} &= \hat{H}_p = f\,, \quad \beta(x(0), x(T)) = 0\,; \\ \dot{p} &= -\hat{H}_x\,, \quad p(0) = -\nabla_1 k - \nabla_1 \beta\,\rho, \\ & \qquad\qquad p(T) = \nabla_2 k + \nabla_2 \beta\,\rho\,, \quad \rho \in I\!R^s; \\ \hat{H}_u &= 0\,; \\ \mu^T g &= 0\,, \qquad \mu \geq 0\,, \quad g \leq 0. \end{aligned}$$

Notice that in the transversality condition given above the subscripts 1, 2 stand for the gradient components corresponding to the initial and to the final state vectors respectively.

For a given admissible pair (x_0, u_0), let $g, f, \ldots$ be evaluated along the state-control trajectory. We will denote by g^σ, $\sigma > 0$, the set of σ-active constraints at t, e.g. the set of g_i such that $0 \geq g_i(t, x_0(t), u_0(t)) \geq -\sigma$.

(C1) Invertibility: For some positive constant σ, the gradients w.r.t. u of the σ- active constraints, $\nabla_u g^\sigma$, are uniformly linearly independent a.e. on $[0, T]$.

(C2) Controllability: There are functions $y \in W^1_\infty$, $v \in L_\infty$ satisfying

$$\nabla_x g^\sigma y + \nabla_u g^\sigma v = 0\,; \quad \dot{y} - \nabla_x f\, y - \nabla_u f\, v = 0 \qquad a.e.$$

together with the boundary condition $\nabla_1 \beta^T y(0) + \nabla_2 \beta^T y(T) = 0$.

In order to formulate second-order conditions, the index set $I_\delta = \{i : \mu_i \geq \delta\}$ and the related tangent spaces

$$\begin{aligned} T_\delta &= \{\zeta : (\nabla_{(x,u)} g^i)^T \zeta = 0 \quad \forall i \in I_\delta\}, \\ T'_\delta &= \{v : (\nabla_u g^i)^T v = 0 \;\forall i \in I_\delta\} \end{aligned}$$

are useful. The conditions then can be given as follows:

(C3) Legendre-Clebsch Condition: For some positive δ, a constant $\alpha > 0$ exists such that the estimate $v^T \hat{H}_{uu}(t) v \geq \alpha |\, v\,|^2$ holds $\forall\, v \in T'_\delta$ uniformly a.e. on the interval $[0, T]$.

(C4) Riccati Inequality: With the abbreviations $R = -(\nabla_u g^\delta)^+ \nabla_x g^\delta$ and $P = (\nabla_u g^\delta)^\perp$ where $(\cdot)^+$ stands for the *pseudoinverse* and $(\cdot)^\perp$ for the *null-space projector* of a matrix $(\cdot)$, introduce

$$\begin{aligned} a^\delta &= f_x + f_u R\,, \\ h^\delta_{xx} &= \hat{H}_{xx} + \hat{H}_{xu}R + R^T\hat{H}_{ux} + R^T\hat{H}_{uu}R\,, \\ \left(h^\delta_{uu}\right)^{(-1)} &= P\left(P\hat{H}_{uu}P\right)^{-1}P\,, \quad h^\delta_{xu} = \hat{H}_{xu} + R^T\hat{H}_{uu}\,. \end{aligned}$$

For some $\gamma > 0$, the matrix differential inequality

$$\begin{aligned} \dot{Q} \succeq\ & (h^\delta_{xu} + Qf_u)\left(h^\delta_{uu}\right)^{(-1)}(h^\delta_{xu} + Qf_u)^T \\ & - (a^\delta)^T Q - Qa^\delta - h^\delta_{xx} + \gamma I \qquad a.e. \text{ on } [0,T] \end{aligned} \tag{5}$$

has a bounded in $[0,1]$ solution satisfying the boundary restrictions

$$\begin{aligned} \xi^T\left(\nabla_1^2 k + \nabla_1^2\beta\,\rho + Q(0)\right)\xi &\geq \gamma\,|\xi|^2 \qquad \forall\,\xi \in I\!R^n \\ \xi^T\left(\nabla_2^2 k + \nabla_2^2\beta\,\rho - Q(1)\right)\xi &\geq \gamma\,|\xi|^2 \qquad \forall\,\xi \in I\!R^n. \end{aligned}$$

It is known that **(C1)** – **(C4)** not only guarantee **(R1)** – **(R3)** (see [7]), but also a) the LIPSCHITZ stability of the solution in L_∞ w.r.t. small data perturbations ([11], [15], [3]), b) (in the case of a continuous control u_0) the convergence of the EULER and related discretization methods ([14], [4], [2]).

2. Bounded-strong minima. Piecewise conditions

are formulated as integral conditions in terms of Ψ. When we are interested in local growth estimations, this fact gives us the chance to combine piecewise changing growth characterizations as they are typical for switches in the optimal control or for the junction of free arcs and arcs where certain constraints are active.

We begin our optimality analysis with the auxiliary functional $\Psi = \Psi(x, u, S)$ from (4), where the integrand has the form

$$R[t] = (H(t, x, u, \nabla_x S) + S_t)\,[t]\,. \tag{6}$$

As it has been shown in [7], using $\Psi(x_0, u_0, S) = 0$ together with an expansion

$$S(t,x) = S_0(t) + p(t)^T(x - x_0(t)) + 0.5\,(x - x_0(t))^T Q(t)\,(x - x_0(t))$$

and the optimal data $\nabla_x S = p_0 + Q(x - x_0)$ and $\dot{S}_0 = -r_0$, one can express R in the following way:

$$\begin{aligned} R[t] &= \hat{H}(x, u, p_0, \mu_0) - \hat{H}(x_0, u_0, p_0, \mu_0) - \hat{H}_x(x_0, u_0, p_0, \mu_0)^T (x - x_0) \\ &\quad +0.5\,(x - x_0)^T \dot{Q}(x - x_0) + (x - x_0)^T Q\,(\,f(x, u) - f(x_0, u_0)\,) \\ &\quad + \mu_0^T\,(\,g(x, u) - g(x_0, u_0)\,) \end{aligned}$$

By rearranging the terms related to variations w.r.t x or u, one can separate two terms $R_{1,2}$ such that $R[t] = R_1[t] + R_2[t]$ and (with $p = p_0 + Q(x - x_0)$)

$$\begin{aligned} R_1[t] &= \hat{H}(x, u_0, p, \mu_0) - \hat{H}(x_0, u_0, p, \mu_0) \\ &\quad - \hat{H}_x(x_0, u_0, p_0, \mu_0)^T (x - x_0) \;+ 0.5\,(x - x_0)^T \dot{Q}(x - x_0) \\ &\approx 0.5\,(x - x_0)^T \left(\hat{H}_{xx} + Q f_x + f_x^T Q + \dot{Q}\right) (x - x_0); \\ R_2[t] &= \hat{H}(x, u, p, \mu_0) - \hat{H}(x, u_0, p, \mu_0) - \mu_0^T g(x, u) \end{aligned}$$

Under condition **(C4)**, in particular $R_1[t]$ will be uniformly positive if $\|x - x_0\|_\infty$ is sufficiently small.

In order to estimate Ψ resp. $R[t]$, in [6] the following general result has been proved (Theor. 5 in the cited paper):

Theorem 2 *Suppose* (x_0, u_0) *to be a weak local minimizer satisfying together with some matrix function* $Q \in W_\infty^1$ *the conditions* **(C1)** - **(C4)**. *Let*

$$R_2[t] \geq c_1 |\,u - u_0(t)|^\nu - c_2 |\,x - x_0(t)| \qquad \forall u : \; |u| \leq M \tag{7}$$

hold almost everywhere on $[0, 1]$ *with* $\nu(t) \in \{1, 2\}$ *and constants not depending on* t. *Then* (x_0, u_0) *is a bounded-strong local minimizer, and positive constants* c, ϵ *exist such that*

$$J(x, u) - J(x_0, u_0) \geq c \,\|\, x - x_0\|_2^2 \tag{8}$$

for all admissible (x, u) *with* $\|\,x - x_0\|_\infty \leq \epsilon$, $\|u\|_\infty \leq M$.

Important special cases are the following:

Case 1: $\hat{H}(x_0, v, p_0, \mu_0)$ is strongly convex w.r.t. v.
In this case, $R_2 \geq c_1 |u - u_0|^2 - O(|\,x - x_0|)$, – a situation which can be often observed when $\hat{H}$ takes its minimum in an inner point of the control set.

Case 2: $u_0 = arg \min \hat{H}(x_0, v, p_0, \mu_0)$, and $-\mu_0^T g(x_0, u) \geq c_1' |u - u_0|$.
Here one can conclude that $R_2 \geq c_1' |\,u - u_0| - O(|\,x - x_0|)$. The situation

occurs when a certain strict complementarity condition holds together with the invertibility assumption **(C1)**.

The criteria given above have been discussed in detail in [6] and tested on a nonlinear example with generically discontinuous optimal control regime in [7]. As a further illustration, we consider here an example from [14]:

Example: The tunneldiode oscillator.

$$\begin{aligned} \min\ J(x,u) &= \int_0^T \left(u^2(t) + x^2(t)\right)\, dt \\ \text{s.t.}\quad \dot{x}_1 &= x_2\,, \qquad x_1(0) = x_2(0) = -5,\ x_1(T) = x_2(T) = 0, \\ \dot{x}_2 &= -x_1 + x_2(1.4 - 0.14x_2^2) + 4u, \\ \text{for}\quad & |\,u\,| \le 1 \qquad a.e.\ \text{on}\ [0,T]. \end{aligned}$$

In the paper [14], the conditions **(C1)** - **(C4)** were checked numerically. In particular, by using a multiple shooting method a bounded matrix function Q was constructed so that the RICCATI condition was satisfied. Thus it is reasonable to assume that a solution to **(C4)** exists with $\|Q\|_\infty < \infty$, and that **(C1)** - **(C3)** hold true for some positive σ and δ. In the example situation for $T = 4.5$ the structure of the optimal control was obtained as $u = 1$ on $[0, \tau_1)$, $u = -1$ on (τ_2, τ_3) and $u = -2p_2$ elsewhere (where $0 < \tau_1 < \tau_2 < \tau_3 < T$ are found approximately from the numerical solution). The points τ_i are the so-called junction points. Consider the term R_2 from our above growth analysis for Ψ:
Using the example data, we get

$$\begin{aligned} R_2 &= u^2 - u_0^2 + 4p_2(u - u_0) + \lambda_{0,1}(1 - u_0) + \lambda_{0,2}(u_0 + 1) \\ &= (2u_0 + 4p_{0,2})(u - u_0) + (u - u_0)^2 + 4(u - u_0)[Q(x - x_0)]_2 \\ &\ge \min\left\{(\lambda_1 + \lambda_2)\,|u - u_0|,\ 0.5\,|u - u_0|^2\right\} - c\,|x - x_0| \end{aligned}$$

whenever $|x - x_0| \le \epsilon < 1$. Therefore, the conditions of Theorem 2 are applicable in the example, and the considered solution turns out to be a bounded-strong local minimum satisfying a local L_2 quadratic growth estimate.

3. Bounded-strong optimality in bang-bang case

Consider a dynamical system with linear in state and control equation and known initial position. We will ask for a control regime which in given time gains the system to a final state as close as possible to the origin.

In general, in an arbitrarily given time the system not always can be terminated. Since a (small) deviation from zero in the final position is allowed, this problem class is also called *soft termination control.* As a model case, consider a problem with box control constraints:

$$
\begin{aligned}
(\mathbf{P}_S) \qquad & min \quad J(x,u) &=& \ \frac{1}{2}\,\|\, x(T)\,\|^2 & & \\
& s.t. \quad \dot{x}(t) &=& \ A(t)x(t) + B(t)u(t) \quad a.e. \text{ in } [0,T]; & & (9)\\
& & & x(0) = a\,; & & (10)\\
& & & |\, u_i(t)\,| \le 1,\ i=1,\ldots,k, \quad a.e. \text{ in } [0,T]\,. & & (11)
\end{aligned}
$$

The HAMILTON function related to $(\mathbf{P}_S)$ has the form

$$H(t,x,u,p) \;=\; p^T A(t)\,x \;+\; p^T B(t)\,u\;,$$

whereas the *augmented* HAMILTONian reads as

$$\hat{H}(t,x,u,p,\mu) \;=\; H \;+\; \mu_1^T(u-e) \;-\; \mu_2^T(u+e)$$

with $\mu_{1,2} \ge 0,\ \ e = (1,1,\ldots,1)^T$.
From PONTRYAGIN's maximum principle, we obtain the *switching function*

$$\sigma(t) \;=\; B(t)^T p(t)\;, \tag{12}$$

where the costate p satisfies the adjoint equation

$$\dot{p}(t) \;=\; -A(t)^T p(t)\;, \quad p(T) = x(T)\;,$$

and the optimal control is given by

$$u_{0,i} \in \begin{cases} \{+1\} & \text{if} \quad \sigma_i(t) \;<\; 0, \\ \{-1\} & \text{if} \quad \sigma_i(t) \;>\; 0, \\ [-1,+1] & \text{if} \quad \sigma_i(t) \;=\; 0, \end{cases} \qquad i=1,\ldots,k\,. \tag{13}$$

Further, the multiplier functions μ_j suffice the relations

$$\mu_1(t) \;=\; \Big(B(t)^T p(t)\Big)_-\;, \quad \mu_2(t) \;=\; \Big(B(t)^T p(t)\Big)_+$$

where the right-hand sides denote the positive resp. negative part of the related vector components.

In the case that $\sigma_i(t) \equiv 0$ on a certain interval $I \subseteq [0,T]$, we say the control u_0 has a *singular arc* (on I). In our optimality analysis, we will restrict ourselves to the case of piecewise constant u_0 without singular arcs:

Assumption 1 *The optimal control has no singular arcs. In addition, the set of switching points* $\Sigma \;=\; \{\, t \in [0,T] \;:\; \exists\, i \in \{1,\ldots,m\}$ *with* $\sigma_i(t) = 0\,\}$ *is finite, i.e.* $\Sigma \;=\; \{\, t_s : \; 1 \le s \le l\,\}$ *for some* $l \in N$.

Remark: It is well known, that the above assumption holds true e.g. in the case that A is time-independent and has exclusively real eigenvalues.

Under the Assumption 1, the solution obviously satisfies **(C1)**, **(C2)**: Indeed, $\nabla_u g^0 = diag(\gamma_j)$ with $\gamma_j \in \{+1, -1\}$, and the controllability assumption **(C2)** in the case of a (linear) initial value problem always trivially holds.

When we try, however, to apply the second order optimality criteria **(C3)** and **(C4)** to our problem, serious difficulties occur:

First of all, the LEGENDRE - CLEBSCH condition **(C3)** becomes singular since $\hat{H}_{uu} = 0$ a.e. on $[0, T]$. It can be fulfilled formally only in the limit sense with $\delta = 0$ where $T_0 = \{0\}$ for all $t \in [0, T] \backslash \Sigma$.

The limit case of the corresponding RICCATI inequality for $\delta \to 0$ gives

$$\begin{aligned} \dot{Q} + AQ + QA^T &\succeq \gamma I \qquad a.e. \\ I - Q(T) &\succeq \gamma I \end{aligned} \tag{14}$$

which obviously can be fulfilled with arbitrary positive γ. In particular, one can choose Q_1 such that in both parts of (14) equality holds with $\gamma = 0.5$. Then for $\gamma < 0.5$, the matrix function $Q_\gamma = 2\gamma Q_1$ solves (14). Moreover, it belongs to $L_\infty(0, T; I\!R^{n \times n})$ with $\|Q_\gamma\|_\infty = 2\gamma \|Q_1\|_\infty \leq c(A)\,\gamma$.

It has to be mentioned, however, that in the case of a singular matrix $\hat{H}_{uu}$ the conditions **(C3)**, **(C4)** with $\delta = 0$ in general are *not* sufficient to show the optimality of the solution even in the weak local sense. For our problem class e.g. the linearization in the (nearly-)active constraints will allow only for zero control variation, which together with the admissibility assumption for the linear state equation case reduces the state variation to zero, too. The local technique of deriving estimates for Ψ (resp. for $J - J^*$) from its TAYLOR expansion (see [21] and [7], [6]) therefore fails in the given situation.

Having in mind these arguments, let us restart the optimality analysis for $\Psi = \Psi(x, u, S)$ (cf. (4)) with the integrand $R = R_1 + R_2$ from (6). Notice that from $f(x, u) = Ax + Bu$ we obtain in particular the following estimate for R_1 near x_0:

$$R_1[t] \approx 0.5\,(x - x_0)^T \left(\dot{Q} + QA + A^TQ + \hat{H}_{xx}\right)(x - x_0)\,.$$

In the case when $(\mathbf{P}_S)$ is considered, we have $\hat{H}_{xx} = 0$. Choosing $Q = Q_\gamma$ and $y = x - x_0$ such that $\|y\|_\infty \leq \epsilon_1$, with sufficiently small $\epsilon_1 > 0$ we get

$$R_1[t] \geq 0.25\,y^T \left(\dot{Q} + QA + A^TQ\right) y \geq \frac{\gamma}{4}\,|y|^2\,, \tag{15}$$

We will further derive appropriate estimates for R_2 and the integrals Ψ under the following regularity assumption on the zeros of the switching function σ:

Assumption 2 *There exist positive constants c_0, $\bar{\delta}$ with the following property: For $\delta \in (0, \bar{\delta})$, denote $\omega_\delta = \bigcup_{1 \le s \le l}(t_s - \delta, t_s + \delta)$. Then,*

$$\min_i \left| \left(B^T p(t) \right)_i \right| \ge c_0 \delta \qquad \forall\, t \in [0,T] \backslash \omega_\delta \,.$$

In the case of problem $(\mathbf{P}_S)$, the term R_2 connected with variations w.r.t. u may be expressed by

$$\begin{aligned} R_2[t] &= \hat{H}(x, u, p_0, \mu_0) - \hat{H}(x, u_0, p_0, \mu_0) + (x - x_0)^T Q B (u - u_0) \\ &\qquad - \mu_{0,1}^T (u - u_0) + \mu_{0,2}^T (u - u_0). \end{aligned}$$

Denoting $v = u - u_0$ with an arbitrary feasible control u, we have:

$$\begin{aligned} \left(B^T p_0\right)_i > 0 \quad &\Rightarrow \quad (u_o)_i = -1 \quad \Rightarrow \qquad v_i \ge 0 \\ \left(B^T p_0\right)_i < 0 \quad &\Rightarrow \quad (u_o)_i = +1 \quad \Rightarrow \qquad v_i \le 0 \,. \end{aligned}$$

Therefore,

$$R_2[t] = \left(B^T p_0\right)^T v + y^T Q B v \ge \sum_{i=1}^m \left| B^T p_0 \right|_i \cdot |v_i| - \left| y^T Q B v \right| .$$

In order to estimate the integral over R_2, each part of the right-hand side will be integrated separately and estimated now:

$$J_1 = \int_0^T \left| y^T Q B v \right| dt \le \|y\|_\infty \|Q\|_\infty \|B\|_\infty \int_0^T |v(t)|\, dt \,.$$

From the state equation it follows that $\|y\|_\infty \le c(A, B)\, \|v\|_1$, thus

$$J_1 \le \|Q\|_\infty \|B\|_\infty \, c(A, B)\, \|v\|_1^2 =: \gamma\, c_2 \|v\|_1^2$$

when $Q = Q_\gamma$, and for a certain constant $c_2 = c_2(A, B) > 0$.

The estimation of

$$J_2 = \int_0^T \sum_{i=1}^m \left| B^T p_0 \right|_i \cdot |v_i|\, dt$$

will make essential use of the Assumption 2 on the switching function behavior near its zeros. Rewriting

$$J_2 = \int_0^T \ldots\, dt = \int_{[0,T] \backslash \omega_\delta} \ldots\, dt + \int_{\omega_\delta} \ldots\, dt \,,$$

it is easy to see that the second part is nonnegative. Neglecting this integral (which for small δ is of small size), from Assumption 2 we conclude

$$J_2 \geq c_0\delta \int_{[0,T]\setminus\omega_\delta} \sum_{i=1}^{m} |\, v_i(t)\,|\; dt \geq c_0\delta \int_{[0,T]\setminus\omega_\delta} |\, v(t)\,|\; dt$$

(where $|v|$ stands for the EUKLIDean vector norm of $v = v(t)$ in $\mathbb{R}^m$). On the other hand side,

$$\begin{aligned} \|v\|_1 &= \int_0^T |v(t)|\, dt = \int_{[0,T]\setminus\omega_\delta} |v(t)|\, dt + \int_{\omega_\delta} |v(t)|\, dt \\ &\leq \int_{[0,T]\setminus\omega_\delta} |v(t)|\, dt + c_1\delta \end{aligned}$$

for $c_1 = 4kM$ if only $\|u\|_\infty$ and $\|u_0\|_\infty$ are bounded by M ($M = 1$ in $(\mathbf{P}_S)$ e.g.). Inserting this estimate into our relation for J_2, we obtain

$$J_2 \geq c_0\delta\, (\, \|v\|_1 - c_1\delta\,)\ .$$

Combining now the estimates for J_1 and J_2, an estimate for $\int R_2\, dt$ results:

$$\int_0^T R_2[t]\, dt \geq c_0\delta\, (\, \|v\|_1 - c_1\delta\,) - c_2\gamma\, \|v\|_1^2\ . \tag{16}$$

LEMMA 1 (Weak local optimality.)
Let the Assumptions 1, 2 hold for the optimal data. Then positive constants ϵ, c_w exist such that

$$\int_0^T R[t]dt \geq c_w \left(\|\, x - x_0\|_2^2 + \|\, u - u_0\|_1^2 \right) \tag{17}$$

for all admisssible $(x,u) \in W_\epsilon$. Therefore, (x_0,u_0) is a strict weak local minimizer (with $J(x,u) - J(x_0,u_0)$ satisfying a local quadratic growth condition similiar to (17)).

Proof: Our previous analysis allows to discuss the estimate (16) for various δ. In particular, choose

$$\delta = \frac{1}{2\,c_1}\, \|v\|_1\ .$$

Then, for $\|v\|_\infty = \|\, u - u_0\|_\infty \leq \epsilon < \epsilon_2 = (2c_1\bar{\delta})/T$,

$$\delta \leq \frac{1}{2\,c_1}\, T\, \|v\|_\infty < \bar{\delta},$$

and we obtain

$$\int_0^T R_2[t]\,dt \ \geq \ \left(\frac{c_0}{4c_1} - c_2\gamma\right) \|v\|_1^2 \ =: \ \tilde{c}(\gamma)\,\|v\|_1^2 \ .$$

If γ is taken from $\left(0, \frac{c_0}{4c_1c_2}\right)$ then $\tilde{c}(\gamma)$ is positive, so that together with (15) the last estimate leads to

$$\int_0^T R[t]\,dt \ \geq \ \frac{\gamma}{4}\,\|\, x - x_0\|_2^2 \ + \ \tilde{c}(\gamma)\,\|\, u - u_0\|_1^2 \ .$$

Thus, the desired conclusion follows by setting $c_w \ = \ \max_{\gamma>0} \min\{\frac{\gamma}{4}, \tilde{c}(\gamma)\}$ and $\epsilon = \min\{\epsilon_1, \epsilon_2\}$. □

LEMMA 2 (Strong local optimality.)
Let the Assumptions 1, 2 hold true. Then positive constants ϵ, c_s exist such that

$$\int_0^T R[t]\,dt \ \geq \ c_s\,\|\, x - x_0\|_\infty^2 \qquad \text{for all admisssible } (x,u) \ \in \ \tilde{W}_\epsilon\,. \qquad (18)$$

Therefore, (x_0, u_0) is a strict strong local minimizer (with $J(x,u) - J(x_0, u_0)$ satisfying a local quadratic growth condition similiar to (18)).

Proof: Consider (16) with

$$\delta \ = \ \min\left\{\frac{1}{2\,c_1}, \frac{\bar{\delta}}{2MT}\right\} \|v\|_1 \ =: \ c_3\|v\|_1 \ .$$

Since $\|v\|_\infty \ \leq \ 2M$, we have $\delta \ \leq \ \frac{\bar{\delta}}{2MT}\,T\,\|v\|_\infty \ \leq \ \bar{\delta}$ so that, for $\gamma < (c_0c_3)/(4c_2)$ and $\|y\|_\infty \leq \epsilon_1$,

$$\int_0^T R[t]\,dt \ \geq \ \frac{c_0c_3}{4}\,\|v\|_1^2 \ . \qquad (19)$$

follows. Notice that due to the state equation we have

$$\|y\|_\infty \leq \ c(A,B)\,\|v\|_1,$$

and consequently (18) holds for $c_s \leq c_0c_3/(4\,c^2(A,B))$. □

Remark: Notice that in the case of a compact control set as in the model problem $(\mathbf{P}_S)$ the definitions of strong and of bounded-strong local optimality coincide.

As a conclusion from the last two Lemmas we easily obtain:

Theorem 3 *Let the Assumptions 1, 2 be satisfied for the solution* (x_0, u_0) *of problem* $(\mathbf{P}_S)$. *Then,* (x_0, u_0) *is a (bounded-)strong local minimizer, and positive constants* c, ϵ *exist such that*

$$J(x,u) - J(x_0,u_0) \geq c\Big(\|x - x_0\|_2^2 + \|u - u_0\|_1^2 \Big) \tag{20}$$

for all admissible pairs with $\|x - x_0\|_\infty \leq \epsilon$.

4. Minimizing sequence stabilization

In this final section, it will be shown how the results of Theorems 2 and 3 can be used to obtain certain preliminary convergence results for minimizing sequences of $(\mathbf{P})$ resp. $(\mathbf{P}_S)$. The results are orientated on [6] (Propos. 6).
For convenience, the following assumption is made on the system dynamics:

Assumption 3 *There exists a constant* $M > 0$ *such that for any piecewise continuous* u *with* $\|u\|_\infty \leq M$ *the state boundary value problem (1), (2) has a bounded solution on* $[0, T]$. *If solutions corresponding to* $u_{1,2}$ *are denoted by* $x_{1,2}$ *resp., then* $\| x_1 - x_2 \|_\infty \leq c_0 \| u_1 - u_2 \|_1$ *holds true for some constant* $c_0 > 0$

Consider first the situation of section 3 where the LEGENDRE-CLEBSCH condition is fulfilled in the sense of **(C3)**, and where Theorem 2 allows for a local quadratic growth estimation of the objective functional.

LEMMA 3 *Let* (x_0, u_0) *be a bounded-strong local minimizer for* $(\mathbf{P})$ *and suppose the Assumptions 1-3 to hold true. Further, assume the estimate (8), i.e.*

$$J(x,u) - J(x_0,u_0) \geq c\,\|x - x_0\|_2^2$$

to be valid for $\|x - x_0\|_\infty \leq \epsilon$. *If* (x_k, u_k) *is a minimizing sequence with uniformly bounded and piecewise continuous controls* u_k, *and if for all* k, $\|u_k - u_0\|_1$ *is sufficiently small, then* $x_k \to x_0$ *in* L_2 *sense.*

The proof is similiar to that of Proposition 6 in [6]. Notice, that in practice a minimizing sequence can be obtained e.g. by the EULER discretization approach ([1], also [2]). The error estimates in the cited papers are given in terms of the maximal deviation in the discretization grid points, a condition which for piecewise continuous controls u_0 is sufficient for their L_1 closeness to the solution.

We add an analogous result for the bang-bang situation, although in this case (without an appropriate coercivity assumption like **(C3)**) the convergence of discretization schemes and their benefit for constructing minimizing sequences theoretically is yet an open question:

LEMMA 4 *Let (x_0, u_0) be a bounded-strong local minimizer for* $(\mathbf{P}_S)$ *and suppose the Assumptions 1-3 to hold true. Further, assume the estimate (20), i.e.*

$$J(x,u) - J(x_0,u_0) \geq c\Big(\|x - x_0\|_2^2 + \|u - u_0\|_1^2 \Big)$$

to be valid for $\|x - x_0\|_\infty \leq \epsilon$. If (x_k, u_k) is a minimizing sequence with uniformly bounded and piecewise continuous controls u_k, and if for all k, $\|u_k - u_0\|_1$ is sufficiently small, then $x_k \to x_o$ in L_2, and moreover, $u_k \to u_0$ w.r.t. L_1 topology.

References

[1] Dontchev, A. L., and Hager, W. W. Lipschitzian stability in nonlinear control and optimization, *SIAM J. Control and Optimization*, 31:569-603, 1993.

[2] Dontchev, A. L., and Hager, W. W. The Euler approximation in state constrained optimal control, *Math. Comp.*, 70(233):173-203, 2001.

[3] Dontchev, A. L., and Malanowski, K. A characterization of Lipschitzian stability in optimal control, In Ioffe et al. [8], pages 62-76.

[4] Felgenhauer, U. *Diskretisierung von Steuerungsproblemen unter stabilen Optimalitätsbedingungen*, Habilitationsschrift, Brandenburgische Technische Universität Cottbus, 1999. (122 p., in German)

[5] Felgenhauer, U. On smoothness properties and approximability of optimal control functions. In D. Klatte, J. Rückmann, and D. Ward, editors, *Optimization with Data Perturbation II*, vol. 101 of *Ann. Oper. Res.* Baltzer Sc. Publ., The Netherlands,vol. 101:23-42, 2001.

[6] Felgenhauer, U. Structural properties and approximation of optimal controls, *Nonlinear Analysis*, (TMA), 47(3):1869-1880, 2001.

[7] Felgenhauer, U. Weak and strong optimality conditions for constrained control problems with discontinuous control, *J. Optim. Theor. Appl.*, 110(2):361-387, 2001.

[8] A. Ioffe, S. Reich, and I. Shafrir, editors. *Calculus of variations and optimal control. Haifa 1998*, vol. 411 of *Chapman & Hall/CRC Res. Notes Math.*, Boca Raton, FL, 2000. Chapman & Hall/CRC.

[9] Klötzler, R. On a general conception of duality in optimal control, In: vol. 703 of *Lect. Notes Math.*, 189-196. Springer-Verlag, New York, Heidelberg, Berlin, 1979.

[10] Malanowski, K. Two-norm approach in stability and sensitivity analysis of optimization and optimal control problems, *Adv. in Math.Sc. and Applic.*, 2:397-443, 1993.

[11] Malanowski, K. Stability and sensitivity analysis of solutions to infinite-dimensional optimization problems, In J. Henry nad J.-P. Yvon, editors, *Proc. 16th IFIP-TC7 Conference on System Modelling and Optimization*, vol. 197 of *Lect. Notes in Control and Inf. Sci.*, 109-127, London, 1994. Springer-Verlag.

[12] Malanowski, K. Stability and sensitivity analysis of solutions to nonlinear optimal control problems, *Appl. Math. and Optim.*, 32:111-141, 1995.

[13] Malanowski, K. Stability analysis of solutions to parametric optimal control problems, In J. Guddat, H. T. Jongen, F.Nožička, G. Still, and F.Twilt, editors, *Proc. IV. Conference on "Parametric Optimization and Related Topics"* Enschede 1995, *Ser. Approximation and Optimization*, 227-244, Frankfurt, 1996. Peter Lang Publ. House.

[14] Malanowski, K., Büskens, C., and Maurer, H. Convergence of approximations to nonlinear control problems, In A. V. Fiacco, editor, *Mathematical Programming with Data Perturbation*, vol. 195 of *Lect. Notes Pure Appl. Mathem.*, 253-284. Marcel Dekker, Inc., New York, 1997.

[15] Malanowski, K., and Maurer, H. Sensitivity analysis for parametric optimal control problems with control-state constraints, *Comput. Optim. Appl.*, 5:253–283, 1996.

[16] Malanowski, K., and Maurer, H. Sensitivity analysis for state constrained optimal control problems, *Discrete Contin. Dynam. Systems*, 4:241 – 272, 1998.

[17] Maurer, H., and Pickenhain, S. Second order sufficient conditions for optimal control problems with mixed control-state constraints, *J. Optim. Theor. Appl.*, 86:649–667, 1995.

[18] Milyutin, A. A., and Osmolovskii, N. P. *Calculus of variations and optimal control*, Amer. Mathem. Soc., Providence, Rhode Island, 1998.

[19] Osmolovskii, N. P. Quadratic conditions for nonsingular extremals in optimal control (A theoretical treatment). *Russian J. of Mathem. Physics*, 2:487-512, 1995.

[20] Osmolovskii, N. P. Second-order conditions for broken extremals. In Ioffe et al. [8], pages 198-216.

[21] Pickenhain, S. Sufficiency conditions for weak local minima in multidimensional optimal control problems with mixed control-state restrictions, *Zeitschr. f. Analysis u. Anwend. (ZAA)*, 11:559–568, 1992.

GRAPH ISOMORPHISM ALGORITHM BY PERFECT MATCHING

Kazuma Fukuda
Department of Internet Media Sysytem
Information Technology R&D Center
MITSUBISHI ELECTRIC Corporation
Kamakura, Kanagawa, JAPAN
kfukuda@isl.melco.co.jp

Mario Nakamori
Department of Computer Science
Tokyo A&T University
Koganei, Tokyo, JAPAN
nakamori@cc.tuat.ac.jp

Abstract No polynomial time algorithm is known for the graph isomorphism problem. In this paper, we determine graph isomorphism with the help of perfect matching algorithm, to limit the range of search of 1 to 1 correspondences between the two graphs: We reconfigure the graphs into layered graphs, labeling vertices by partitioning the set of vertices by degrees. We prepare a correspondence table by means of whether labels on 2 layered graphs match or not. Using that table, we seek a 1 to 1 correspondence between the two graphs. By limiting the search for 1 to 1 correspondences between the two graphs to information in the table, we are able to determine graph isomorphism more efficiently than by other known algorithms. The algorithm was timed with on experimental data and we obtained a complextity of $O(n^4)$.

Keywords: Graph Isomorphism, Regular Graph

1. Introduction

The graph isomorphism problem is to determine whether two given graphs are isomorphic or not. It is not known whether the problem belongs to the class P or the class NP–complete. It has been shown, however, that the problem can be reduced to a group theory problem (van Leeuwen, 1990).

Most studies of graph isomorphism (Hopcroft and Wong, 1974; Lueker, 1979; Babai et al., 1980; Galil et al., 1987; Hirata and Inagaki, 1988; Akutsu, 1988) restrict graphs by their characteristics. Some studies are undertaken based on group theory. Most studies are concerned on the existence of algorithms (Filotti and Mayer, 1980; Babai et al., 1982; Luks, 1982; Babai and Luks, 1983; Agrawal and Arvind, 1996), and a few papers report the implementation of algorithms (Corneil and Gotlieb, 1970) and experimental results.

At present the best computational complexity by worst case analysis (Babai and Luks, 1983; Kreher and Stinson, 1998) is $O\left(c^{n^{1/2+o(1)}}\right)$. This algorithm makes use of the unique certification of a graph.

In the present paper, we consider the graph isomorphism problem for non-oriented connected regular graphs whose vertices and edges have no weight. We seek graph isomorphism by means of perfect matching to limit the range of 1-to-1 correspondences between the two graphs as follows.

First, we choose one vertex as root for each graph and reconfigure the graphs into layered graphs corresponding to the chosen vertices. Next, we label those vertices by partitioning the set of vertices by the distance from the root vertex. We construct a correspondence table which reflects whether labels on 2 layered graphs are the same or not. Then, referring to that table, we search for a 1-to-1 correspondence between the two graphs.

In other words, we create a bipartite graph between V_1 and V_2 and find a perfect matching in this bipartite graph.

In the worst case, we might enumerate all the combinations of vertices among the two graphs, which would be of exponential order. However, we have been successful in determining the isomorphism of graphs within a reasonable time using experimental data; these results are also reported in the present paper.

We consider only regular graphs. Since the general graph isomorphism problem can be reduced to the regular graph isomorphism problem in polynomial time (Booth, 1978), this restriction does not lose generality.

1.1. Perfect Matching Problem

The matching problem on a bipartite graph is a problem that of finding a set of edges such that any two edges do not share the same vertex (Iri, 1969). If the set covers all the vertices, the set is called *perfect matching.*

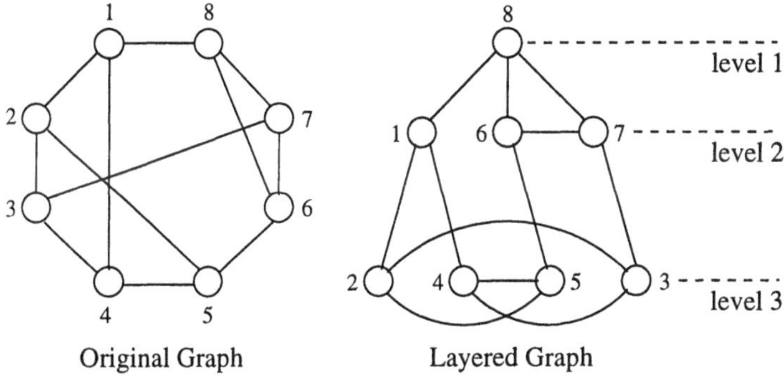

Figure 1. Layered Graph

It is known that there exist polynomial algorithms of finding a perfect matching. (Micali and Vazirani, 1980 etc.)

1.2. Preliminaries

Let the two given regular graphs be $G_1 = (V_1, E_1)$, $G_2 = (V_2, E_2)$, where $|V_1| = |V_2| = |V| = n$, $|E_1| = |E_2| = |E|$ $(= O(n^2))$. Each vertex is uniquely labeled and is stored in an array of size n. Graph isomorphism is defined as follows.

Definition 1 *Two graphs $G_1 = (V_1, E_1)$ and $G_2 = (V_2, E_2)$ are isomorphic, if there is a 1-to-1 correspondence $f : V_1 \to V_2$, such that $(v, v') \in E_1$ iff $(f(v), f(v')) \in E_2$ for any $(v, v') \in E_1$. This function f is called an isomorphism between G_1 and G_2.*

Similarly we could define graph isomorphism in the case where one vertex is fixed in each graph.

We consider only regular graphs for which the vertex degree satisfies $3 \le d \le \lfloor \frac{n-1}{2} \rfloor$, because of the relation between a graph and its complement.

2. Reconfiguring Graphs to Layered Graphs

In the present paper, we make use of layered graphs to determine isomorphism.

2.1. Layered Graphs

Given a graph G and a vertex $r \in V$, the layered graph $L(G, r)$ with root r consists of

- vertices of G,

- edges of G,
- $level(u)$ for each vertex u,

where $level(u)$ is the shortest distance (or the depth) from r to u (Figure 1). Transforming a graph with n vertices to a layered graph can be done in $O(n^2)$ time.

2.2. Characteristics of Layered Graphs

We divide the set of vertices adjacent to v into 3 subsets, $D_u(v)$, $D_s(v)$, and $D_d(v)$, as follows:

- $D_u(v) = \{v' \mid (v, v') \in E \text{ and } level(v') = level(v) - 1\}$,
- $D_s(v) = \{v' \mid (v, v') \in E \text{ and } level(v') = level(v)\}$,
- $D_d(v) = \{v' \mid (v, v') \in E \text{ and } level(v') = level(v) + 1\}$.

Let the number of vertices of each subset be d_u , d_s, and d_d :

- $d_u(v) = |D_u(v)|$, (upper degree)
- $d_s(v) = |D_s(v)|$, (same level degree)
- $d_d(v) = |D_d(v)|$. (lower degree)

It follows that the degree of v, $d(v)$, is equal to $d_u(v) + d_s(v) + d_d(v)$.

It is trivial to derive at the following:

- $d_u(r) = d_s(r) = 0, \ d_d(r) = d(r)$,
- each vertex v except the root vertex satisfies $d_u(v) \geq 1$,
- all vertices adjacent to the vertices in *level* i have *level* i or $(i \pm 1)$.

Given these assumptions, we propose the following.

Proposition 1 *Two graphs $G_1 = (V_1, E_1)$ and $G_2 = (V_2, E_2)$ are isomorphic if and only if there are vertices $v_1 (\in V_1)$ and $v_2 (\in V_2)$ and the two layered graphs $L(G_1, v_1)$ and $L(G_2, v_2)$ are isomorphic.*

Each vertex $v (\in V)$ has a label[1] ($level(v)$, $d_u(v)$, $d_s(v)$, $d_d(v)$). Let the label be denoted by $M(v)$. We call the set of vertices that have the same labels a "class," which we denote by B_i ($1 \leq i \leq k$, where k is the number of classes). For example, data from Figure 1 are shown in Table 1 sorted by label. We denote by $\mathcal{L}(G, v)$ the vertices of G partitioned into classes.

[1] A label for a general vertex is constructed by graph appending each vertix's degree $d(v)$ to the level.

Table 1. Example of Labeling. Data are from graph shown in Figure 1

	8	1	6	7	2	4	5	3
level	1	2	2	2	3	3	3	3
$d(v)$	3	3	3	3	3	3	3	3
$d_u(v)$	0	1	1	1	1	1	1	1
$d_s(v)$	0	0	1	1	2	2	2	2
$d_d(v)$	3	2	1	1	0	0	0	0
class	1	2	3	3	4	4	4	4

3. Finding a 1-to-1 correspondence between two graphs

In this section, we consider how to make use of perfect matching algorithm in order to determine the isomorphism of graphs.

3.1. Correspondence between 2 Layered Graphs

For two given graphs, we consider all layered graphs for which a vertex of the graph is the root.

For $v_i \in V_1$ and $v_j \in V_2$, we set $c_{ij} = 1$ if $\mathcal{L}(G_1, v_i)$ and $\mathcal{L}(G_2, v_j)$ have the same labels and partitions, otherwise $c_{ij} = 0$. Thus, we have a correspondence table as shown in Table 2.

It is easy to see that each table entry's value is unique and does not depend on expressions of the two graphs.

Table 2. Table of Layered Graphs

		G_1						
		1	2	$\cdots$	i	$\cdots$	$n-1$	n
	1	0	1	$\cdots$	0	$\cdots$	1	0
	2	0	1	$\cdots$	0	$\cdots$	1	0
	$\vdots$				$\vdots$			
G_2	j	1	0	$\cdots$	1	$\cdots$	0	1
	$\vdots$				$\vdots$			
	$n-1$	1	0	$\cdots$	1	$\cdots$	0	1
	n	0	0	$\cdots$	0	$\cdots$	1	0

The entries with a value of 1 are candidates for a 1-to-1 correspondence between vertices the two graphs. As a result, we could take that correspondence, by finding perfect matchings according to the table. (Figure 2)

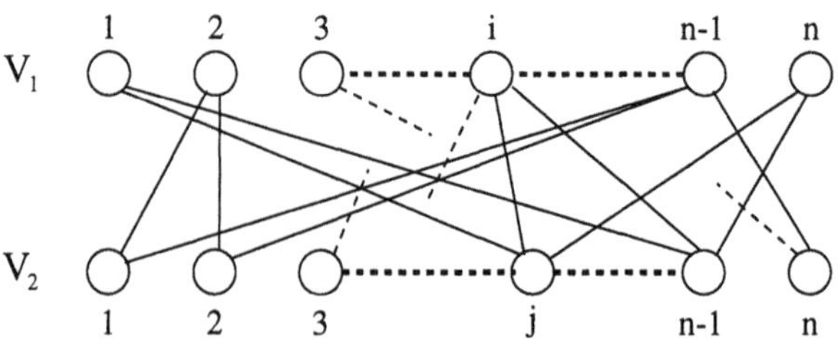

Figure 2. 1-to-1 correspondences as perfect matchings

In the graph isomorphism problem, we have to determine whether there exists a 1-to-1 correspondence between vertices in two graphs checking all possible perfect matchings[2] (in the correspondence table). Of course, the possible perfect matchings do not always indicate isomorphism, so we have to enumerate all perfect matchings and to test for isomorphism. However, the table limits the range searched for a 1-to-1 correpondence.

If there is no perfect matching between two graphs based on this table, they are not isomorphic.

3.2. Solutions and Issues

We have implemented the above algorithm and in Section 4 applied it experimentally to determine isomorphism. We test for 1-to-1 correspondence between vertices in two graphs as follows.

- Construct a 1-to-1 correspondence table as preprocessing.
- Test for 1-to-1 correspondence between vertices in the two graphs.

Next, we enumerate 1-to-1 correspondences one by one until we find a perfect matching between vertices in graphs.

The program based on our algorithm and described in the next section has not adopted stronger methods to bound recursion, because we want to make it easier to understand effectiveness by using a table.

However, if all entries in the table are 1's, we have to enumerate all perfect matchings. This results in many combinations of 1-to-1 correspondence to test. This might be the worst situation for our algorithm. In such situation, however, we could consider 2 cases whether 2 graphs are isomorphic or not.

[2]In practice, it is not necessary to enumerate those perfect matchings to determine isomorphism.

In the former case, since both graphs would have much symmetry, we could find a 1-to-1 correspondence earlier. In the latter case, we do not need to enumerate all perfect matchings as follows :

- Consider the two layered graph which both root vertices are corresponding in the table.
- Within **each corresponding class** between the two layered graphs, test 1-to-1 correspondences.
  - Examine the number of same vertices adjacent to 2 vertices in each corresponding class.
  - If there is no 1-to-1 correspondence for at least one class, they are not isomorphic.

Thus, we could reduce complexity of enumeration.

Also we need to consider what features of graphs indicate the worst complexity.

Among other known algorithms the best complexity in the worst case analysis is of time $O\left(c^{n^{1/2+o(1)}}\right)$ (Babai and Luks, 1983; Kreher and Stinson, 1998). That algorithm determines isomorphism by certifying graphs uniquely. Though it certifies by partitioning the set of vertices recursively, the basic idea in partitioning is as follows : "which partitioned set contains vertices adjacent to a certain vertex?" To prevent unnecessary recursions, it takes advantage of certifications results. The complexity of certification is of exponential order.

4. Experiments

We have implemented the program described above and experimented on various regular graphs.

4.1. Environment and Graph Data

Our experiment was carried out with a Celeron 450MHz, 128 MB memory (and 128 MB swaps) and C (gcc-2.91.66) on Linux (2.2.14). We measured running time using a UNIX like OS command "time."

We have constructed various regular graphs for input using a program that was implemented according to Matsuda et al., 1992. Those graphs have numbers of vertices from 20 to 120 with vertex degree of 10. We constructed not only various isomorphic graphs that have the same number of vertices and degree but also non-isomorphic ones.

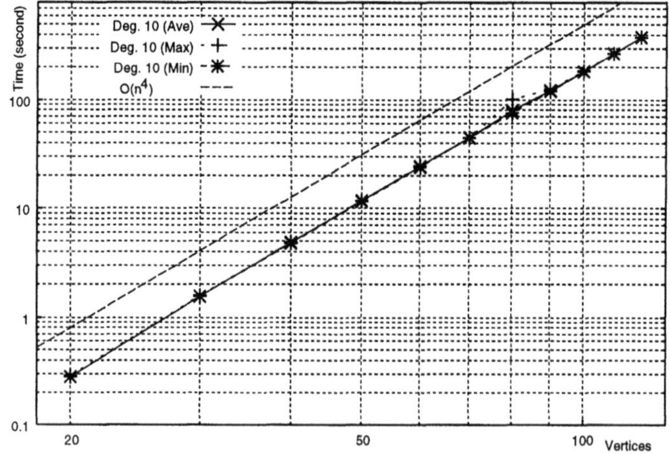

Figure 3. Isomorphic case

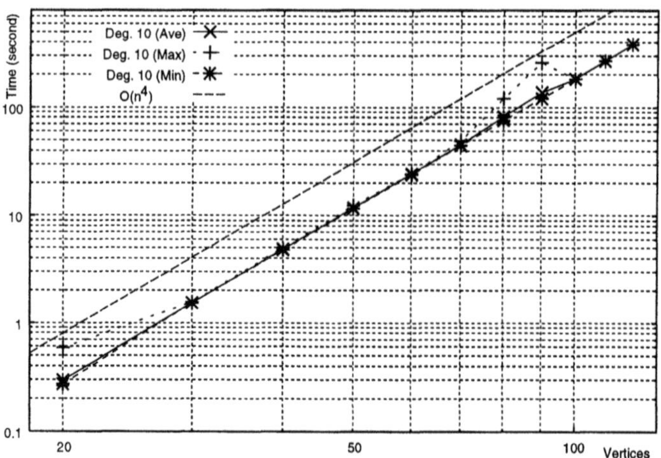

Figure 4. Non-isomorphic case

4.2. Results and Estimation

Computational results are shown in Figures 3 and 4. In these figures, we show the average and the maximum time versus the number of vertices in a graph, and depict the resulting curve[3].

[3]That was multiplied by adequate constants to be easily able to compare.

As a result, we conclude that the experimental time complexity is proportional to $O(n^4)$ regardless of whether the graphs are isomorphic or not. These results tend to be closer to the complexity of making a correspondence table than of examining 1 to 1 correspondences (perfect matchings) between the two graphs. Besides, we have seen almost the same results in both cases isomorphic and non isomorphic.

We anticipated that complexity might be larger as all the perfect matchings might be enumerated in the non-isomorphic case, but the result of our experiment showed to be much more efficient.

Differences between average time, maximum time and minimum time in the number of vertices and degree are very small, so the program is quite stable. Standard deviations in the results are also very small (though not shown here) and didn't have any result over 1 second. Furthermore, in the non-isomorphic case, we could determine lack of isomorphism by testing only the table (in the graphs used at least).

5. Conclusions

In the present paper, targeting nonweighted, undirected and connected regular graphs, we considered graph isomorphism by means of perfect matching to limit the range of 1 to 1 correspondence between two graphs as follows. First, we reconfigured the given graph as a layered graph, labeled vertices by partitioning the set of vertices by distance from a root vertex, and prepared a correspondence table by means of whether labels on 2 layered graphs matched or not. Using that table, we find 1 to 1 correspondences between the two graphs. In our experiments, we could determine isomorphism within a practical and stable time.

For further research, we have to examine other types of graphs, and analyse complexity of the program for them. Also, we wish to compare our results with practical running results of the best algorithm described in Babai and Luks, 1983 and Kreher and Stinson, 1998 whose worst complexity are known to have exponential time.

References

Agrawal, M. and Arvind, V. (1996). A note on decision versus search for graph automorphism. *Information and Computation*, 131:179–189.

Akutsu, T. (1988). A polynomial time algorithm for subgraph isomorphism of tree-like graphs. *IPSJ 90-AL-17-2.*

Babai, L., Erdös, P., and Selkow, S. M. (1980). Random graph isomorphism. *SIAM J. Comput.*, 9:628–635.

Babai, L., Grigoryev, D. Y., and Mount, D. M. (1982). Isomorphism of graphs with bounded eigenvalue multiplicity. *Proc. 14th Annual ACM Symp. Theory of Computing*, pages 310–324.

Babai, L. and Luks, E. M. (1983). Canonical labeling of graphs. *Proc. 14th Annual ACM Symp. on Theory of Computing, Boston*, pages 171–183.

Babel, L., Baumann, S., Ludecke, M., and Tinhofer, G. (1997). Stabcol: Graph isomorphism testing based on the weisfeiler-leman algorithm. Technical Report Preprint TUM-M9702, Munich.

Barrett, J. W. and Morton, K. W. (1984). Approximate symmetrization and Petrov-Galerkin methods for diffusion-convection problems. *Comput. Methods Appl. Mech. Engrg.*, 45:97–122.

Booth, K. S. (1978). Isomorphism testing for graphs, semigroups, and finite automata are polynomially equivalent problems. *SIAM J. Comput.*, 7:273–279.

Corneil, D. G. and Gotlieb, C. C. (1970). An efficient algorithm for graph isomorphism. *J. ACM*, 17:51–64.

Cull, P. and Pandy, R. (1994). Isomorphism and the n-queens problem. *ACM SIGCSE Bulletin*, 26:29–36.

Filotti, I. S. and Mayer, J. N. (1980). A polynomial-time algorithm for determining the isomorphism of graphs of fixed genus. *Proc. 12th Annual ACM Symp. Theory of Computing*, pages 236–243.

Galil, Z., Hoffmann, C. M., Luks, E. M., Schnorr, C. P., and Weber, A. (1987). An $o(n^3 \log n)$ deterministic and an $o(n^3)$ las vegas isomorphism test for trivalent graphs. *J. ACM*, 34:513–531.

Hirata, T. and Inagaki, Y. (1988). Tree pattern matching algorithm. *IPSJ 88-AL-4-1.*

Hopcroft, J. and Wong, J. (1974). Linear time algorithms for isomorphism of planar graphs. *Proc. 6th Annual ACM Symp. Theory of Computing*, pages 172–184.

Iri, M. (1969). *Network Flow, Transportation and Scheduling.* Academic Press.

Köbler, J., Schöning, U., and Torán, J. (1992). Graph isomorphism is low for pp. *J. of Computer Complexity*, 2:301–330.

Kreher, D. L. and Stinson, D. R. (1998). *Combinational Algorighms: Generation, Enumeration and Search.* CRC.

Lueker, G. S. (1979). A linear time algorithm for deciding interval graph isomorphism. *J. ACM*, 26:183–195.

Luks, E. M. (1982). Isomorphism of graphs of bounded valence can be tested in polynomial time. *J. Computer and System Sciences*, 25:42–65.

Matsuda, Y., Enohara, H., Nakano, H., and Horiuchi, S. (1992). An algorithm for generating regular graphs. *IPSJ 92-AL-25-3.*

Micali, S. and Vazirani, V. V. (1980). An $o(\sqrt{V} \cdot e)$ algorithm for finding maximum matching in general graphs. *Proc. 21st Ann. IEEE Symp. Foundations of Computer Science*, pages 17–27.

Torán, J. (2000). On the hardness of graph isomorphism. *Proc. 41st Annual Symposium on Foundations of Computer Science, California*, pages 180–186.

van Leeuwen, J. (1990). *Handbook of Theoretical Computer Science, Vol. A: Algorithm and Complexity.* Elseveir.

A REDUCED SQP ALGORITHM FOR THE OPTIMAL CONTROL OF SEMILINEAR PARABOLIC EQUATIONS

Roland Griesse
Lehrstuhl für Ingenieurmathematik
Universität Bayreuth, Germany
roland.griesse@uni-bayreuth.de

Abstract This paper deals with optimal control problems for semilinear time-dependent partial differential equations. Apart from the PDE, no additional constraints are present. Solving the necessary conditions for such problems via the Newton-Lagrange method is discussed. Motivated by issues of computational complexity and convergence behavior, the Reduced Hessian SQP algorithm is introduced. Application to a system of reaction-diffusion equations is outlined, and numerical results are given to illustrate the performance of the reduced Hessian algorithm.

Keywords: optimal control, parabolic equation, semilinear equation, reduced SQP method, reaction-diffusion equation

Introduction

There exist two basic classes of algorithms for the solution of optimal control problems governed by partial differential equations (PDEs). They both are of an iterative fashion and are different in that *Newton-type* methods require the repeated solution of the (non-linear) PDE while the algorithms of *SQP-type* deal with the linearized PDE only. Newton-type methods have been successfully applied, e.g., to control problems for the Navier-Stokes equations in [4] and will not be discussed here.

The main focus of this paper is on SQP-type methods which basically use Newton's algorithm in order to solve the first order necessary conditions. This scheme leads to a linear boundary value problem for the state and adjoint variables. It is the size of the *discretized* linear boundary value problem that motivates a variant of this approach in the first place: The reduced SQP method, which has been the subject of the following papers: [5] introduces reduced Hessian methods in Hilbert spaces. [4] studies various second-order methods for optimal control of the time-dependent Navier-Stokes equations. [2] and [3] discuss algorithms based on inexact factorization of the full Hessian step (11) which involve the reduced Hessian (or approximations thereof) in the factors. [1] examines preconditioners for the KKT matrices arising in interior point methods, also using reduced Hessian techniques.

This paper is organized as follows: In Section 1, the class of semilinear second order parabolic partial differential equations is introduced with control provided in distributed fashion. Section 2 covers optimal control problems for these PDEs and establishes the first order necessary conditions. Section 3 describes the basic SQP method in function spaces (also called the *Newton-Lagrange method* in this context), that can be used to solve these conditions. The reduced Hessian method is derived as a variant thereof. It will be seen that this method is applicable only if the *linearized* PDE is uniquely solvable with continuous dependence on the right hand side data. The purpose of the reduced Hessian method is to significantly decrease the size of the discretized SQP steps. The associated algorithm which requires the repeated solution of the linearized state equation and of the corresponding adjoint is presented in detail. In Section 4, this procedure is applied to a system of reaction-diffusion PDEs. Finally, numerical results are given in Section 5.

While the ideas and algorithm are worked out for distributed control problems throughout this paper, boundary and mixed control problems can be treated in the very same manner with only minor modification of notation.

1. Semilinear Parabolic Equations

Let Ω be a bounded domain in $\mathbb{R}^2$ with sufficiently smooth boundary Γ and $Q = \Omega \times (0,T)$, $\Sigma = \Gamma \times (0,T)$ with given final time $T > 0$. We consider semilinear parabolic initial-boundary value problems of the following type:

$$\begin{aligned} y_t(x,t) + A(x)y(x,t) + n(x,t,y(x,t),u(x,t)) &= 0 \quad \text{in} \quad Q \\ \partial_n y(x,t) + b(x,t,y(x,t)) &= 0 \quad \text{on} \quad \Sigma \\ y(x,0) - y_0(x) &= 0 \quad \text{on} \quad \Omega. \end{aligned} \tag{1}$$

The elliptic differential operator $A(x)y = -\sum_{i,j=1}^{2} D_j(a_{ij}(x)D_i y)$ is represented by the matrix $\bar{A}(x) = (a_{ij}(x)) \in \mathbb{R}^{2\times 2}$ which is assumed to be symmetric, and $\partial_n y(x,t) = n(x)^T \bar{A}(x)\nabla y(x,t) = \sum_{i,j=1}^{2} a_{ij} n_i(x) D_j y(x,t)$ is the so-called *co-normal derivative* along the boundary Γ. When A is the negative Laplace operator $-\Delta$, $\bar{A}$ gives the identity matrix and $\partial_n y(x,t)$ is simply the normal derivative or Neumann trace of $y(x,t)$.

Questions of solvability, uniqueness and regularity for non-linear PDEs shall not be answered here. Please refer to [7] and the references cited therein. We assume that there exist Banach spaces Y for the state, U for the control and Z for the adjoint variable such that the semilinear parabolic problem (1) is well-posed in the abstract form

$$e(y,u) = 0 \quad \text{with} \quad e : Y \times U \to Z' \tag{2}$$

where Z' is the dual space of Z. The operator e may represent a *strong* or *weak* form of the state equation (1). Casting the PDE in this convenient form will allow us later to view the control problem as a PDE-constrained optimization problem and hence support a solution approach based on the Lagrange functional. However, in the detailed presentation of the algorithms, we will return to interpreting the operator e and its linearization e_y as time-dependent PDEs.

2. Optimal Control Problems

In the state equation (1), the function u defined on Q is called the *distributed* control function. A *Neumann boundary control* problem arises when, instead of u, a control function v is present in the boundary nonlinearity $b(x,t,y(x,t),v(x,t))$. Other possibilities include *Dirichlet boundary control* or even combinations of all of the above. Examples of boundary control problems can be found, e.g., in [3] and [1]. Everything presented in this paper can be and in fact has been applied to boundary control problems with only minor modifications.

The core of optimal control problems is to choose the control function $u \in U$ in order to minimize a given objective function. In practical terms, the objective can, e.g., aim at energy minimization or tracking a given desired state.

We shall use the *objective* for the distributed control case from [7]:

$$f(y,u) = \int_\Omega \varphi(x, y(x,T))\, dx + \int_Q g(x,t,y,u)\, dx\, dt \tag{3}$$

where φ asseses the terminal state and g evaluates the distributed control effort and the state trajectory in $(0,T)$.

The abstract optimal control problem considered throughout the rest of this paper can now be stated:

$$\begin{aligned} \text{Minimize} \quad & f(y,u) \quad && \text{over} \quad (y,u) \in Y \times U \\ \text{s.t.} \quad & e(y,u) = 0 \quad && \text{holds.} \end{aligned} \tag{4}$$

A particularly simple situation arises when the state equation (1) is in fact linear in (y,u) and the objective (3) is convex or even quadratic positive definite. However, in the general case, our given problem (4) to find an optimal control u and a corresponding optimal state y minimizing (3) while satisfying the state equation $e(y,u) = 0 \in Z'$ is a non-convex problem. We will not address the difficult question of global optimal solutions but rather assume that a local optimizer $(\hat{y},\hat{u})$ exists. The following *first order necessary conditions* involving the adjoint variable λ are well-known, see, e.g., [7] (with $-\lambda$ instead of λ):

$$\begin{aligned}
-\lambda_t + A(x)^*\lambda + n_y(x,t,y,u)\lambda + g_y(x,t,y,u) &= 0 \quad \text{in} \quad Q \\
\partial_n\lambda + b_y(x,t,y)\lambda &= 0 \quad \text{on} \quad \Sigma \\
\lambda(T) + \varphi_y(x,y(T)) &= 0 \quad \text{in} \quad \Omega \\
g_u(x,t,y,u) + n_u(x,t,y,u)\lambda &= 0 \quad \text{in} \quad Q \\
y_t + A(x)y + n(x,t,y,u) &= 0 \quad \text{in} \quad Q \\
\partial_n y + b(x,t,y) &= 0 \quad \text{on} \quad \Sigma \\
y(0) - y_0(x) &= 0 \quad \text{on} \quad \Omega.
\end{aligned} \tag{5}$$

These can be derived by constructing the *Lagrangian*

$$L(y,u,\lambda) = f(y,u) + \langle e(y,u),\lambda\rangle_{Z',Z} \tag{6}$$

and evaluating the conditions

$$L_y(y,u,\lambda) = 0 \quad \text{in} \quad Y' \quad \textit{(adjoint equation)} \tag{7}$$

$$L_u(y,u,\lambda) = 0 \quad \text{in} \quad U' \quad \textit{(optimality condition)} \tag{8}$$

$$L_\lambda(y,u,\lambda) = e(y,u) = 0 \quad \text{in} \quad Z' \quad \textit{(state equation)} \tag{9}$$

in their strong form.

Triplets $(\hat{y},\hat{u},\hat{\lambda})$ that satisfy the first order necessary conditions are called *stationary points*. Obviously, the conditions (5) or (7)–(9) constitute a non-linear two-point boundary value problem involving the non-linear *forward equation* (initial values given) for the state y and the linear *backward equation* (terminal conditions given) for the adjoint λ. In the next section we introduce an algorithm to solve this problem.

3. SQP Algorithms

As we have seen in the previous section, finding stationary points $(\hat{y},\hat{u},\hat{\lambda})$ and thus candidates for the optimal control problem requires the solution of the non-linear operator equation system (7)–(9). This task can be attacked by Newton's method that is commonly used to find zeros of non-linear differentiable functions.

Suppose that we are given a triplet (y^k, u^k, λ^k), the current iterate. The Newton step to compute updates $(\delta y, \delta u, \delta\lambda)$ reads

$$\begin{bmatrix} L_{yy}(y^k, u^k, \lambda^k) & L_{yu}(y^k, u^k, \lambda^k) & e_y(y^k, u^k)^* \\ L_{uy}(y^k, u^k, \lambda^k) & L_{uu}(y^k, u^k, \lambda^k) & e_u(y^k, u^k)^* \\ e_y(y^k, u^k) & e_u(y^k, u^k) & 0 \end{bmatrix} \begin{bmatrix} \delta y \\ \delta u \\ \delta\lambda \end{bmatrix} = - \begin{bmatrix} L_y(y^k, u^k, \lambda^k) \\ L_u(y^k, u^k, \lambda^k) \\ e(y^k, u^k) \end{bmatrix}$$

$$= - \begin{bmatrix} f_y(y^k, u^k) + e_y(y^k, u^k)^*\lambda^k \\ f_u(y^k, u^k) + e_u(y^k, u^k)^*\lambda^k \\ e(y^k, u^k) \end{bmatrix} \text{ in } \begin{bmatrix} Y' \\ U' \\ Z' \end{bmatrix}. \tag{10}$$

This method is referred to as the *Newton-Lagrange* algorithm. It falls under the category of SQP solvers since (10) are also the necessary conditions of an auxiliary QP problem, see, e.g., [6]. Note that in contrast to the so-called *Newton approach* (cf. [4]), the iterates (y^k, u^k) of the SQP method are *infeasible* w.r.t. the non-linear state equation, i.e. the method generates control/state pairs that satisfy the PDE (1) only in the limit.

The operators appearing in the matrix on the left hand side (the *Hessian of the Lagrangian*) deserve some further explanation. First it is worth recalling that the first partial Fréchet derivative of a mapping $g : X_1 \times X_2 \to Y$ between normed linear spaces $X = X_1 \times X_2$ and Y at a given point $x = (x_1, x_2) \in X$ is a bounded linear operator, e.g., $g_{x_1}(x) \in \mathcal{L}(X_1, Y)$. Consequently, the second partial Fréchet derivatives at x are $g_{x_1x_1}(x) \in \mathcal{L}(X_1, \mathcal{L}(X_1, Y))$, $g_{x_1x_2}(x) \in \mathcal{L}(X_2, \mathcal{L}(X_1, Y))$, etc. They can equivalently be viewed as bi-linear bounded operators, e.g., the latter taking its first argument from X_2 and its second from X_1 and mapping this pair to an element of Y.

The *adjoint operators* (or, precisely speaking, conjugate operators) appearing in the equation (10) can most easily be explained by their property of switching the arguments' order in bilinear maps:

$$\begin{aligned} e_y(y^k, u^k) &\in \mathcal{L}(Y, Z') \\ e_y(y^k, u^k)^* &\in \mathcal{L}(Z'', Y') \hookrightarrow \mathcal{L}(Z, Y') \quad \text{since} \quad Z \hookrightarrow Z'' \\ e_y(y^k, u^k)^*(z, y) &= e_y(y^k, u^k)(y, z) \qquad \text{for all} \quad y \in Y, z \in Z. \end{aligned}$$

Exploiting the fact that the adjoint variable λ appears linearly in the Lagrangian L, the Newton step (10) can be rewritten in terms of the new iterate λ^{k+1} rather than the update $\delta\lambda$. For brevity, the arguments (y^k, u^k, λ^k) will be omitted from now on:

$$\begin{bmatrix} L_{yy} & L_{yu} & e_y^* \\ L_{uy} & L_{uu} & e_u^* \\ e_y & e_u & 0 \end{bmatrix} \begin{bmatrix} \delta y \\ \delta u \\ \lambda^{k+1} \end{bmatrix} = - \begin{bmatrix} f_y \\ f_u \\ e \end{bmatrix}. \tag{11}$$

As can be expected, this system (obtained by linearization of (7)–(9)) represents a *linear* two-point boundary value problem whose solution is now the main focus.

To render problem (11) amenable for computer treatment, some discretization has to be carried out. Inevitably, its discretized version will be a large system of linear equations since it ultimately contains the values of the state, control and adjoint at all discrete time steps and all nodes of the underlying spatial grid. Thus, one seeks to minimize the dimension of the system by decomposing it into smaller parts. The *reduced Hessian algorithm* is designed just for this purpose:

Roughly speaking, it solves the linear operator equation (11) for δu first, using Gaussian elimination on the symbols in the matrix. A prerequisite to this procedure is the bounded invertibility of $e_y(y,u)$ for all (y,u) which are taken as iterates in the course of the algorithm. In other words, the linearized state equation $e_y(y,u)h = f$ has to be uniquely solvable for h (with continuous dependence on the right hand side $f \in Z'$) at these points (y,u). One obtains the reduced Hessian step

$$\begin{aligned}\left(e_u^* e_y^{-*} L_{yy} e_y^{-1} e_u + L_{uu} - L_{uy} e_y^{-1} e_u - e_u^* e_y^{-*} L_{yu}\right) \delta u & \\ = e_u^* e_y^{-*} \left[f_y - L_{yy} e_y^{-1} e\right] - f_u + L_{uy} e_y^{-1} e & \qquad (12)\\ e_y\, \delta y = \; - e - e_u \delta u & \qquad (13)\\ e_y^*\, \lambda^{k+1} = \; - f_y - L_{yy}\delta y - L_{yu}\delta u. & \qquad (14)\end{aligned}$$

The operator preceding δu is called the *reduced Hessian* $H_{\delta u}$ in contrast to the *full Hessian* matrix H appearing in (11). Note that both the full and the reduced Hessian are self-adjoint operators. After discretization, the reduced Hessian will be small and dense, whereas the full Hessian will be large and sparse. Aiming at solving a discretized version of (12) using an iterative solver, the action of the reduced Hessian on given elements $\delta u \in U$ has to be computed, plus the right hand side of (12). It can be shown that once an approximate solution δu to (12) is found, the remaining unknowns δy and λ^{k+1} obeying (13) and (14) can be expressed in terms of quantities already calculated. The overall procedure to solve (7)–(9) applying the reduced Hessian method on the inner loop decomposes nicely into the steps described in figure 1 using the auxiliary variables $h_1, h_3 \in Y$ and $h_2, h_4 \in Z$.

In many practical cases, the objective and the PDE separate as

$$f(y,u) = f_1(y) + f_2(u) \quad \text{and} \quad e(y,u) = e_1(y) + e_2(u) \qquad (15)$$

which entails $L_{uy} = L_{yu} = 0$.

We observe that for the computation of the right hand side b as well as for every evaluation of $H_{\delta u}\Box$, it is required to solve one equation in-

Reduced SQP Algorithm

1 Set $k = 0$ and initialize (y^0, u^0, λ^0).

2 Solve

(a) $e_y h_1 = e$

(b) $e_y^* h_2 = f_y - L_{yy} h_1$

and set $b := e_u^* h_2 - f_u + L_{uy} h_1$.

3 For every evaluation of $H_{\delta u}\Box$ inside some iterative solver of $H_{\delta u}\delta u = b$, solve

(a) $e_y h_3 = e_u \Box$

(b) $e_y^* h_4 = L_{yy} h_3 - L_{yu}\Box$

and set $H_{\delta u}\Box := e_u^* h_4 - L_{uy} h_3 + L_{uu}\Box$.

4 Set $u^{k+1} := u^k + \delta u$.

5 Set $\delta y := -h_1 - h_3$ and $y^{k+1} := y^k + \delta y$.

6 Set $\lambda^{k+1} := -h_2 + h_4$.

7 Set $k := k + 1$ and go back to step 2.

Figure 1. Reduced SQP Algorithm

volving e_y and another involving e_y^*. It will be seen in the sequel that in our case of e representing a time-dependent PDE these are in fact solutions of the linearized forward (state) equation and the corresponding backward (adjoint) equation, see figure 2 in the following section.

Note that the linear system involving the reduced Hessian $H_{\delta u}$ is significantly reduced in size as compared to the full Hessian of the Lagrangian, the more so as in practical applications, there are many more state than control variables.

4. Example

As an example, distributed control of a semilinear parabolic system of reaction-diffusion equations will be discussed. The PDE system describes a chemical reaction $C_1 + C_2 \to C_3$ where the three substances are subject to diffusion and a simple non-linear reaction law.

While in the discussion so far only *one* (scalar) PDE appears, the generalization to *systems* of PDEs is straightforward. In the example, the state $y = (c_1, c_2, c_3)^T$ as well as the adjoint $\lambda = (\lambda_1, \lambda_2, \lambda_3)^T$ now have three scalar components while the control is still one-dimensional. The linearized systems occuring in the computation of the auxiliary variables $h_1, \ldots, h_4$ feature a coupling between their components which is generated by the non-linearity in the state equation (16). Also note that this example satisfies the separation condition (15).

The reaction-diffusion system under consideration is given by

$$\begin{aligned} c_{1_t} &= D_1 \Delta c_1 - k_1 c_1 c_2 & \partial_n c_1 &= 0 & c_1(0) &= c_{10} \\ c_{2_t} &= D_2 \Delta c_2 - k_2 c_1 c_2 + u & \partial_n c_2 &= 0 & c_2(0) &= c_{20} \\ c_{3_t} &= D_3 \Delta c_3 + k_3 c_1 c_2 & \partial_n c_3 &= 0 & c_3(0) &= c_{30} \end{aligned} \tag{16}$$

where the control acts only through component two. The boundary conditions simply mean that the boundary of the reaction vessel is impermeable. The constants D_i and k_i are all non-negative and denote diffusion and reaction coefficients, respectively.

The objective in this case is a standard least-squares-type functional

$$f(y, u) = \int_\Omega [c_1(x, T) - c_{1d}]^2 \, dx + \gamma \int_Q u(x, t)^2 \, dx \, dt$$

in order to minimize the distance of component one's terminal state $c_1(x, T)$ to a given desired state c_{1d} while taking control cost into account, weighted by a factor $\gamma > 0$. In case one is interested in maximum product yield, the term $-\int_\Omega c_3(x, T) \, dx$ can be inserted into the objective.

The individual steps in the reduced Hessian algorithm for this particular example are given in figure 2. There the vector $(c_1^k, c_2^k, c_3^k, u^k, \lambda_1^k, \lambda_2^k, \lambda_3^k)^T$ denotes the current iterate. It stands out that the linear systems for $h_1, \ldots, h_4$ can equivalently be written as

$$h_{i_t} + \hat{K} h_i = f_i \quad \text{for} \quad i \in \{1, 3\} \tag{17}$$

$$-h_{j_t} + \hat{K}^T h_j = g_j \quad \text{for} \quad j \in \{2, 4\} \tag{18}$$

where the operator matrix

$$\hat{K} = \begin{bmatrix} -D_1\Delta - k_1 c_2^k & -k_1 c_1^k & 0 \\ -k_2 c_2^k & -D_2\Delta - k_2 c_1^k & 0 \\ k_3 c_2^k & k_3 c_1^k & -D_3\Delta \end{bmatrix} \tag{19}$$

is *non-symmetric*. Please notice that this phenomenon does *not* occur in *scalar* PDE control problems.

Reduced Hessian steps for the reaction-diffusion example

Solve for $h_1 = (h_{11}, h_{12}, h_{13})^T$:

$$\begin{aligned}
&h_{11_t} - D_1\Delta h_{11} - k_1 c_2^k h_{11} - k_1 c_1^k h_{12} = c_{1_t}^k - D_1\Delta c_1^k - k_1 c_1^k c_2^k \\
&h_{12_t} - D_2\Delta h_{12} - k_2 c_1^k h_{12} - k_2 c_2^k h_{11} = c_{2_t}^k - D_2\Delta c_2^k - k_2 c_1^k c_2^k + u^k \\
&h_{13_t} - D_3\Delta h_{13} + k_3 c_2^k h_{11} + k_3 c_1^k h_{12} = c_{3_t}^k - D_3\Delta c_3^k + k_3 c_1^k c_2^k \\
&\partial_n h_{11} = 0 \qquad \partial_n h_{12} = 0 \qquad \partial_n h_{13} = 0 \\
&h_{11}(0) = c_1^k(0) - c_{10} \qquad h_{12}(0) = c_2^k(0) - c_{20} \qquad h_{13}(0) = c_3^k(0) - c_{30}
\end{aligned}$$

Solve for $h_2 = (h_{21}, h_{22}, h_{23})^T$:

$$\begin{aligned}
&-h_{21_t} - D_1\Delta h_{21} - k_1 c_2^k h_{21} - k_2 c_2^k h_{22} + k_3 c_2^k h_{23} = g_{21} \\
&-h_{22_t} - D_2\Delta h_{22} - k_2 c_1^k h_{22} - k_1 c_1^k h_{21} + k_3 c_1^k h_{23} = g_{22} \\
&-h_{23_t} - D_3\Delta h_{23} = 0 \\
&\qquad g_{21} = -k_1 h_{12}\lambda_1^k - k_2 h_{12}\lambda_2^k - k_3 h_{12}\lambda_3^k \\
&\qquad g_{22} = -k_1 h_{11}\lambda_1^k - k_2 h_{11}\lambda_2^k - k_3 h_{11}\lambda_3^k \\
&\partial_n h_{21} = 0 \qquad \partial_n h_{22} = 0 \qquad \partial_n h_{23} = 0 \\
&h_{21}(T) = 2[c_1^k(T) - c_{1d} - h_{11}(T)] \qquad h_{22}(T) = 0 \qquad h_{23}(T) = 0
\end{aligned}$$

Set $b = -h_{22} - 2\gamma u^k$.
Solve for $h_3 = (h_{31}, h_{32}, h_{33})^T$:

$$\begin{aligned}
&h_{31_t} - D_1\Delta h_{31} - k_1 c_2^k h_{31} - k_1 c_1^k h_{32} = 0 \\
&h_{32_t} - D_2\Delta h_{32} - k_2 c_1^k h_{32} - k_2 c_2^k h_{31} = -\square \\
&h_{33_t} - D_3\Delta h_{33} - k_3 c_2^k h_{31} - k_3 c_1^k h_{32} = 0 \\
&\partial_n h_{31} = 0 \qquad \partial_n h_{32} = 0 \qquad \partial_n h_{33} = 0 \\
&h_{31}(0) = 0 \qquad h_{32}(0) = 0 \qquad h_{33}(0) = 0
\end{aligned}$$

Solve for $h_4 = (h_{41}, h_{42}, h_{43})^T$:

$$\begin{aligned}
&-h_{41_t} - D_1\Delta h_{41} - k_1 c_2^k h_{41} - k_2 c_2^k h_{42} - k_3 c_2^k h_{43} = g_{41} \\
&-h_{42_t} - D_2\Delta h_{42} - k_2 c_1^k h_{42} - k_1 c_1^k h_{41} - k_3 c_1^k h_{43} = g_{42} \\
&-h_{43_t} - D_3\Delta h_{43} = 0 \\
&\qquad g_{41} = k_1 h_{32}\lambda_1^k + k_2 h_{32}\lambda_2^k + k_3 h_{32}\lambda_3^k \\
&\qquad g_{42} = k_1 h_{31}\lambda_1^k + k_2 h_{31}\lambda_3^k + k_3 h_{31}\lambda_3^k \\
&\partial_n h_{41} = 0 \qquad \partial_n h_{42} = 0 \qquad \partial_n h_{43} = 0 \\
&h_{41}(T) = 2h_{31}(T) \qquad h_{42}(T) = 0 \qquad h_{43}(T) = 0
\end{aligned}$$

Set $H_{\delta u}\square := -h_{42} + 2\gamma\square$.

Figure 2. Reduced SQP Algorithm for the Reaction-Diffusion Example

5. Numerical Results

In this section, results obtained from an implementation of the reduced Hessian algorithm will be presented. All coding has been done in Matlab 6.0 using the PDE toolbox to generate the spatial mesh and the finite element matrices. The performance of the reduced Hessian algorithm will be demonstrated in comparison to an iterative algorithm working on the *full* Hessian of the Lagrangian H given in (11).

To this end the convergence behavior over iteration count of *one* particular SQP step (corresponding to steps 2 and 3 in the algorithm) will be shown. For the tests we chose

$$\begin{array}{lll}
c_1^k(x,t) = 0.5 & c_{10}(x) = 0.1 + \chi_{\{x_1>0.3\}}(x) & \lambda_1^k(x,t) = 0 \\
c_2^k(x,t) = 0.5 & c_{20}(x) = 0.1 + \chi_{\{x_2>0.3\}}(x) & \lambda_2^k(x,t) = 0 \\
c_3^k(x,t) = 0.5 & c_{30}(x) = 0 & \lambda_3^k(x,t) = 0 \\
c_{1d}(x) = 0 & D_1 = 0.01 & k_1 = 0.5 \\
u^k(x,t) = 0 & D_2 = 0.05 & k_2 = 1.5 \\
\gamma = 1 & D_3 = 0.15 & k_3 = 2.5
\end{array}$$

on some finite element discretization of the unit circle $\Omega \subset \mathbb{R}^2$, where χ_A denotes the indicator function of the set $A \cap \Omega$. The final time was $T = 10$.

As was seen earlier in equation (11), there are three block rows in H, corresponding to the linearizations of the adjoint equation, the optimality condition and the state equation, respectively. For our tests, these have been semi-discretized using piecewise linear triangular finite elements in space. The ODE systems obtained by the method of lines are of the following form:

$$M\dot{y} + \;\; Ky = f \quad \text{(forward equations)} \tag{20}$$
$$-M\dot{\lambda} + K^T\lambda = g \quad \text{(backward equations)} \tag{21}$$

They were treated by means of the implicit Euler scheme with constant step size. Of course, suitable higher order integrators can be used as well. Using this straightforward approach yields one drawback that becomes apparent in figure 3: The discretized full Hessian matrix H is no longer symmetric, although the continuous operator H is self-adjoint. The same holds for the discretized reduced Hessian $H_{\delta u}$.

This is due to the treatment of initial and terminal conditions in the linearized state and the adjoint equation. Nevertheless, there are methods that reestablish symmetry, but these will not be pursued in the course of this paper since qualitatively, the convergence results remain unchanged. For that reason, the non-symmetry will be approved,

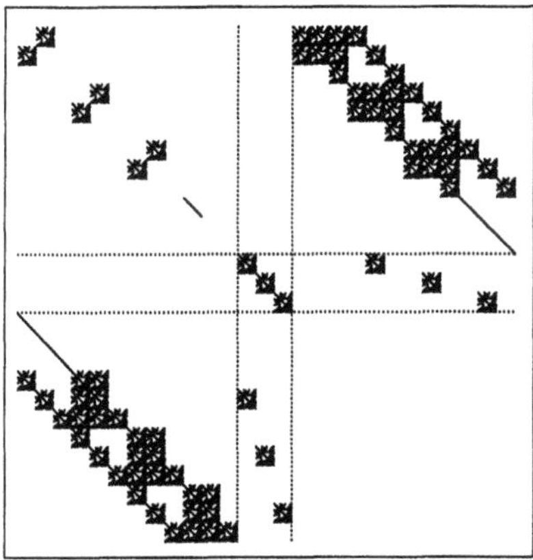

Figure 3. Non-symmetry of discretized full Hessian, $nt = 4$ time steps, implicit Euler, dotted lines indicate blocks corresponding to (11)

thereby waiving the possibility to use, e.g., a conjugate gradient method to solve the reduced problem but relying on iterative solvers capable of non-symmetric problems. In the tests, `GMRES` has proved quite efficient on the full Hessian problem while `CGS` and `BICGSTAB` failed to generate reasonably better iterates than the initial all-zero guess. For the reduced Hessian, all three algorithms found the solution to high accuracy, and `CGS` needed the fewest iterations to do so. As a common basis, `GMRES` with no restarts was used for both the full and the reduced Hessian problem.

Note that while the discretized state and adjoint allocate nt (equal to 4 in figure 3) discrete time steps, the discretized control needs only $nt - 1$. This is attributed to the use of the Euler method where, after discretization, $u(t = 0)$ does not appear in any of the equations.

In order to illustrate the convergence properties, it is convenient to have the *exact* discretized solution $(\delta y, \delta u, \lambda^{k+1})$ of the full SQP step (11) at hand. To that end, the full Hessian matrix was set up explicitly for a set of relatively coarse discretizations, and the exact solution was computed using a direct solver based on Gaussian elimination (Matlab's backslash operator). The exact solution δu of (12) was obtained in the same way after setting up the reduced Hessian matrix, where the corresponding δy and λ^{k+1} were calculated performing the forward/backward integration given by (13) and (14). These two reference solution triplets differ only by entries of order 1E-15 and will be considered equal.

It has to be mentioned that for these low-dimensional examples (cf. table 1), a direct solver is a lot faster than any iterative algorithm. However, setting up the exact reduced Hessian matrix of course is not an option for fine discretizations.

Figures 4–6 illustrate the convergence behavior of `GMRES` working on the reduced versus the full Hessian matrix: For δu_{ref} denoting the exact discretized solution, the graphs show the relative error history

$$e^j(t) = \frac{||\delta u^j(t) - \delta u_{\text{ref}}(t)||}{||\delta u_{\text{ref}}(t)||} \tag{22}$$

in the L^2 norm, where $\delta u^j(t)$ denotes the approximate solution generated by the iterative solver after j iterations, taken at the time grid point $t \in [0, T]$. The same relative errors can be defined for δu substituted by $\delta c_1, \ldots, \delta c_3$ or $\lambda_1^{k+1}, \ldots, \lambda_3^{k+1}$ which are the components of the state update δy and the new adjoint estimate λ^{k+1}.

Each figure shows the relative error history $e_j(t)$ of either δu or δc_1 obtained using GMRES with no restarts after $j = 4, 8, \ldots, 28$ iterations on the reduced problem and after $j = 100, 200, \ldots, 600$ iterations on the full problem. The figures for δc_2, δc_3 and $\lambda_1^{k+1}, \ldots, \lambda_3^{k+1}$ look very much the same and are not shown here. The discretization level is characterized by the number of discrete time steps nt and the number of grid points in the finite element mesh poi. Table 1 lists the number of optimization variables in the full and reduced case for the individual discretizations used.

nt	*poi*	*# of vars (reduced)*	*# of vars (full)*
9	25	200	1550
9	81	648	5022
19	81	1458	10692

Table 1. Number of optimization variables for different discretizations

It can clearly be seen that the iterative solver works very well on the reduced system while it needs many iterations on the full matrix. This was to be expected since it is a well-known fact (see, e.g., [1] and [2]) that iterative solvers working on the full Hessian require preconditioning. Although the evaluation of $H_{\delta u}$ times a vector is computationally more expensive than H times a vector, the reduced Hessian algorithm is by far the better choice over the unpreconditioned full algorithm. To give some idea why the reduced Hessian algorithm outperforms the full Hessian version, let us define

$$P = \begin{bmatrix} -e_u^* e_y^{-*} & I & e_u^* e_y^{-*} L_{yy} e_y^{-1} \\ 0 & 0 & I \\ I & 0 & 0 \end{bmatrix} \tag{23}$$

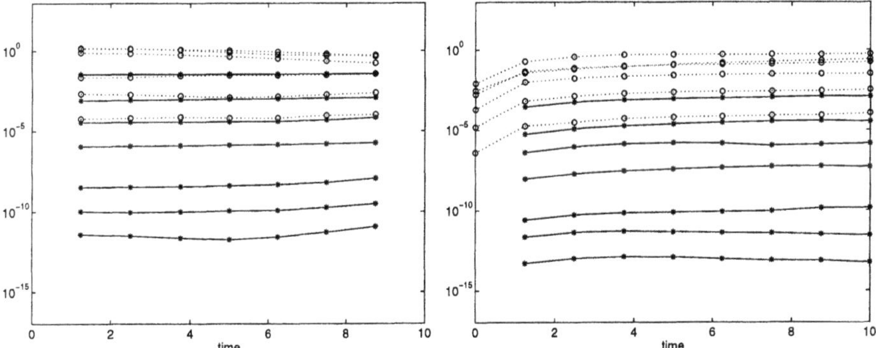

Figure 4. Relative error history for δu (left) and δc_1 (right) on the reduced (solid lines) problem for $j = 4, 8, \ldots, 28$ iterations and on the full (dotted lines) problem for $j = 100, 200, \ldots, 600$ iterations at discretization level $nt = 9$, $poi = 25$

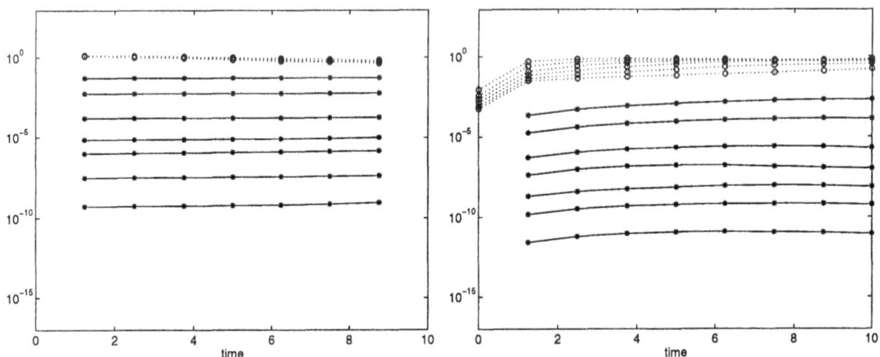

Figure 5. Relative error history for δu (left) and δc_1 (right) on the reduced (solid lines) problem for $j = 4, 8, \ldots, 28$ iterations and on the full (dotted lines) problem for $j = 100, 200, \ldots, 600$ iterations at discretization level $nt = 9$, $poi = 81$

as the *left preconditioner* for the full Hessian problem (11) with the first two columns permuted (for simplicity, the separation condition (15) is assumed to hold): From (11), we get

$$P \begin{bmatrix} & L_{yy} & e_y^* \\ L_{uu} & & e_u^* \\ e_u & e_y & \end{bmatrix} \begin{bmatrix} \delta u \\ \delta y \\ \lambda^{k+1} \end{bmatrix} = -P \begin{bmatrix} f_y \\ f_u \\ e \end{bmatrix}, \tag{24}$$

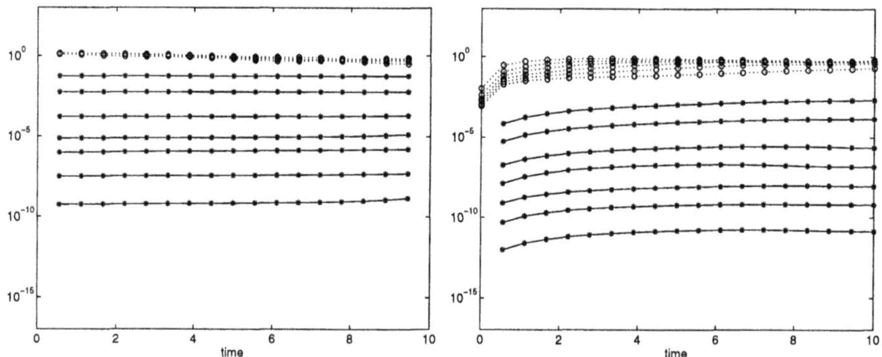

Figure 6. Relative error history for δu (left) and δc_1 (right) on the reduced (solid lines) problem for $j = 4, 8, \ldots, 28$ iterations and on the full (dotted lines) problem for $j = 100, 200, \ldots, 600$ iterations at discretization level $nt = 19$, $poi = 81$

which is equivalent to the block-triangular system

$$\begin{bmatrix} H_{\delta u} & & \\ e_u & e_y & \\ & L_{yy} & e_y^* \end{bmatrix} \begin{bmatrix} \delta u \\ \delta y \\ \lambda^{k+1} \end{bmatrix} = \begin{bmatrix} e_u^* e_y^{-*} \left[f_y - L_{yy} e_y^{-1} e \right] - f_u \\ -e \\ -f_y \end{bmatrix} \tag{25}$$

whose rows are just the equations (12)–(14). Hence the reduced Hessian problem is nothing else than the full problem after preconditioning with P. Comparing (11) to (25), it turns out that the preconditioning actually provides the iterative solver with some insight into the interdependence of the unknown variables. While in the full Hessian system, the solver takes *all* variables as degrees of freedom, in the reduced system only the *true* free variables (i.e. the controls) appear and the state and the adjoint are calculated consistently. From this point of view, the reduced Hessian method resembles what is usually called a *direct single shooting* approach, applied to a linear-quadratic model.

The necessity to have the full and reduced Hessian matrix explicitly available for the numerical tests limits the discretization levels to very coarse ones throughout this paper. In practice, however, control problems for time-dependent PDEs with about 275 000 unknowns (including 40 000 control variables) have been successfully solved on a desktop PC within 2 hours using the reduced Hessian SQP algorithm.

References

[1] Battermann, A. & Heinkenschloss, M.: *Preconditioners for Karush-Kuhn-Tucker Matrices Arising in the Optimal Control of Distributed Systems*, in: W. Desch,

F. Kappel, K. Kunisch (eds.), Optimal Control of Partial Differential Equations, Vorau 1997, Birkhäuser Verlag, Basel, Boston, Berlin, 1998, pp. 15-32.

[2] Biros, G. & Ghattas, O.: *Parallel Lagrange-Newton-Krylov-Schur Methods for PDE-Constrained Optimization. Part I: The Krylov-Schur Solver*, Technical Report, Laboratory for Mechanics, Algorithms, and Computing, Carnegie Mellon University, 2000.

[3] Biros, G. & Ghattas, O.: *Parallel Lagrange-Newton-Krylov-Schur Methods for PDE-Constrained Optimization. Part II: The Lagrange-Newton Solver, and its Application to Optimal Control of Steady Viscous Flows*, Technical Report, Laboratory for Mechanics, Algorithms, and Computing, Carnegie Mellon University, 2000.

[4] Hinze, M. & Kunisch, K.: *Second Order Methods for Optimal Control of Time-dependent Fluid Flow*, Bericht Nr. 165 des Spezialforschungsbereichs F003 Optimierung und Kontrolle, Karl-Franzens-Universität Graz (1999), to appear in SIAM J. Control Optim.

[5] Kupfer, F.-S.: *An infinite-dimensional convergence theory for reduced SQP methods in Hilbert space*, SIAM J. Optimization 6, 1996.

[6] Nocedal, J. & Wright, S.: *Numerical Optimization*, Springer, 1999.

[7] Tröltzsch, F.: *On the Lagrange–Newton–SQP Method for the Optimal Control of Semilinear Parabolic Equations*, SIAM J. Control Optim. 38, No. 1, pp. 294–312, 1999.

ON NUMERICAL PROBLEMS CAUSED BY DISCONTINUITIES IN CONTROLS

Christian Großmann, Antje Noack, Reiner Vanselow *
Dresden University of Technology
Institute of Numerical Mathematics
D - 01062 Dresden, Germany
{grossm, noack, vanselow}@math.tu-dresden.de

Abstract The regularity of solutions of parabolic initial-boundary value problems directly depends upon the regularity of the boundary data. Reduced regularity of boundary data arise e.g. in optimal boundary control problems governed by evolution equations by a discretization of the control by piecewise constant functions and results in refined grids if automatic step size procedures in time are applied. In the present study effects to numerical methods for solving the state equations are illustrated. Moreover, an appropriate splitting of the solution is used to improve the numerical behavior of the discretization technique as well as of the optimization method applied to the control problem itself.

Keywords: Boundary control, parabolic equation, discretization.

1. Introduction

Smoothness properties of solutions of parabolic initial-boundary value problems directly depend upon the smoothness of initial and boundary data. As a consequence, discretizing the boundary control by piecewise given functions generically results in a reduced smoothness of solutions of the related state equations. However, the efficiency of numerical methods for partial differential equations depends on the regularity of the desired solution. This yields specific effects like severe local grid refinements in time when standard discretization techniques are applied. In the present paper we investigate such effects and in case of piecewise constant Dirichlet controls we use a splitting of the solution to improve

*Partial funding of this research provided by DFG grant GR1777/2-2.

the numerical behavior of the discretization technique as well as of the optimization method applied to the control problem.

Throughout the paper we consider spatial one-dimensional boundary heat control problems

$$J(u) := \frac{1}{2}\int_0^1 [w(x,T;u) - q(x)]^2\, dx + \frac{\alpha}{2}\int_0^T [u(t) - p(t)]^2\, dt \to \min_{u\in U} ! \quad (1.1)$$

subject to the state equations

$$\begin{aligned} \frac{\partial w}{\partial t} - \sigma^2 \frac{\partial^2 w}{\partial x^2} &= f \quad \text{in} \quad Q := (0,1)\times(0,T],\\ w &= 0 \quad \text{on} \quad \Gamma_l := \{0\}\times(0,T],\\ \gamma_D\, w + \gamma_N \frac{\partial w}{\partial x} &= u \quad \text{on} \quad \Gamma_r := \{1\}\times(0,T],\\ w &= g \quad \text{on} \quad Q_0 := [0,1]\times\{0\}, \end{aligned} \quad (1.2)$$

with control u and the state $w(\cdot,\cdot;u)$ in the weak sense (cf. [11], [12]). Here γ_D, $\gamma_N \geq 0$ are given coefficients satisfying $\gamma_D + \gamma_N > 0$ and $\alpha > 0$ is a fixed regularization parameter. Further, $U \neq \emptyset$, $U \subset L_\infty(0,T)$ denotes a set of admissible controls, $q \in L_2(0,1)$ is the given target temperature and $p \in U$ denotes some fixed reference control.

For controls $u \in U$ here we restrict ourselves to discretizations $u^\tau \in U^\tau$ defined over a given time grid

$$0 = t^0 < t^1 < \cdots < t^{M^c-1} < t^{M^c} = T \quad (1.3)$$

by piecewise constant functions, i.e.

$$u^\tau \in U^\tau \iff u^\tau(t) = u^k \in \mathbb{R},\ \forall t \in (t^{k-1}, t^k],\ k = 1,\ldots,M^c. \quad (1.4)$$

Here M^C denotes the number of time intervals for control discretization. To distinguish between discretizations of control and states we indicate the first ones by upper scripts (as above) and the second ones by lower scripts.

In case of Dirichlet controls, i.e. $\gamma_D = 1$, $\gamma_N = 0$, jumps of u^τ at inner grid points t^k, $k = 1,\ldots,M^c - 1$, cause discontinuities of the related solution $w(\cdot,\cdot;u^\tau)$. In literature, there are several results on the numerical treatment of the heat equation with irregular solutions, where the irregularities result from the functions f or g (cf. [2], [6], [9], [10]).

In literature one can find three approaches to overcome such difficulties. In the first one fitted methods are constructed with coefficients which are adapted to the singularities (cf.[6]). In the second approach a standard method is chosen but with specifically refined meshes in the neighborhood of singularities (cf. [2] , [9], [10]). The third approach splits off the singularities. In the present paper we apply this splitting. Related details are discussed in the following section.

2. Numerical treatment of the state equations

2.1 Splitting in case of Dirichlet conditions

In the present subsection we consider the case $\gamma_D = 1$, $\gamma_N = 0$. At first, we assume compatibility at $x = 0$, i.e. $g(0) = 0$, and (for the sake of a uniform description) we extend the piecewise constant function u^τ to $t = 0$ by $u^\tau(0) := u^0 := g(1)$.

Let us now introduce functions $w^k : Q \to \mathbb{R}$, $k = 1, \ldots, M^c$, by

$$w^k(x,t) := \begin{cases} 1 - \operatorname{erf}\left(\dfrac{1-x}{2\sigma\sqrt{t - t^{k-1}}}\right) & , \text{ if } x \in [0,1],\ t > t^{k-1} \\ 0 & , \text{ if } x \in [0,1],\ t \le t^{k-1} \end{cases} \tag{2.1}$$

with the error function

$$\operatorname{erf}(\xi) := \frac{2}{\sqrt{\pi}} \int_0^\xi e^{-s^2} ds, \quad \xi \in \mathbb{R}. \tag{2.2}$$

Definition (2.1), (2.2) yields $w^k \in C^\infty(\bar{Q} \backslash \{(1, t^{k-1})\})$ and

$$\frac{\partial w^k}{\partial t} - \sigma^2 \frac{\partial^2 w^k}{\partial x^2} = 0 \qquad \text{in } Q.$$

Further, w^k has a jump w.r.t. t at $(1, t^{k-1})$. Hence, occurring discontinuities of the solution of (1.2) at the points $(1, t^k)$, $k = 0, \ldots, M^c - 1$, originated by jumps in u, can be captured by the functions w^k. Namely, using superposition, the solution $w(\cdot,\cdot;u^\tau)$ of (1.2) can be written as

$$w(x,t;u^\tau) = \hat{w}(x,t;u^\tau) + v(x,t;u^\tau), \qquad (x,t) \in \bar{Q} \tag{2.3}$$

for any given $u^\tau \in U^\tau$, where $\hat{w}(\cdot,\cdot;u^\tau)$ is defined by

$$\hat{w}(x,t;u^\tau) := \sum_{k=1}^{M^c} \left(u^k - u^{k-1}\right) w^k(x,t), \qquad (x,t) \in \bar{Q} \tag{2.4}$$

and $v(\cdot,\cdot;u)$ denotes the solution of the related parabolic problem

$$\begin{aligned} \frac{\partial v}{\partial t} - \sigma^2 \frac{\partial^2 v}{\partial x^2} &= f \quad \text{in } Q, \\ v = -\hat{w} \quad \text{on } \Gamma_l, \qquad v &= 0 \quad \text{on } \Gamma_r, \\ v &= g \quad \text{on } Q_0. \end{aligned} \tag{2.5}$$

Due to $\hat{w}(0,0;u^\tau) = g(0) = 0$ for any $u^\tau \in U^\tau$, the smoothness of $\hat{w}$ at $x = 0$ and sufficiently smooth functions f and g, the discontinuities of w are completely captured by $\hat{w}$. Hence, problem (2.5) allows a better numerical treatment than the original PDE.

2.2 Discretization of the state equations

In the preceeding section we described the principle impact of piecewise discretizations of controls to the smoothness of the solutions of the state equations. Now, we sketch consequences of reduced regularity to numerical methods applied to (1.2) with discretized boundary data.

Among the variety of methods let us consider semi-discretization by standard method of lines (MOL) as well as full discretization schemes. The major difference of both approaches is that in the first one standard ODE solvers with efficient step size control can be applied while the full scheme provides a direct access to the time grid which will later be advantageous in evaluating adjoint states for the optimal control problem.

Consider some spatial grid $\{x_i\}_{i=0}^N$ over the interval $[0,1]$, i.e.

$$0 = x_0 < x_1 < \cdots < x_{N-1} < x_N = 1. \tag{2.6}$$

Using simple finite differences we obtain a spatial semi-discretization of the PDE by

$$\begin{aligned} h_{i+1/2}\frac{dw_i}{dt}(t) - \sigma^2\left[\frac{w_{i+1}(t)-w_i(t)}{h_{i+1}} - \frac{w_i(t)-w_{i-1}(t)}{h_i}\right] \\ = h_{i+1/2}\, f(x_i,t), \qquad i = 1,\dots,N-1 \end{aligned} \tag{2.7}$$

with $h_i := x_i - x_{i-1}$, $i = 1,\dots N$ and $h_{i+1/2} := (h_i + h_{i+1})/2$. Here and in the sequel w_i denote functions which approximate $w(x_i,\cdot;u^\tau)$. In addition to (2.7) the boundary conditions from (1.2) at $x = 1$ are taken into account by

$$\gamma_D\, w_N(t) = u^\tau(t), \quad t \in (0,T] \tag{2.8a}$$

and

$$\begin{aligned} \frac{h_N}{2}\frac{dw_N}{dt}(t) &= \frac{\sigma^2}{\gamma_N}\left[u^\tau(t) - \gamma_D w_N(t)\right] \\ &-\sigma^2\frac{w_N(t)-w_{N-1}(t)}{h_N} + \frac{h_N}{2} f(x_N,t), \quad t \in (0,T] \end{aligned} \tag{2.8b}$$

for $\gamma_N = 0$ and $\gamma_N \neq 0$, respectively, while at $x = 0$ we have in both cases $w_0(t) = 0$. If we consider splitting then instead of (1.2) we apply semi-discretization to problem (2.5) and we have $v_0(t) = -\hat{w}(0,t;u^\tau)$, $v_N(t) = 0$.

Together with the initial conditions

$$w_i(0) = g(x_i), \quad i = 1, \ldots, N \tag{2.9}$$

we obtain an IVP system for the functions w_i. Notice that in case of Dirichlet control the number of unknowns is $N-1$ otherwise N. We will not explicitly distinguish these cases and write for simplicity in the sequel just N.

In our first approach we treat the IVP (2.7)-(2.9) by standard ODE codes for stiff IVPs. In particular, in our study we applied BDF-codes and trapezoidal rule with automatic step size control.

Alternatively to semi-discretization and standard ODE codes, to which in the sequel we refer shortly as semi-discretization, in a second approach we apply implicit Euler method with a fixed time step T/M to (2.7) - (2.9), which we denote in the sequel as full discretization.

In both approaches discrete states are denoted by $w_{i,j}$, $i = 0, 1, \ldots, N$, $j = 0, 1, \ldots, M$, where M is the number of time steps.

3. Numerical treatment of the control problem

3.1 Gradient evaluation

Discretization of the controls and the state equations leads to an approximation of the original optimal control problem (1.1), (1.2) by a finite dimensional quadratic programming problem. Let us consider the case that no constraints are imposed upon the controls.

The state equations result in an affine mapping transferring discrete controls $u^\tau \in U^\tau$ into discrete terminal states $w_{\cdot,M}$, i.e. we have

$$w_{\cdot,M} = A_{h,\tau}\, u^\tau + a_{h,\tau} \tag{3.1}$$

with some matrix $A_{h,\tau} \in \mathcal{L}(\mathbb{R}^{M^c}, \mathbb{R}^N)$ and some vector $a_{h,\tau} \in \mathbb{R}^N$. With discrete scalar products $\langle \cdot, \cdot \rangle$ in $\mathbb{R}^N$ and $\mathbb{R}^{M^c}$, respectively, we obtain problems of the type

$$J_{h,\tau}(u^\tau) \to \min ! \qquad \text{s.t.} \qquad u^\tau \in U^\tau \tag{3.2}$$

with

$$J_{h,\tau}(u^\tau) := \frac{1}{2}\langle A_{h,\tau}\, u^\tau - \overline{q}_{h,\tau}, A_{h,\tau}\, u^\tau - \overline{q}_{h,\tau} \rangle + \frac{\alpha}{2}\langle u^\tau - p^\tau, u^\tau - p^\tau \rangle. \tag{3.3}$$

Here $\overline{q}_{h,\tau} := q_h - a_{h,\tau}$, and $p^\tau \in U^\tau$ denotes some approximation of p. Further, the necessary optimality conditions are given by

$$J'_{h,\tau}(u^\tau) = A^*_{h,\tau}\,(A_{h,\tau} u^\tau - \overline{q}_{h,\tau}) + \alpha\,(u^\tau - p^\tau) = 0.$$

It should be noticed that in case of full discretization $A_{h,\tau}$, $a_{h,\tau}$ are known, but will not be constructed explicitly because of the dynamic nature of the discrete state equations. However, in case of semi-discretization where some ODE software code is applied to (2.7)-(2.9) then $A_{h,\tau}$, $a_{h,\tau}$ depend on various additional features, like built-in automatic step size controls. In case of semi-discretization as well as full discretization is applied the image $A_{h,\tau}u^\tau$ can be determined for any $u^\tau \in U^\tau$ by discrete time integration. Moreover, adjoint equations provide an efficient tool for gradient evaluations replacing the calculation of $A^*_{h,\tau}(A_{h,\tau}u^\tau - \overline{q}_{h,\tau})$. For the optimal control problem (1.1), (1.2) the corresponding adjoint problem is defined by (cf. [1], [4], [7], [11])

$$
\begin{aligned}
\frac{\partial z}{\partial t} + \sigma^2 \frac{\partial^2 z}{\partial x^2} &= 0 \quad \text{in} \quad Q, \\
z = 0 \quad \text{on} \quad \Gamma_l, \quad \gamma_D\, z + \gamma_N \frac{\partial z}{\partial x} &= 0 \quad \text{on} \quad \Gamma_r, \\
z &= w - q \quad \text{on} \quad [0,1] \times \{T\}
\end{aligned}
\tag{3.4}
$$

and the reduced gradient of the objective at $u \in U$ in direction $s \in L_\infty(0,T)$ is given by

$$
J'(u)\, s = \frac{\sigma^2}{\gamma_D + \gamma_N} \left\langle z(1,\cdot\,;u) - \frac{\partial z}{\partial x}(1,\cdot\,;u) + \alpha(u-p), s \right\rangle_{(0,T)} . \tag{3.5}
$$

Notice that after reversing the time orientation the adjoint problem (3.4) is of parabolic type as the state equation (1.2). However, unlike in the state equation in the adjoint equation we meet incompatibility only at one time level, namely $t = T$.

For the remaining part of this section we restrict ourselves to the case $\gamma_D = 1$, $\gamma_N = 0$. Further, for simplicity in the sequel we consider equidistant spatial grids and denote its step size by $h > 0$.

When applying standard ODE solvers to the related semi-discrete IVP (2.7)-(2.9) and an appropriate discretization to the scalar product in (3.5) we obtain the following approximation of the discrete directional derivative

$$
J'(u^\tau)s^\tau \approx \sum_{j=1}^{M^c} \left[\frac{\sigma^2}{h} \sum_{k \in K_j} (\vartheta_k - \vartheta_{k-1}) z_{N-1,k-1} + \alpha\, \tau^j \,(u^j - p^j) \right] s^j, \tag{3.6}
$$

where $u^j, p^j, s^j \in \mathbb{R}$, $j = 1,\ldots, M^c$ are the coefficients of $u^\tau, p^\tau, s^\tau \in U^\tau$, $(z_{i,j})$ is the discrete solution of the adjoint problem (3.4), $\{\vartheta_k\}_{k=0}^M$ denotes the time grid generated by the applied ODE solver and

$$
K_j := \{k \in \{1,\ldots,M\} \colon \vartheta_k \in (t^{j-1}, t^j]\}, \; \tau^j := t^j - t^{j-1}, \; j = 1,\ldots,M^c.
$$

To obtain (3.6) from (3.5) besides simple integration, the derivative of $\frac{\partial z}{\partial x}$ at $x = 1$ is approximated by one sided finite differences where we take into account the boundary condition $z(1, \cdot) = 0$. Equation (3.6) provides the representation of the discrete gradient $J'(u^\tau)$ via the adjoints.

In case of full discretization, the discrete gradient can be evaluated directly via the corresponding discrete adjoint system. Similarly to the continuous adjoint system after time reversal it turns out to be an implicit Euler scheme again. The obtained formula for the discrete gradient (3.6) can be also interpreted as an approximation of the continuous one.

Our numerical experiments confirmed the fact that the discrete adjoints of the full discretization lead to exact gradients as generated in automatic differentiation tools (see [3]). However, if software tools are applied to semi-discretization of (2.5) and of the adjoint equations (3.4) then only an approximation of the gradients is obtained. One reason for that deviation is that applications of ODE solvers with time step control lead to discretizations of the states and adjoint states with different time grids. Thus the discretization of the adjoint states is not adjoint to the discrete states in the sense of the discrete L_2-norm but only an approximation. Moreover, the summation in the formula for the discrete gradient (3.6) causes a further amplification of the error. Hence, to guarantee convergence of optimization techniques based on this approach a sufficiently high order of accuracy in the applications of ODE software is required which becomes rather expensive for fine discretizations.

3.2 Selected minimization techniques

Since the gradient can be obtained quite easily via adjoint states conjugate gradient methods as well as quasi-Newton techniques (e.g. Broyden's symmetric update, DFP-method, ...) are appropriate for solving the discrete quadratic minimization problem (3.2).

To make the paper self contained we describe briefly the major steps of methods used in our tests for solving (3.2). Let us denote the elements of a sequence $\{\mathbf{u}^l\}$ of discrete controls by $\mathbf{u}^l := u^{\tau,l} \in U^\tau$. In the considered piecewise constant approximation we can represent $\mathbf{u}^l$ by its coefficients $u^{k,l} \in \mathbb{R}$, $k = 0, 1, \ldots, M^c$, $l = 0, 1, \ldots$.

As one of the methods of choice we applied conjugate gradient methods. Starting with some $\mathbf{u}^0 \in U^\tau$ and $\beta_0 := 0$, these methods generate a minimizing sequence $\{\mathbf{u}^l\} \subset U^\tau$ recursively by

$$\begin{aligned} \mathbf{s}^l &:= -J'(\mathbf{u}^l) + \beta_l \, \mathbf{s}^{l-1}, \qquad \beta_{l+1} := \frac{\|J'(\mathbf{u}^l)\|_2^2}{\|J'(\mathbf{u}^{l-1})\|_2^2} \\ \mathbf{u}^{l+1} &:= \mathbf{u}^l + \alpha_l \mathbf{s}^l, \quad \text{with Cauchy step size} \quad \alpha_l > 0. \end{aligned} \tag{3.7}$$

CG methods terminate with the optimal control given by the final minimizer in a finite number of steps provided exact function and gradient evaluations are applied and no rounding errors occur. This is, however, unrealistic in the problems under consideration but the convergence can be accelerated by appropriate preconditioning (cf. [5], [8]). We detected that in the case of Dirichlet control the analytic solution (2.4) which captures jumps in boundary data serves for preconditioning.

In case of unconstrained controls the Cauchy (i.e. minimizing) step size α is easily obtained. However, penalty methods for the treatment of constraints require additional step size procedures.

As other methods of choice we included quasi-Newton methods into our study. Their basic idea is to define the search direction $\mathbf{s}^l$ at $\mathbf{u}^l$ by

$$H_l\,\mathbf{s}^l = -J_l', \tag{3.8}$$

where $J_l' := J'(\mathbf{u}^l)$ and H_l, $l = 0, 1, \ldots$, denote matrices satisfying the related quasi-Newton equation

$$H_l\,(\mathbf{u}^l - \mathbf{u}^{l-1}) = J_l' - J_{l-1}', \quad l = 1, 2, \ldots. \tag{3.9}$$

Starting with the identity $H_0 := I$ the matrices H_l are updated by appropriate formulas. In particular, we considered Broyden's symmetric update. Let

$$\mathbf{r}^{l+1} := J_{l+1}' - J_l' - H_l\,(\mathbf{u}^{l+1} - \mathbf{u}^l).$$

Then the new matrix H_{l+1} is defined by

$$H_{l+1} := H_l + \frac{\mathbf{r}^{l+1}\,(\mathbf{r}^{l+1})^T}{(\mathbf{r}^{l+1})^T(\mathbf{u}^{l+1} - \mathbf{u}^l)}. \tag{3.10}$$

In the evaluation $\mathbf{u}^{l+1} := \mathbf{u}^l + \alpha_l \mathbf{s}^l$ the step size $\alpha_l > 0$ has been selected according to a simplified Armijo rule. For a detailed description of CG-methods and quasi-Newton methods we refer e.g. to [5], [8].

Occurring constraints

$$|u^j| \leq 1, \quad j = 1, \ldots, M^c$$

on controls have been included by the penalty term ($\rho > 0$)

$$P_\rho(u^\tau) := \frac{c}{2} \sum_{j=1}^{M^c} \tau^j \left[\sqrt{(u^j+1)^2 + \rho} + \sqrt{(u^j-1)^2 + \rho} - 2\right]. \tag{3.11}$$

For $\rho \to 0+$ this tends uniformly to the well-known non-smooth penalty

$$P_0(u^\tau) := c \sum_{j=1}^{M^c} \tau^j \left[\max\{0, -u^j - 1\} + \max\{0, u^j - 1\}\right],$$

which is exact for sufficiently large constant $c > 0$. For $\rho > 0$ the penalty P_ρ is infinitely often differentiable. This forms an advantage in comparison with loss functions. Further, unlike for barriers the values $P_\rho(u^\tau)$ are finite for any discrete control u^τ. For the first derivative and the Hessian we have

$$P'_\rho(u^\tau)\, s^\tau \;=\; \frac{c}{2}\sum_{j=1}^{M^c} \tau^j \left[\frac{u^j+1}{\sqrt{(u^j+1)^2+\rho}} + \frac{u^j-1}{\sqrt{(u^j-1)^2+\rho}}\right] s^j$$

and

$$P''_\rho(u^\tau) \;=\; \frac{c\,\rho}{2}\,\mathrm{diag}\left(\tau^j\left[\frac{1}{((u^j+1)^2+\rho)^{3/2}} + \frac{1}{((u^j-1)^2+\rho)^{3/2}}\right]\right),$$

respectively. These derivatives have been used directly in the quasi-Newton methods, i.e. only components related to $J(\cdot)$ are taken into consideration by the quasi-Newton update. On the other hand, in Armijo's step size rule only penalty terms have to be repeatedly evaluated due to the quadratic nature of $J(\cdot)$. This accelerates the code compared to an application of an all-purpose minimization routine.

4. Numerical experiments

4.1 Preliminaries

In our numerical experiments we tested the performance of different techniques applied to IBVPs (1.2) with discontinuous boundary data as well as studied effects in connection with boundary control problems of tracking type. All experiments are implemented in MATLAB. The focus in Examples 1, 2 was directed towards the behavior of automatic step size procedures in ODE codes and to an improvement of the efficiency of such codes by using the splitting described in subsection 2.1. In connection with optimal control in Examples 3, 4 we studied the influence of discontinuities in boundary data on the convergence of minimization techniques.

In all examples we choose equidistant grids $x_i = i/N$ and $t^k = (Tk)/M^c$. Further, in the first two examples we choose $\sigma = 1/2$, $T = 1$, but in the last two $\sigma = 1$, $T = 0.1$.

The following tables and figures report on numerical results obtained by the BDF-code **ode15s** (option BDF=on) using several maximal orders of consistency (option MaxOrder) and trapezoidal rule **ode23t**, respectively. If not written otherwise, the default values of the relative and absolute error tolerance RelTol=1e−3 and AbsTol=1e−6, respectively, are used. Further, in the Dirichlet case ($\gamma_D = 1$, $\gamma_N = 0$) we split the experiments in direct solving problem (1.2) by the method of line (named 'direct' in the following tables) and in applying superposition (2.3) to treat occurring jumps in boundary data. In the latter case we solve numerically the remaining smooth problem (2.5). All described effects depend on M^c and the height of the jumps.

4.2 Example 1 (state equations)

For the first example we choose $f \equiv 0, ; g \equiv 0, \quad M^c = 3, \quad N = 50$ with boundary data u^τ in (1.4) according to $\{u^k\} = (1, -2, 3)^T$. Fig. 1 shows the obtained solution $w(x, t; u^\tau)$ for Dirichlet and Neumann boundary conditions, respectively. The number of required time steps is

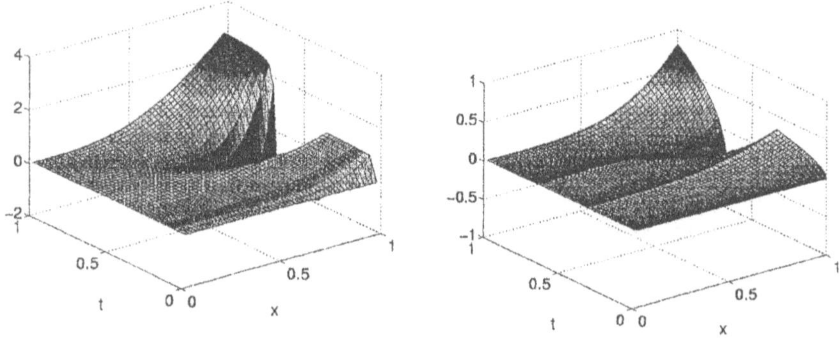

Figure 1. $\gamma_D = 1, \ \gamma_N = 0$ and $\gamma_D = 0, \ \gamma_N = 1$

reported in Tab. 1. For the trapezoidal rule code the related results are marked with T instead of the order as done for BDF code.

treatment	direct				superposition			
maximal order	1	2	5	T	1	2	5	T
obtained time steps	2331	518	322	377	364	111	58	81

Table 1. Comparison of different approaches

The left two graphs in Fig. 2 illustrate the behavior of the automatic step size control when applied directly or after splitting in case of Dirichlet boundary conditions. Further, in the right graph step size results in case of Neumann boundary conditions are reported.

The numerical experiments show (see Fig. 2) that each jump in the control u^τ reduces the time step size drastically. On the other hand, splitting-off the discontinuities (in case of Dirichlet-boundary conditions) in advance avoids these time step size reductions and, hence, yields a more effective numerical procedure.

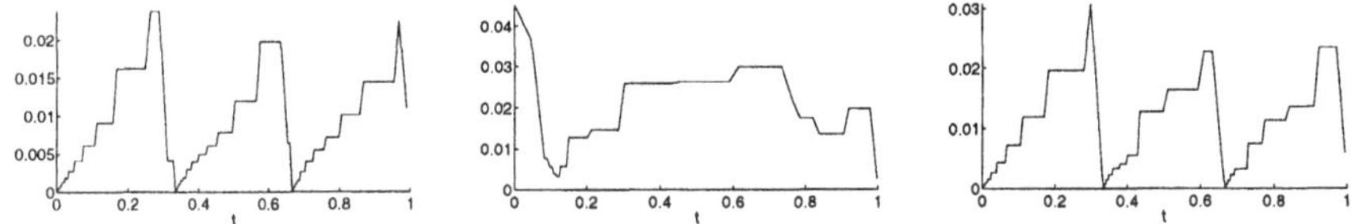

Figure 2. Time step sizes for order 5

4.3 Example 2 (state equation, known exact solution)

In this example we consider a problem with Dirichlet boundary conditions where the exact solution is known. The required discontinuous boundary data are generated by means of the function $\hat{w}$ introduced in Section 2. Unlike in the previous tests here we concentrate on the error behavior. Let the exact solution be given by

$$w(x,t;u^\tau) = g(x) - \left[\hat{w}(x,t;u^\tau) - (1-x)\,\hat{w}(0,t;u^\tau)\right]$$

with $g(x) = 10x^2(1-x)^2$. To study one internal jump only we choose $M^c = 2$ and u^τ in (1.4) according to $\{u^k\} = (1,-1)^T$.

Fig. 3 shows the obtained solution $w(x,t;u^\tau)$ and together with Fig. 4 the error of the BDF-code with MaxOrder=5 for superposition and the direct approach, respectively. In the right picture of Fig. 4 the neighborhood of the point $(x,t) = (1,0)$, where a jump is located, is cut off.

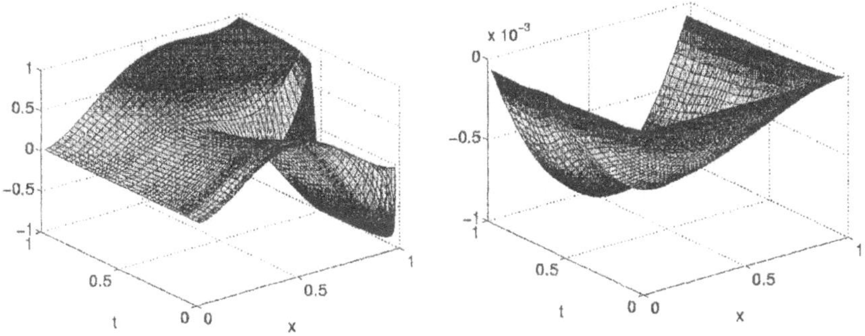

Figure 3. Solution $w(x,t;u^\tau)$ and Error for superposition

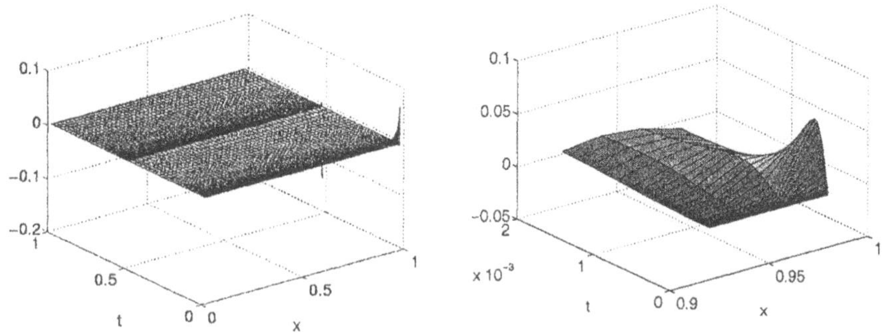

Figure 4. Error in case of the direct solution

Choosing different numbers N of spatial grid points with fixed accuracy RelTol=AbsTol=1e−8 we obtain

N	1600	800	200	50
obtained time steps	1274/133	1138/135	894/140	666/150
error at $t = 1$	9e-07	3e-06	6e-05	9e-04

Table 2. Comparison of required time steps for different N

where in the second row of Tab. 2 the first number is related to direct treatment, the second to superposition.

The numerical experiments reflect (see Tab. 2 and Fig. 3,4), that the step size reduction is the more severe the larger N is.

Finally, we notice that the numerical solution converges at $t = T$, although there is no convergence locally near jumps.

4.4 Example 3 (unconstrained control problem)

We consider the optimal control problem (1.1), (1.2) with

$$p \equiv 0, \ f \equiv 0, \ g \equiv 0 \quad \text{and} \quad q(x) = 0.05\, x^2 \sin(4\pi x).$$

The convergence behavior of a CG-algorithm as well as a quasi-Newton method with Broyden's update is compared for both the approaches discussed in Subsection 3.1, i.e. that the calculation of the discrete gradient (3.5) is based on semi-discretization with discretized continuous adjoints and full discretization with discrete adjoints, respectively. In case of semi-discretization superposition is used for the solution of state as well as for the adjoint state equations. The remaining regular problems were treated by the BDF-code of MATLAB with MaxOrder=5. In Fig. 5 we

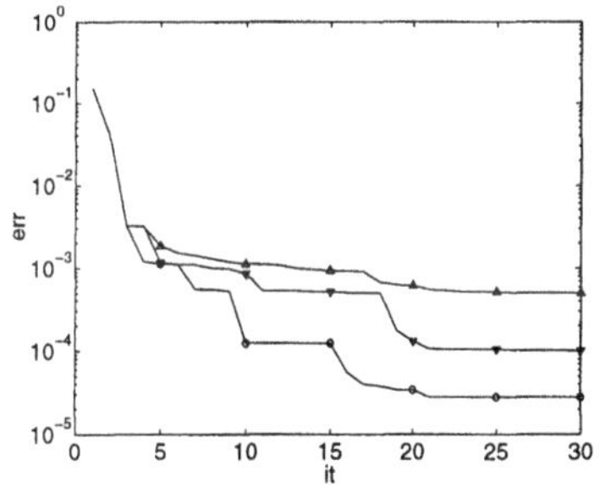

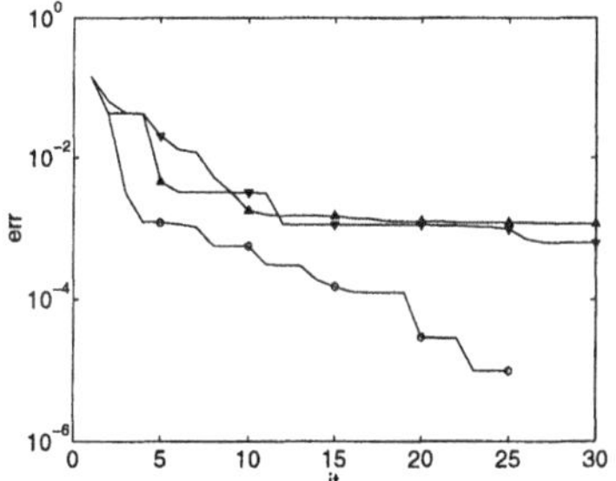

Figure 5. CG-algorithm and Broyden's method

included results from semi-discretization, RelTol=AbsTol=1e−5, semi-discretization, RelTol=AbsTol=1e−12 and full discretization, $M = 500$. The related curves are marked by $\Delta-$, $\nabla-$ and $\circ-$, respectively.

Further, in Fig. 6 the corresponding optimal controls received by the CG-algorithm are reported. We indicate that further slow improvements were obtained beyond the iteration steps plotted in Fig. 6.

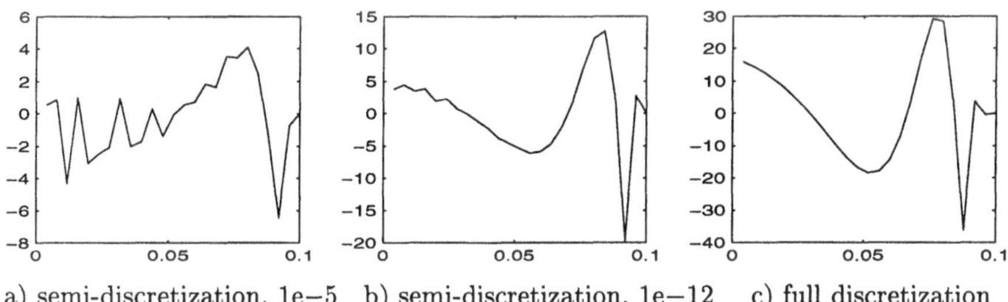

a) semi-discretization, 1e−5 b) semi-discretization, 1e−12 c) full discretization

Figure 6. Optimal control obtained by CG-algorithm

In Tab. 3 the influence of the control grid is given for full discretization. Semi-discretizations with sufficiently high accuracy in the ODE solvers show a similar behavior. In general we remark that additionally

M^c	CG method	Broyden's update
10	4.38e-03	4.84e-03
25	2.59e-03	2.78e-03
50	1.42e-03	1.87e-03
100	1.40e-03	1.73e-03

Table 3. Comparison of convergence behavior for different control grids

to slower convergence semi-discretization in both cases of accuracy is more expensive, i.e. consumes significantly more computer time, than the full discretization.

4.5 Example 4 (constrained control problem)

We choose $f \equiv 0$, $g \equiv 0$. Further, we start with a control problem (1.1), (1.2) which possesses the optimal solution

$$u_{ref}(t) = 1.5 \sin\left(\frac{4\pi t}{T}\right), \quad t \in [0,T]$$

if no constraints are given for the controls. Using this the functions q and p are defined by $q(x) := w(x,T;u_{ref})$ and $p := u_{ref}$, respectively, with the solution $w(\cdot,\cdot;u_{ref})$ of the state equation (1.2) for $u = u_{ref}$.

M^c	semi-discretization	full discretization	clipping
10	1.41e-03	1.36e-03	6.93e-03
25	8.83e-04	2.71e-04	2.30e-0.3
50	3.23e-04	1.06e-05	1.10e-03
100	7.28e-04	9.84e-06	6.60e-04

Table 4. Obtained objective values for different control grids

In Tab. 4 the achieved optimal values are reported for the two approaches. In addition, we show in the last column the objective value for the discrete control which is obtained from the unconstrained optimal one by simple clipping along the constraints.

The following Fig. 7 shows discrete optimal controls obtained by Broyden's update (3.10) to the quadratic part (from the state equations) and by direct use of up to second order derivatives of the penalties as given in Section 3. Further, in Fig. 8 the approximation of the tracked target and a comparison between the constrained and the unconstrained optimal controls are given.

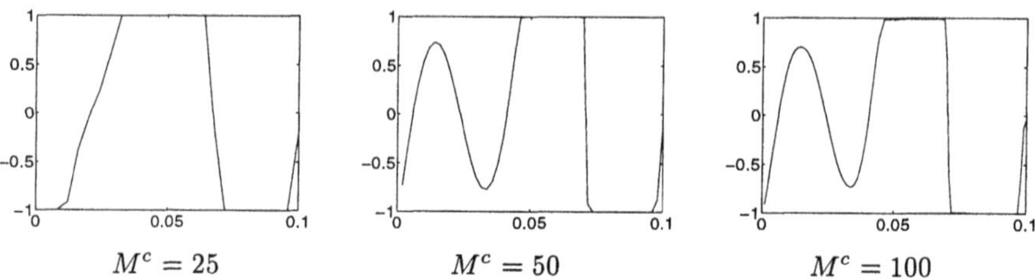

Figure 7. Discrete optimal controls, full discretization

The computational experiments showed a very similar behavior as in the unconstrained case. In semi-discretization the state as well as the

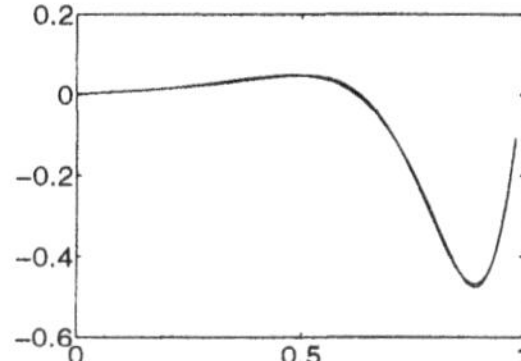

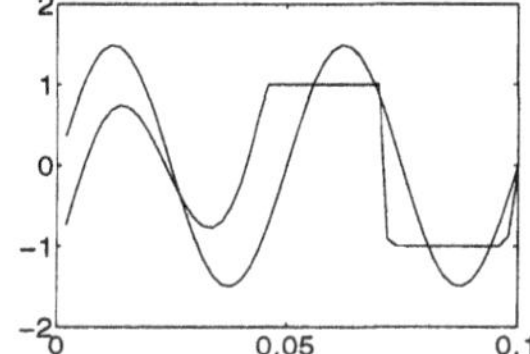

Figure 8. Approximation of the target Constrained, unconstrained control

adjoint system have to be solved with a sufficiently high accuracy to ensure a good approximation of the gradient. This, however, results in high time consumption in the applied ODE solver. On the other hand, full discretization in more general cases (in particular in higher spatial dimensions) requires additional preparatory work compared with the use of available software codes.

5. Conclusions

Piecewise constant discretization of boundary controls yields a reduced smoothness of the solutions of state equations. In all our considered examples this resulted in locally small step sizes if ODE solvers were applied to a semi-discretization of the state equations. These problems could be avoided by considering in advance a specific splitting of the state equations.

In the examples of optimal control problems semi-discretization was only used in connection with a separation of the discontinuities. Hence, the ODE solvers were, in fact, applied to the regular subproblem. Nevertheless, this approach turned out to be more time consuming then full discretization combined with discrete adjoints. In addition, full discretization often yielded better values of the objectives and proved to be faster for comparable accuracy. Further, if lower accuracies were applied to speed up the ODE codes in semi-discretization then the optimization became slow due to the fact that discretizations of continuous adjoint problems lead to only rough approximations of gradients.

References

[1] Casas, E. (1997). Pontryagin's principle for state-constraint boundary control problems of semilinear parabolic equations. *SIAM J. Control Optim.* 35:1297-1327.

[2] Crouzeix, M. and Thomee, V. (1987). On the discretization in time of semilinear equations with nonsmooth initial data. *Math. Comput.* 49:359-377.

[3] Griewank, A. (2000). *Evaluating derivatives: Principles and techniques of algorithmic differentiation.* SIAM Publ., Philadelphia.

[4] Grossmann, C. and Noack, A. (2001). Linearizations and adjoints of operator equations – constructions and selected applications. *TU-Preprint MATH-NM-08-01.*

[5] Grossmann, C. and Terno, J. (1993). *Numerik der Optimierung.* Teubner, Stuttgart.

[6] Hemker, P.W. and Shishkin, G.I. (1993). Approximation of parabolic PDEs with a discontinuous initial condition. *East-West J. Numer. Math.* 1:287-302.

[7] Kelley, C.T. and Sachs, E.W. (1999). A trust region method fo parabolic boundary control problems. *SIAM J. Optim.* 9:1064-1081.

[8] Nocedal, J. and Wright, S.J. (1999). *Numerical optimization.* Springer, New York.

[9] Rannacher, R. (1984). Finite element solution of diffusion problems with irregular data. *Numer. Math.* 43:309-327.

[10] Sammon, P. (1983). Fully discrete approximation methods for parabolic problems with nonsmooth initial data. *SIAM J. Numer. Anal.* 20:437-470.

[11] Tröltzsch, F. (1984). *Optimality conditions for parabolic control problems and applications.* Teubner, Leipzig.

[12] Tröltzsch, F. (1994). Semidiscrete Ritz-Galerkin approximation of non-linear parabolic boundary control problems - strong convergence of optimal controls. *Appl. Math. Optim.* 29:309-329.

[13] Tychonoff, A.N. and Samarsky, A.A. (1959). *Differentialgleichungen der Mathematischen Physik.* Verlag d. Wissenschaft, Berlin.

SOLUTIONS DIFFERENTIABILITY OF PARAMETRIC OPTIMAL CONTROL FOR ELLIPTIC EQUATIONS

Kazimierz Malanowski
Systems Research Institute
Polish Academy of Sciences
ul. Newelska 6, 01-447 Warszawa, Poland
kmalan@ibspan.waw.pl

Abstract A family of parameter dependent elliptic optimal control problems with nonlinear boundary control is considered. The control function is subject to amplitude constraints. It is shown that under standard coercivity conditions the solutions to the problems are Bouligand differentiable (in L^s, $s < \infty$) functions of the parameter. The differentials are characterized as the solutions of accessory linear-quadratic problems.

Keywords: Parametric optimal control, elliptic equation, nonlinear boundary control, control constraints, Bouligand differentiability of the solutions

1. Introduction

In this paper, we analyse differentiability, with respect to the parameter, of solutions to a nonlinear boundary optimal control problem for an elliptic equation. Our aim is to show that, under a standard coercivity condition, the solutions to the optimal control problem are *Bouligand differentiable* functions of the parameter. Let us recall this concept of differentiability (see [3, 8, 9]).

Definition 1 *A function ϕ, from an open set $\mathcal{G}$ of a normed linear space H into another normed linear space X, is called Bouligand differentiable (or B-differentiable) at a point $h_0 \in \mathcal{G}$ if there exists a positively homogeneous mapping $D_h\phi(h_0) : \mathcal{G} \to X$, called B-derivative, such that*

$$\phi(h_0 + \Delta h) = \phi(h_0) + D_h\phi(h_0)\Delta h + o(\|\Delta h\|_H). \qquad (1)$$

Clearly, if $D_h\phi(h_0)$ is linear, it becomes Fréchet derivative.

As in [4] and in [5], the sensitivity, i.e., differentiability analysis for the original nonlinear problem is reduced to the same analysis for the accessory linear-quadratic problem. The starting point of the analysis is the Lipschitz stability result for the solutions to the linear-quadratic elliptic problems due to A.Unger [11]. Using this result, B-differentiability is proved in two steps. First, passing to the limit in the difference quotient, we show the directional differentiability and characterize the directional differential as the solution of an auxiliary linear-quadratic optimal control problem. Using this characterization, in the second step we show that an estimate of the form (1) holds, i.e., the solutions are Bouligand differentiable. This result can be considered as a generalization of that obtained in [1], where a different methodology was used to prove the directional differentiability of the solutions to parametric elliptic problem, under the assumption that the cost functional is quadratic with respect to the control.

2. Preliminaries

Let $\Omega \subset I\!R^n$ denote a bounded domain with boundary Γ. As usually, by Δy and $\partial_\nu y$ we denote the Laplace operator and the co-normal derivative of y at Γ, respectively. Moreover, let H be a Banach space of parameters and $G \subset H$ an open and bounded set of feasible parameters.

For any $h \in G$ consider the following elliptic optimal control problem:

$$(\mathrm{O}_h) \quad \text{Find } (y_h, u_h) \in Z^\infty := (W^{1,2}(\Omega) \cap C(\bar{\Omega})) \times L^\infty(\Gamma) \text{ such that}$$

$$\begin{aligned} F(y_h, u_h, h) = & \min\{F(y,u,h) \\ := & \int_\Omega \varphi(y(x),h)dx + \int_\Gamma \psi(y(x),u(x),h)dS_x\} \end{aligned} \tag{2}$$

subject to

$$\begin{aligned} -\Delta y(x) + y(x) &= 0 && \text{in } \Omega, \\ \partial_\nu y(x) &= b(y(x),u(x),h) && \text{on } \Gamma, \end{aligned} \tag{3}$$

$$u \in \mathcal{U} := \{v \in L^\infty(\Gamma) \mid m_1 \le v(x) \le m_2 \text{ a.e. in } \Gamma\}. \tag{4}$$

In this setting, $m_1 < m_2$ are fixed real numbers, dS_x denotes the surface measure induced on Γ, and the subscript x indicates that the integration is performed with respect to x. We assume:

(A1) The domain Ω has $C^{1,1}$-boundary Γ.

(A2) For any $h \in G$, the functions $\varphi(\cdot,h) : I\!R \to I\!R$, $\psi(\cdot,\cdot,h) : I\!R \times I\!R \to I\!R$ and $b(\cdot,\cdot,h) : I\!R \times I\!R \to I\!R$ are of class C^2. Moreover, for any fixed $u \in I\!R$ and $h \in G$, $b(\cdot,u,h) : I\!R \to I\!R$ is monotonically decreasing. There is a bound $c_G > 0$ such that

$$|b(0,0,h)| + |D_{(y,u)}b(0,0,h)| + |D^2_{(y,u)}b(0,0,h)| \le c_G \quad \forall h \in G.$$

Moreover, for any $K > 0$ there exists a constant $l(K)$ such that

$$|D^2_{(y,u)}b(y_1,u_1,h) - D^2_{(y,u)}b(y_2,u_2,h)| \le l(K)(|y_1 - y_2| + |u_1 - u_2|)$$

for all y_i, u_i such that $|y_i| \le K, |u_i| \le K$, and all $h \in G$. The same conditions as above are also satisfied by φ and ψ.

(A3) The functions $b(y,u,\cdot)$, $D_y b(y,u,\cdot)$ and $D_u b(y,u,\cdot)$ are Fréchet differentiable in h. Similar properties posses functions φ and ψ.

By the following lemma, proved in [6], problem (O_h) is well posed.

Lemma 1 *If* **(A1)** - **(A3)** *hold, then for any $u \in \mathcal{U}$ and any $h \in G$ there exists a unique weak solution $y(u,h) \in W^{1,2}(\Omega) \cap C(\bar{\Omega})$ of (3). Moreover, there exists $c > 0$ such that*

$$\|y(u',h') - y(u'',h'')\|_{C(\bar{\Omega})} \le c(\|u' - u''\|_{L^\infty(\Gamma)} + \|h' - h''\|_H). \tag{5}$$

Define the following Hamiltonian and Lagrangian

$$\mathcal{H} : I\!R^3 \times G \to I\!R, \quad \mathcal{L} : W^{1,2}(\Omega) \times L^\infty(\Gamma) \times W^{1,2}(\Omega) \times G \to I\!R,$$

$$\mathcal{H}(y,u,p,h) := \psi(y,u,h) + pb(y,u,h), \tag{6}$$

$$\begin{aligned} \mathcal{L}(y,u,p,h) &:= F(y,u,h) - \int_\Omega p(-\Delta y + y)dx \\ &= \int_\Omega [\varphi(y,h) - (\nabla p, \nabla y) - (p,y)]dx + \int_\Gamma \mathcal{H}(y,u,p,h)dS_x. \end{aligned} \tag{7}$$

We assume:

(A4) For a given reference value $h_0 \in G$ of the parameter, there exists a local solution $(y_0, u_0) \in Z^\infty$ of (O_{h_0}) and an associated state $p_0 \in W^{1,2}(\Omega) \cap C(\bar{\Omega})$, such that the following first-order necessary optimality conditions hold

$$D_y\mathcal{L}(y_0,u_0,p_0,h_0)z = 0 \qquad \text{for all } z \in W^{1,2}(\Omega), \tag{8}$$

$$D_u\mathcal{L}(y_0,u_0,p_0,h_0)(u-u_0) \ge 0 \qquad \text{for all } u \in \mathcal{U}. \tag{9}$$

In a standard way, conditions (8) and (9) yield the adjoint equation and the pointwise stationarity of the Hamiltonian:

$$\left.\begin{aligned} -\Delta p_0(x) + p_0(x) &= D_y\varphi(y_0(x),h_0), && \text{in } \Omega, \\ \partial_\nu p_0(x) &= D_y\mathcal{H}(y_0(x),u_0(x),p_0(x),h_0), && \text{on } \Gamma, \end{aligned}\right\} \tag{10}$$

$$\left.\begin{array}{c} D_u\mathcal{H}(y_0(x),u_0(x),p_0(x),h_0)(u-u_0(x))\geq 0 \\ \text{for all } u\in[m_1,m_2] \text{ and a.a. } x\in\Gamma. \end{array}\right\} \tag{11}$$

Conditions (10) and (11) together with the state equation (3) constitute the optimality system for (O_{h_0}). It will be convenient to rewrite this optimality system in the form of a generalized equation. To do that, define the spaces

$$\begin{array}{l} X^s := W^{1,s}(\Omega)\times L^s(\Gamma)\times W^{1,s}(\Omega), \\ \Delta^s = L^s(\Omega)\times L^s(\Gamma)\times L^s(\Omega)\times L^s(\Gamma)\times L^s(\Gamma),\ s\in[2,\infty] \end{array} \tag{12}$$

and the following set-valued mapping with closed graph:

$$\mathcal{N}(u)=\left\{\begin{array}{ll} \lambda\in\{L^\infty(\Gamma)\mid \int_\Gamma \lambda(v-u)\,dS_x\leq 0\quad \forall v\in\mathcal{U}\} & \text{if } u\in\mathcal{U}, \\ \emptyset & \text{if } u\notin\mathcal{U}. \end{array}\right. \tag{13}$$

Denote $\xi=(y,u,p)\in X^\infty$. Let the function $\mathcal{F}:\ X^\infty\times G\to\Delta^\infty$, as well as the multivalued mapping $\mathcal{T}:\ X^\infty\to 2^{\Delta^\infty}$ be defined as follows

$$\mathcal{F}(\xi,h)=\left[\begin{array}{ll} -\Delta y+y & \text{in } \Omega \\ \partial_\nu y-b(y,u,h) & \text{on } \Gamma \\ -\Delta p+p-D_y\varphi(y,h) & \text{in } \Omega \\ \partial_\nu p-D_y\mathcal{H}(y,u,p,h) & \text{on } \Gamma \\ D_u\mathcal{H}(y,u,p,h) & \text{on } \Gamma \end{array}\right],\quad \mathcal{T}(\xi)=\left[\begin{array}{c} \{0\} \\ \{0\} \\ \{0\} \\ \{0\} \\ \mathcal{N}(u) \end{array}\right]. \tag{14}$$

Then the optimality system (3), (10), (11) for (O_{h_0}) can be expressed in the form of the following generalized equation:

$$0\in\mathcal{F}(\xi_0,h_0)+\mathcal{T}(\xi_0). \tag{15}$$

3. Application of abstract theorems for generalized equations

We are going to investigate conditions under which there exists a neighborhood G_0 of h_0 such that, for each $h\in G_0$, the generalized equation

$$0\in\mathcal{F}(\xi,h)+\mathcal{T}(\xi) \tag{16}$$

has a locally unique solution $\xi_h=(y_h,u_h,p_h)$, which is Bouligand differentiable function of h. We will follow the same scheme as in [4, 5]. Namely the proof will be in two steps. First, we show existence, local uniqueness and Lipschitz continuity of the solutions to (16). In the second step, we use these properties, to show differentiability of the solutions. In both steps we need the following auxiliary generalized equation, obtained from (16) by linearization of $\mathcal{F}(\cdot, h_0)$ at the reference solution and by perturbation:

$$\delta \in \mathcal{F}(\xi_0, h_0) + D_\xi \mathcal{F}(\xi_0, h_0)(\zeta - \xi_0) + \mathcal{T}(\zeta), \tag{17}$$

where $\delta \in \Delta^\infty$ is the perturbation. Clearly, for $\delta = 0$, ξ_0 is a solution to (17). We will denote by $\mathcal{B}_\rho^X(x_0) := \{x \in X \mid \|x - x_0\|_X \leq \rho\}$ the closed ball of radius ρ centered at x_0 in a Banach space X.

The following Robinson's implicit function theorem (see, Theorem 2.1 and Corollary 2.2 in [7]) allows to deduce existence and local Lipschitz continuity of the solutions to the nonlinear generalized equation (16), from the same properties of the solutions to the linearized equation (17).

Theorem 1 *If there exist $\rho_1 > 0$ and $\rho_2 > 0$ such that, for each $\delta \in \mathcal{B}_{\rho_1}^{\Delta^\infty}(0)$ there is a unique solution ζ_δ in $\mathcal{B}_{\rho_2}^X(\xi_0)$ of (17), which is Lipschitz continuous in δ, then there exist $\sigma_1 > 0$ and $\sigma_2 > 0$ such that, for each $h \in \mathcal{B}_{\sigma_1}^H(h_0)$ there is a unique solution ξ_h in $\mathcal{B}_{\sigma_2}^X(\xi_0)$ of (16), which is Lipschitz continuous in h.*

Similarly, the following theorem due to Dontchev (see, Theorem 2.4 and Remark 2.6 in [2]) allows to reduce differentiability analysis for the solutions to (16) to the same analysis for the solutions to (17).

Theorem 2 *If the assumptions of Theorem 1 are satisfied and, in addition, the solutions $\zeta_\delta \in \mathcal{B}_{\rho_2}^X(\xi_0)$ of (17) are Bouligand differentiable functions of δ in a neighborhood of the origin, with the differential $(D_\delta \zeta_0; \eta)$, then the solutions ξ_h of (16) are Bouligand differentiable in a neighborhood of h_0. For a direction $g \in H$, the differential at h_0 is given by*

$$(D_h \xi_0; g) = (D_\delta \zeta_0; -D_h \mathcal{F}(\xi_0, h_0) g). \tag{18}$$

Remark 1 In Theorem 1, Lipschitz continuity of ζ and ξ is understood in the sense of that norm in the space X, in which $\mathcal{F}(\cdot, h)$ is differentiable. On the other hand, Theorem 2 remains true, if B-differentiability of ζ_δ is satisfied in a norm in the image space X *weaker* than that in which Lipschitz continuity in Theorem 1 holds (see, Remark 2.11 in [2]); e.g., in L^s, $(s < \infty)$, rather than in L^∞. This property will be used in Section 4.

In order to apply Theorems 1 and 2 to (O_h), we have to find the form of the linearization (17) of the optimality system (16), for $\mathcal{F}$ and $\mathcal{T}$ given in (14). To simplify notation, the functions evaluated at the reference point will be denoted by subscript "0", e.g., $\varphi_0 := \varphi(y_0, h_0)$, $\mathcal{H}_0 := \mathcal{H}(y_0, u_0, p_0, h_0)$. Moreover, we denote $\xi_0 := (y_0, u_0, p_0)$.

Let $\delta = (\delta^1, \delta^2, \delta^3, \delta^4, \delta^5) \in \Delta^\infty$ be a vector of perturbations. By simple calculations we obtain the following form of (17):

$$(\mathrm{LO}_\delta) \qquad \left.\begin{array}{l} -\Delta z + z = e^1 + \delta^1, \\ \partial_\nu z - D_y b_0 z = e^2 + \delta^2 + D_u b_0 v, \end{array}\right\} \tag{19}$$

$$\left.\begin{array}{l} -\Delta q + q = e^3 + \delta^3 + D^2_{yy}\varphi_0 z, \\ \partial_\nu q - D_y b_0 q = e^4 + \delta^4 + D^2_{yy}\mathcal{H}_0 z + D^2_{yu}\mathcal{H}_0 v, \end{array}\right\} \tag{20}$$

$$D^2_{uy}\mathcal{H}_0 z + D^2_{uu}\mathcal{H}_0 v + D_u b_0 q - e^5 - \delta^5 \in -\mathcal{N}(v), \tag{21}$$

where $e = (e^1, e^2, e^3, e^4, e^5) \in \Delta^\infty$ is a given vector.

Note that

$$(z_0, v_0, q_0) = (y_0, u_0, p_0) \tag{22}$$

is a solution to (LO_δ) for $\delta = 0$. An inspection shows that (LO_δ) can be treated as an optimality system for the following linear-quadratic *accessory problem*:

$$(\mathrm{LP}_\delta) \quad \begin{array}{l} \text{Find } (z_\delta, v_\delta) \in Z^\infty \text{ such that} \\ \mathcal{I}(z_\delta, v_\delta, \delta) = \min \mathcal{I}(z, v, \delta) \\ \text{subject to} \end{array}$$

$$\begin{array}{rll} -\Delta z(x) + z(x) & = \delta^1(x) & \text{in } \Omega, \\ \partial_\nu z(x) & = D_y b_0(x) z(x) + D_u b_0(x) v(x) & \\ & \quad + e^2(x) + \delta^2(x) & \text{on } \Gamma, \end{array} \tag{23}$$

$$v \in \mathcal{U},$$

where

$$\begin{aligned} \mathcal{I}(z, v, \delta) := {} & \tfrac{1}{2}((z, v), D^2\mathcal{L}_0(z, v)) + \textstyle\int_\Omega (e^3 + \delta^3) z\,dx + \\ & + \textstyle\int_\Gamma [(e^4 + \delta^4) z + (e^5 + \delta^5) v]\,dS_x, \end{aligned}$$

with the quadratic form

$$\begin{aligned} ((z, v), D^2\mathcal{L}_0(z, v)) := {} & \int_\Omega D^2_{yy}\varphi(y_0, h_0) z^2\,dx \\ & + \int_\Gamma [z, v] \begin{bmatrix} D^2_{yy}\mathcal{H}_0 & D^2_{yu}\mathcal{H}_0 \\ D^2_{uy}\mathcal{H}_0 & D^2_{uu}\mathcal{H}_0 \end{bmatrix} \begin{bmatrix} z \\ v \end{bmatrix} dS_x. \end{aligned} \tag{24}$$

To verify assumptions of Theorems 1 and 2, we have to show that there exist constants $\rho_1, \rho_2 > 0$ such that for each $\delta \in \mathcal{B}^{\Delta^\infty}_{\rho_1}(0)$ there is a unique stationary point $\zeta_\delta := (z_\delta, v_\delta, q_\delta)$ in $\mathcal{B}^{X^\infty}_{\rho_2}(\xi_0)$ of (LP_δ), which is a Lipschitz continuous and Bouligand differentiable function of δ.

4. Differentiability of solutions to accessory problems

As in [4], the starting point in the proof of differentiability of the solutions to (LP_δ) is the Lipschitz continuity property for these solutions. To this end, we will need a coercivity assumption (see, [6]). Let us define the sets of those points, at which the reference control is active:

$$I = \{x \in \Gamma \mid u_0(x) = m_1\}, \quad J = \{x \in \Gamma \mid u_0(x) = m_2\}. \tag{25}$$

Moreover, for any $\alpha \geq 0$ define the sets

$$\begin{aligned} I^\alpha &= \{x \in \Gamma \mid D_u\mathcal{H}(y_0, u_0, p_0, h_0)(x) > \alpha\}, \\ J^\alpha &= \{x \in \Gamma \mid -D_u\mathcal{H}(y_0, u_0, p_0, h_0)(x) > \alpha\}. \end{aligned} \tag{26}$$

As in [6], we assume:

(AC) *(coercivity)* There exist $\alpha > 0$ and $\gamma > 0$ such that

$$((z, v), D^2\mathcal{L}_0(z, v)) \geq \gamma\|v\|^2_{L^2(\Gamma)} \tag{27}$$

for all pairs (z, v) satisfying

$$\begin{aligned} -\Delta z(x) + z(x) &= 0 && \text{in } \Omega, \\ \partial_\nu z(x) - D_y b_0(x) z(x) - D_u b_0(x) v(x) &= 0 && \text{in } \Gamma, \end{aligned}$$

and such that $v \in \{L^2(\Gamma) \mid v(x) = 0 \text{ for a.a. } x \in I^\alpha \cup J^\alpha\}$.

Note that **(AC)** implies the following pointwise coercivity condition (see, e.g., Lemma 5.1 in [10]).

$$D^2_{uu}\mathcal{H}_0(x) \geq \gamma \quad \text{for a.a. } x \in \Gamma \setminus (I^\alpha \cup J^\alpha). \tag{28}$$

By a slight modification of Satz 18 in [11] we get the following Lipschitz continuity result for (LP_δ):

Proposition 1 *If* **(AC)** *holds, then there exist constants $\rho_1 > 0$ and $\rho_2 > 0$ such that, for all $\delta \in \mathcal{B}^{\Delta^\infty}_{\rho_1}(0)$ there is a unique stationary point $\zeta_\delta := (z_\delta, v_\delta, q_\delta)$ in $\mathcal{B}^{X^\infty}_{\rho_2}(\xi_0)$ of (LP_δ). Moreover, there exists a constant $\ell > 0$ such that*

$$\|z_{\delta'} - z_{\delta''}\|_{W^{1,s}(\Omega)}, \|v_{\delta'} - v_{\delta''}\|_{L^s(\Gamma)}, \|q_{\delta'} - q_{\delta''}\|_{W^{1,s}(\Omega)}, \leq \ell\,\|\delta' - \delta''\|_{\Delta^s}$$
for all $\delta', \delta'' \in \mathcal{B}^{\Delta^\infty}_{\rho_1}(0)$ and all $s \in [2, \infty]$

(29)

The proof of B-differentiability of the stationary points of (LP_δ) is in two steps. In the first step, directional differentiability is proved and the directional differential is characterized. This characterization is used in the second step to show that the differential is actually Bouligand. Let us start with the directional differentiability. The proof of the following result is very similar to that of Proposition 4.3 in [5].

Proposition 2 *Let* **(A1)-(A3)** *as well as* **(AC)** *be satisfied and let* $\rho_1, \rho_2 > 0$ *be as in Proposition 1. Then the mapping*

$$\zeta_\delta := (z_\delta, v_\delta, q_\delta) : \mathcal{B}_{\rho_1}^{\Delta^\infty}(0) \to X^2,$$

where $\zeta_\delta \in \mathcal{B}_{\rho_2}^{X^\infty}(\xi_0)$ *denote a unique stationary point of* (LP_δ)*, is directionally differentiable. The directional differential at* $\delta = 0$*, in a direction* $\eta \in \Delta^\infty$*, is given by* $(\varpi_\eta, w_\eta, r_\eta)$*, where* (ϖ_η, w_η) *is the solution and* r_η *the associated adjoint state of the following linear-quadratic optimal control problem:*

(LQ_η) *Find* $(\varpi_\eta, w_\eta) \in W^{1,2}(\Omega) \times L^2(\Gamma)$ *that minimizes*

$$\mathcal{J}_\eta(\varpi, w) = \tfrac{1}{2}((\varpi, w), D^2\mathcal{L}_0(\varpi, w)) + \int_\Omega \eta^3 \varpi \, dx + \int_\Gamma (\eta^4 \varpi + \eta^5 w) \, dS_x \tag{30}$$

subject to

$$\begin{aligned} -\Delta\varpi + \varpi &= \eta^1 && \text{in } \Omega, \\ \partial_\nu \varpi &= D_y b_0 \varpi + D_u b_0 w + \eta^2 && \text{on } \Gamma, \end{aligned} \tag{31}$$

$$w(x) \begin{cases} = 0 & \text{for } x \in (I^0 \cup J^0), \\ \geq 0 & \text{for } x \in (I \setminus I^0), \\ \leq 0 & \text{for } x \in (J \setminus J^0), \\ \text{free} & \text{for } x \in \Gamma \setminus (I \cup J). \end{cases} \tag{32}$$

Note that, by the same argument as in Proposition 1, we find that the stationary points of (LQ_η) are Lipschitz continuous functions of η. Since $(\varpi_0, w_0, r_0) = (0, 0, 0)$, we have

$$\|\varpi_\eta\|_{W^{1,s}(\Omega)}, \ \|w_\eta\|_{L^s(\Gamma)}, \ \|r_\eta\|_{W^{1,s}(\Omega)} \leq \ell \|\eta\|_{\Delta^s}, \quad s \in [2, \infty]. \tag{33}$$

We are now going to show that (ϖ_η, w_η) and r_η are actually B-differentials at $\delta = 0$ of (z_δ, v_δ) and q_δ, respectively.

Theorem 3 *Let* **(A1)-(A3)** *as well as* **(AC)** *be satisfied and let* $\rho_1, \rho_2 > 0$ *be as in Proposition 1. Then the mapping*

$$\zeta_\delta := (z_\delta, v_\delta, q_\delta) : \mathcal{B}_{\rho_1}^{\Delta^\infty}(0) \to X^s, \tag{34}$$

where $\zeta_\delta \in \mathcal{B}_{\rho_2}^{X^\infty}(\xi_0)$ *denote a unique stationary point of* (LP_δ), *is B-differentiable for any* $s \in [2,\infty)$. *The B-differential at* $\delta = 0$ *in a direction* $\eta \in \Delta$ *is given by* $\vartheta_\eta := (\varpi_\eta, w_\eta, r_\eta)$, *where* (ϖ_η, w_η) *is the solution and* r_η *the associated adjoint state of problem* (LQ_η).

Proof The optimality system for (LQ_η) takes the form:

$$\left.\begin{aligned} &-\Delta\varpi + \varpi = \eta^1, \\ &\partial_\nu\varpi - D_y b_0 \varpi = \eta^2 + D_u b_0 w, \end{aligned}\right\} \tag{35}$$

$$\left.\begin{aligned} &-\Delta r + r = \eta^3 + D^2_{yy}\varphi_0\varpi, \\ &\partial_\nu r - D_y b_0 r = \eta^4 + D^2_{yy}\mathcal{H}_0\varpi + D^2_{yu}\mathcal{H}_0 w. \end{aligned}\right\} \tag{36}$$

$$\left.\begin{aligned} &(D^2_{uy}\mathcal{H}_0\,\varpi + D^2_{uu}\mathcal{H}_0 w \; - D_u b_0\, r - \eta^5, v - w) \ge 0 \\ &\text{for all } v \in L^2(Q) \text{ satisfying (32).} \end{aligned}\right\} \tag{37}$$

We have to show that the solution $(\varpi_\eta, w_\eta, r_\eta)$ of (35)-(37) are B-differentials of the solution to (LO_δ). Clearly, $(\varpi_\eta, w_\eta, r_\eta)$ is a positively homogeneous function of η, so, by Definition 1, it is enough to show that

$$z_\eta = z_0 + \varpi_\eta + \sigma_1(\eta), \quad v_\eta = v_0 + w_\eta + \sigma_2(\eta), \quad q_\eta = q_0 + r_\eta + \sigma_1(\eta),$$

$$\text{where } \frac{\|\sigma_1(\eta)\|_{W^{1,s}(\Omega)}}{\|\eta\|_{\Delta^\infty}} \to 0, \quad \frac{\|\sigma_2(\eta)\|_{L^s(Q)}}{\|\eta\|_{\Delta^\infty}} \to 0, \quad \text{as } \|\eta\|_{\Delta^\infty} \to 0,$$

for any $s \in [2,\infty)$. (38)

Denote

$$(z_\eta - z_0) = \tilde{\varpi}_\eta, \quad (v_\eta - v_0) = \tilde{w}_\eta, \quad (q_\eta - q_0) = \tilde{r}_\eta. \tag{39}$$

It follows from (19) and (20) that $(\tilde{\varpi}_\eta, \tilde{w}_\eta, \tilde{r}_\eta)$ satisfies equations identical with (35) and (36):

$$\left.\begin{aligned} &-\Delta\tilde{\varpi} + \tilde{\varpi} = \eta^1, \\ &\partial_\nu\tilde{\varpi} - D_y b_0 \tilde{\varpi} = \eta^2 + D_u b_0 w, \end{aligned}\right\} \tag{40}$$

$$\left.\begin{aligned} &-\Delta\tilde{r} + \tilde{r} = \eta^3 + D^2_{yy}\varphi_0\tilde{\varpi}, \\ &\partial_\nu\tilde{r} - D_y b_0 \tilde{r} = \eta^4 + D^2_{yy}\mathcal{H}_0\tilde{\varpi} + D^2_{yu}\mathcal{H}_0 w. \end{aligned}\right\} \tag{41}$$

To characterize $(\tilde{\varpi}_\eta, \tilde{w}_\eta, \tilde{r}_\eta)$, we still need a condition analogous to (37). To this end, let us choose $\beta \in (0,\alpha)$, where α is given in **(AC)**. Define the sets

$$\begin{aligned} K_1^\beta &= \{x \in I^0 \mid D^2_{uu}\mathcal{H}_0(x) \in (0,\beta)\}, \\ K_2^\beta &= \{x \in J^0 \mid -D^2_{uu}\mathcal{H}_0(x) \in (0,\beta)\}, \\ L^\beta &= \{x \in \Gamma \mid u_0(x) \in (m_1, m_1+\beta) \cup (m_2-\beta, m_2)\}. \end{aligned} \tag{42}$$

Note that

$$\text{meas } (K_1^\beta \cup K_2^\beta \cup L^\beta) \to 0 \quad \text{as } \beta \to 0. \tag{43}$$

Let us split up the set Γ into the following subsets

$$\mathcal{A} = \Gamma \setminus (I \cup J \cup L^\beta), \qquad \mathcal{B} = (I^0 \setminus K_1^\beta) \cup (J^0 \setminus K_2^\beta),$$
$$\mathcal{C} = (I \setminus I^0) \cup (J \setminus J^0), \qquad \mathcal{D} = K_1^\beta \cup K_2^\beta \cup L^\beta.$$

We will analyze conditions analogous to (37) on each of these subsets successively.

Subset $\mathcal{A}$ Choose $\varrho(\beta) = \ell^{-1}\beta$. Then by (22) and (25), as well as by Proposition 1, for all $\eta \in \mathcal{B}^{\Delta^\infty}_{\varrho(\beta)}(0)$ we get

$$v_\eta(x) \in (m_1, m_2) \qquad \text{for a.a. } x \in \mathcal{A}, \tag{44}$$

i.e., by (21)

$$\begin{aligned} D^2_{uy}\mathcal{H}_0(x)\, z_\eta(x) + D^2_{uu}\mathcal{H}_0(x) v_\eta(x) \, + D_u b_0(x,t)\, q_\eta(x,t) \\ -e^5(x) - \eta^5(x) = 0 \qquad \text{for a.a. } x \in \mathcal{A}. \end{aligned} \tag{45}$$

Subtracting from (45) the analogous equation for (z_0, v_0, q_0) and using notation (39), we obtain

$$\begin{aligned} D^2_{uy}\mathcal{H}_0(x)\, \widetilde{\varpi}_\eta(x) + D^2_{uu}\mathcal{H}_0(x)\widetilde{w}_\eta(x) \, - D_u a_0(x)\, \widetilde{r}_\eta(x) \\ -\eta^5(x) = 0 \qquad \text{for a.a. } x \in \mathcal{A}. \end{aligned} \tag{46}$$

Subset $\mathcal{B}$ It follows from Proposition 1 that, shrinking $\varrho(\beta) > 0$ if necessary, for all $\eta \in \mathcal{B}^{\Delta^\infty}_{\varrho(\beta)}(0)$ we obtain

$$D^2_{uy}\mathcal{H}_0(x)\, z_\eta(x) + D^2_{uu}\mathcal{H}_0(x) v_\eta(x) \, + D_u b_0(x)\, q_\eta(x)$$
$$-e^5(x) - \eta^5(x) \begin{cases} > 0 & \text{for a.a. } x \in I^0 \setminus K_1^\beta, \\ < 0 & \text{for a.a. } x \in J^0 \setminus K_2^\beta, \end{cases} \tag{47}$$

which, by (21) implies

$$v_\eta(x) = \begin{cases} m_1(x) & \text{for a.a. } x \in I^0 \setminus K_1^\beta, \\ m_2(x) & \text{for a.a. } x \in J^0 \setminus K_2^\beta, \end{cases}$$

i.e.,

$$\widetilde{w}_\eta(x) = 0 \qquad \text{for a.a. } x \in \mathcal{B}. \tag{48}$$

Subset $\mathcal{C}$ By (22) and (25) we have

$$v_0(x) = u_0(x) = \begin{cases} m_1(x) & \text{for a.a. } x \in I \setminus I^0, \\ m_2(x) & \text{for a.a. } x \in J \setminus J^0 \end{cases} \tag{49}$$

and

$$D^2_{uy}\mathcal{H}_0(x)\, z_0(x) + D^2_{uu}\mathcal{H}_0(x) v_0(x) \, + D_u b_0(x)\, q_0(x) = 0 \quad \text{for a.a. } x \in (I \setminus I^0) \cup (J \setminus J^0). \tag{50}$$

Proposition 1, together with (49) implies that, shrinking $\varrho(\beta)$ if necessary, for any $\eta \in \mathcal{B}^{\Delta^\infty}_{\varrho(\beta)}(0)$ we get

$$v_\eta(x) \in \begin{cases} [m_1, m_2) & \text{for a.a. } x \in I \setminus I^0, \\ (m_1, m_2] & \text{for a.a. } x \in J \setminus J^0. \end{cases} \tag{51}$$

Hence, in view of (21) we have

$$D^2_{uy}\mathcal{H}_0(x)\, z_\eta(x) + D^2_{uu}\mathcal{H}_0(x)\, v_\eta(x) \, + D_u b_0(x)\, q_\eta(x) \\ -\eta^5(x) \begin{cases} \geq 0 & \text{for a.a. } x \in I \setminus I^0, \\ \leq 0 & \text{for a.a. } x \in J \setminus J^0. \end{cases} \tag{52}$$

Conditions (49)–(52) imply:

$$\widetilde{w}_\eta(x) \begin{cases} \geq 0 & \text{for a.a. } x \in I \setminus I^0, \\ \leq 0 & \text{for a.a. } x \in J \setminus J^0 \end{cases}, \tag{53}$$

$$D^2_{uy}\mathcal{H}_0(x)\, \widetilde{\varpi}_\eta(x) + D^2_{uu}\mathcal{H}_0(x)\, \widetilde{w}_\eta(x) \, + D_u b_0(x)\, \widetilde{r}_\eta(x) \\ -\eta^5(x) \begin{cases} \geq 0 & \text{for a.a. } x \in I \setminus I^0, \\ \leq 0 & \text{for a.a. } x \in J \setminus J^0, \end{cases} \tag{54}$$

and

$$(D^2_{uy}\mathcal{H}_0(x)\, \widetilde{\varpi}_\eta(x) + D^2_{uu}\mathcal{H}_0(x)\, \widetilde{w}_\eta(x) \, + D_u b_0(x)\, \widetilde{r}_\eta(x) \\ -\eta^3(x))(w - \widetilde{w}_\eta(x)) \geq 0 \begin{cases} \text{for all } w \geq 0 & \text{on } I \setminus I^0, \\ \text{for all } w \leq 0 & \text{on } J \setminus J^0. \end{cases} \tag{55}$$

<u>Subset $\mathcal{D}$</u> The analysis of subset $\mathcal{D}$ is the most difficult, because we do not know a priori if for $x \in \mathcal{D}$ the constraints are active or not at v_η, no matter how small η is chosen. Without this information, we can say very few about $\widetilde{w}_\eta(x) = v_\eta(x) - v_0(x)$. Let us denote

$$(\widetilde{\eta}^5)'(x) = D^2_{uy}\mathcal{H}_0(x)\, (z_\eta(x) - z_0(x)) + D^2_{uu}\mathcal{H}_0(x)(v_\eta(x) - v_0(x)) \\ +D_u b_0(x)\, (q_\eta(x) - q_0(x)) \text{ for a.a. } x \in \mathcal{D}. \tag{56}$$

By definition (39) we have

$$D^2_{uy}\mathcal{H}_0(x)\, \widetilde{\varpi}_\eta(x) + D^2_{uu}\mathcal{H}_0(x)\widetilde{v}_\eta(x) \, + D_u b_0(x)\, \widetilde{r}_\eta(x) \\ -(\widetilde{\eta}^5)'(x) = 0 \quad \text{for a.a. } x \in \mathcal{D}. \tag{57}$$

Denote $\eta' = (\eta^1, \eta^2, \eta^3, \eta^4, (\eta^5)')$, where

$$(\eta^5)'(x) = \begin{cases} (\widetilde{\eta}^5)'(x) & \text{for } x \in \mathcal{D}, \\ \eta^5(x) & \text{otherwise.} \end{cases} \tag{58}$$

It is easy to see that (40) and (41) together with (46), (48), (53)–(55) and (57) can be interpreted as an optimality system for the optimal control problem $(\widetilde{\mathrm{LQ}}_{\eta'})$, where $(\widetilde{\mathrm{LQ}}_{\eta})$ is the following slight modification of (LQ_{η}):

$(\widetilde{\mathrm{LQ}}_{\eta})$ Find $(\widetilde{\varpi}_\eta, \widetilde{w}_\eta) \in W^2 \times L^2(\Gamma)$ that minimizes $\mathcal{F}_\eta(\varpi, w)$ subject to

$$-\Delta\varpi(x) + \varpi(x) = \eta^1(x) \quad \text{in}\,\Omega,$$

$$\partial_\nu \varpi + D_y b_0\, \varpi = D_y b_0(x)\varpi(x) + D_u b_0(x) w(x) + \eta^2 \quad \text{on}\,\Gamma,$$

$$w(x) \begin{cases} = 0 & \text{for } x \in (I^0 \setminus K_1^\beta) \cup (J^0 \setminus K_2^\beta), \\ \geq 0 & \text{for } x \in (I \setminus I^0), \\ \leq 0 & \text{for } x \in (J \setminus J^0), \\ \text{free} & \text{for } x \in \Gamma \setminus (I \cup J)) \cup (K_1^\beta \cup K_2^\beta). \end{cases}$$

Similarly $(\varpi_\eta, w_\eta, r_\eta)$ can be interpreted as a stationary point of $(\widetilde{\mathrm{LQ}}_{\eta''})$, where $\eta'' = (\eta^1, \eta^2, \eta^3, \eta^4, (\eta^5)'')$, with

$$(\eta^5)''(x) = \begin{cases} (\widetilde{\eta}^5)''(x) & \text{for } x \in \mathcal{D}, \\ \eta^5(x) & \text{otherwise,} \end{cases}$$

$$(\widetilde{\eta}^5)''(x) = D^2_{uy}\mathcal{H}_0(x)\,\varpi_\eta(x) + D^2_{uu}\mathcal{H}_0(x) w_\eta(x) + D_u b_0(x)\, r_\eta(x). \tag{59}$$

It can be easily checked that, as in the case of (LQ_η), the stationary points of $(\widetilde{\mathrm{LQ}}_{\eta})$ are Lipschitz continuous functions of η. Hence, in view of (58) and (59), we have

$$\|\widetilde{\varpi}_\eta - \varpi_\eta\|_{W^{1,s}(\Omega)}, \|\widetilde{w}_\eta - w_\eta\|_{L^s(\Gamma)}, \|\widetilde{r}_\eta - r_\eta\|_{W^{1,s}(\Omega)}$$

$$\leq \ell\, \|\eta' - \eta''\|_{\Delta^s} = \ell \left\{ \int\limits_{K_1^\beta \cup K_2^\beta \cup L^\beta} |(\widetilde{\eta}^5)'(x) - (\widetilde{\eta}^5)''(x)|^s dS_x \right\}^{\frac{1}{s}} \tag{60}$$

Using the definitions (56), (59) and taking advantage of (29) and of (33) we get

$$\begin{aligned}
&|(\tilde{\eta}^5)'(x) - (\tilde{\eta}^5)''(x)| \le |(\tilde{\eta}^5)'(x)| + |(\tilde{\eta}^5)''(x)| \\
&= |D^2_{uy}\mathcal{H}_0(x)\,(z_\eta(x) - z_0(x)) + D^2_{uu}\mathcal{H}_0(x)(v_\eta(x) - v_0(x)) \\
&\qquad + D_u b_0(x)\,(q_\eta(x) - q_0(x))| \\
&\qquad + |D^2_{uy}\mathcal{H}_0(x)\,\varpi_\eta(x) + D^2_{uu}\mathcal{H}_0(x) v_\eta(x) + D_u b_0(x)\, r_\eta(x)| \\
&\qquad \le c\,\|\eta\|_{\Delta^\infty} \quad \text{for a.a. } x \in K_1^\beta \cup K_2^\beta \cup L^\beta.
\end{aligned} \tag{61}$$

Substituting (61) to (60) we obtain

$$\begin{aligned}
&\|\tilde{\varpi}_\eta - \varpi_\eta\|_{W^{1,s}(\Omega)}, \|\tilde{w}_\eta - w_\eta\|_{L^s(\Gamma)}, \|\tilde{r}_\eta - r_\eta\|_{W^{1,s}(\Omega)} \\
&\qquad \le c\|\eta\|_{\Delta^\infty} \left\{\text{meas } (K_1^\beta \cup K_2^\beta \cup L^\beta)\right\}^{\frac{1}{s}}.
\end{aligned} \tag{62}$$

In view of (39) and(43), we find that for any $\epsilon > 0$ and any $s \in [2, \infty)$ we can choose $\beta(\epsilon, s) > 0$ and the corresponding $\varrho(\beta(\epsilon, s))$, so small that

$$\begin{aligned}
&\|z_\eta - z_0 - \varpi_\eta\|_{W^{1,s}(\Omega)}, \|v_\eta - v_0 - w_\eta\|_{L^s(\Gamma)}, \|q_\eta - q_0 - r_\eta\|_{W^{1,s}(\Omega)} \\
&\qquad \le \epsilon\,\|\eta\|_{\Delta^\infty} \quad \text{for all } \eta \in \mathcal{B}^{\Delta^\infty}_{\varrho(\beta(\epsilon,s))}(0).
\end{aligned}$$

This shows that (38) holds and completes the proof of the theorem.

Remark 2 The proof of Theorem 3 cannot be repeated for $s = \infty$ and the counterexample in [4] shows that B-differentiability of (34) cannot be expected for $s = \infty$.

5. Differentiability of the solutions to nonlinear problems

By Theorems 2 and 3, for any h in a neighborhood of h_0, (O_h) has a unique stationary point (y_h, u_h, p_h), which is a B-differentiable function of h. On the other hand, by Theorem 3.7 in [6], for h sufficiently close to h_0, condition **(AC)** implies that (y_δ, u_δ) is a solution to (O_h). Thus, we obtain the following principal result of this paper:

Theorem 4 *If* **(A1)**–**(A7)** *and* **(AC)** *hold, then there exist constants* $\sigma_1, \sigma_2 > 0$ *such that, for any* $h \in \mathcal{B}^H_{\sigma_1}(h_0)$*, there is a unique stationary point* (y_h, u_h, p_h) *in* $\mathcal{B}^{X^\infty}_{\sigma_2}(\xi_0)$ *of* (O_h)*, where* (y_h, u_h) *is a solution of* (O_h)*. The mapping*

$$(y_h, u_h, p_h) :\ \mathcal{B}^H_{\sigma_1}(h_0) \to X^s, \quad s \in [2, \infty) \tag{63}$$

is B-differentiable, and the B-differential evaluated at h_0 in a direction $g \in H$ is given by the solution and adjoint state of the following linear-quadratic optimal control problem

$$(\mathrm{L}_g) \qquad \text{Find } (z_g, v_g) \in W^2 \times L^2(\Gamma) \text{ that minimizes}$$

$$\mathcal{K}_g(z,v) = \tfrac{1}{2}((z,v), D^2\mathcal{L}_0(z,v)) + \int_\Omega D^2_{yh}\varphi_0 g z\, dx + \int_\Gamma D^2_{yh}\mathcal{H}_0 g z\, dS_x + \int_\Gamma D^2_{uh}\mathcal{H}_0 g v\, dS_x$$

subject to

$$\begin{aligned} -\Delta z + z &= 0 && \text{in } \Omega, \\ \partial_\nu z &= D_y b_0\, z + D_u b_0\, v + D_h b_0\, g && \text{on } \Gamma, \end{aligned}$$

and

$$v(x) \begin{cases} = 0 & \text{for } x \in (I^0 \cup J^0), \\ \geq 0 & \text{for } x \in (I \setminus I^0), \\ \leq 0 & \text{for } x \in (J \setminus J^0), \\ \text{free} & \text{for } x \in \Gamma \setminus (I \cup J)). \end{cases}$$

As it was noticed in Introduction, Bouligand differential becomes Fréchet if it is linear. Hence from the form of (L_g), we obtain immediately:

Corollary 1 *If* meas $(I \setminus I^0)$ = meas $(J \setminus J^0) = 0$, *then the mapping (63) is Fréchet differentiable.*

In sensitivity analysis of optimization problems an important role is played by the so-called optimal value function, which on $\mathcal{B}^H_{\sigma_1}(h_0)$ is defined by:

$$\mathcal{F}^0(h) := \mathcal{F}_h(y_h, u_h),$$

i.e., to each $h \in \mathcal{B}^H_{\sigma_1}(h_0)$, $\mathcal{F}^0$ assigns the (local) optimal value of the cost functional. In exactly the same way as in Corollary 5.3 in [5], we obtain the following result showing that Bouligand differentiability of the solutions implies the second order expansion of $\mathcal{F}_0$, *uniform* in a neighborhood of h_0.

Corollary 2 *If assumptions of Theorem 4 hold, then for each $h = h_0 + g \in \mathcal{B}^H_{\sigma_1}(h_0)$*

$$\begin{aligned} \mathcal{F}^0(h) &= \mathcal{F}^0(h_0) + (D_h\mathcal{L}_0, g) \\ &+ \tfrac{1}{2}\left((z_g, v_g, g), \begin{pmatrix} D^2_{yy}\mathcal{L}_0 & D^2_{yu}\mathcal{L}_0 & D^2_{yh}\mathcal{L}_0 \\ D^2_{uy}\mathcal{L}_0 & D^2_{uu}\mathcal{L}_0 & D^2_{uh}\mathcal{L}_0 \\ D^2_{hy}\mathcal{L}_0 & D^2_{hu}\mathcal{L}_0 & D^2_{hh}\mathcal{L}_0 \end{pmatrix} (z_g, v_g, g)\right) \\ &+ o(\|g\|^2_H), \end{aligned} \qquad (64)$$

where (z_g, v_g) *is the B-differential of* (y_h, u_h) *at* h_0 *in the direction* g, *i.e., it is given by the solution to* (L_g).

References

[1] Bonnans, J.F. (1998). "Second order analysis for control constrained optimal control problems of semilinear elliptic systems", *Appl. Math. Optim.*, **38**, 303–325.

[2] Dontchev, A.L. (1995). "Characterization of Lipschitz stability in optimization", In: R.Lucchetti, J.Revalski eds., *Recent Developments in Well-Posed Variational Problems*, Kluwer, pp. 95–116.

[3] Dontchev A.L. (1995). "Implicit function theorems for generalized equations", *Math. Program.,* **70** , 91-106.

[4] Malanowski, K. (2001). "Bouligand differentiability of solutions to parametric optimal control problems", *Num. Funct. Anal. and Optim*, **22**, 973–990.

[5] Malanowski, K. (2002). "Sensitivity analysis for parametric optimal control of semilinear parabolic equations", *J.Convex Anal.,* **9**, 543–561 .

[6] Malanowski, K. and Tröltzsch, F. (2000). "Lipschitz stability of solutions to parametric optimal control problems for elliptic equations", *Control Cybern.,* **29**, 237–256.

[7] Robinson, S.M. (1980). "Strongly regular generalized equations", *Math. Oper. Res.*, **5**, 43–62.

[8] Robinson, S.M. (1987). "Local structure of feasible sets in nonlinear programming, Part III: Stability and sensitivity", *Math. Program. Study,* **30**, 97-116.

[9] Shapiro, A. (1990). "On concepts of directional differentiability", *J. Math. Anal. Appl.,* **66**, 477-487.

[10] Tröltzsch, F. (2000). "Lipschitz stability of solutions to linear-quadratic parabolic control problems with respect to perturbations", *Discr. Cont. Dynam. Systems* **6**, 289–306.

[11] Unger, A. (1997). *Hinreichende Optimalitätsbedingungen 2. Ordnung und Konvergenz des SQP-Verfahrens für semilineare elliptische Randsteuerprobleme.* Ph. D. Thesis, Technische Universität Chemnitz–Zwickau.

SHAPE OPTIMIZATION FOR DYNAMIC CONTACT PROBLEMS WITH FRICTION

A. Myśliński
System Research Institute
Polish Academy of Sciences
01 - 447 Warsaw, ul. Newelska 6, Poland
myslinsk@ibspan.waw.pl

Abstract The paper deals with shape optimization of dynamic contact problem with Coulomb friction for viscoelastic bodies. The mass nonpenetrability condition is formulated in velocities. The friction coefficient is assumed to be bounded. Using material derivative method as well as the results concerning the regularity of solution to dynamic variational inequality the directional derivative of the cost functional is calculated and necessary optimality condition is formulated.

Keywords: Dynamic unilateral problem, shape optimization, sensitivity analysis, necessary optimality condition

1. Introduction

This paper deals with formulation of a necessary optimality condition for a shape optimization problem of a viscoelastic body in unilateral dynamic contact with a rigid foundation. It is assumed that the contact with given friction, described by Coulomb law [2], occurs at a portion of the boundary of the body. The contact condition is described in velocities. This first order approximation seems to be physically realistic for the case of small distance between the body and the obstacle and for small time intervals. The friction coefficient is assumed to be bounded. The equilibrium state of this contact problem is described by an hyperbolic variational inequality of the second order [2, 3, 5, 7, 17].

The shape optimization problem for the elastic body in contact consists in finding, in a contact region, such shape of the boundary of the domain occupied by the body that the normal contact stress is minimized. It is assumed that the volume of the body is constant.

Shape optimization of static contact problems was considered, among others, in [3, 8, 9, 10, 11, 16]. In [3, 8] the existence of optimal solutions and convergence of finite–dimensional approximation was shown. In [9, 10, 11, 16] necessary optimality conditions were formulated using the material derivative approach (see [16]). Numerical results are reported in [3, 11].

In this paper we shall study this shape optimization problem for a viscoelastic body in unilateral dynamical contact. The essential difficulty to deal with the shape optimization problem for dynamic contact problem is regularity of solutions to the state system. Assuming small friction coefficient and suitable regularity of data it can be shown [6, 7] that the solution to dynamic contact problem is enough regular to differentiate it with respect to parameter. Using the material derivative method [16] as well as the results of regularity of solutions to the dynamic variational inequality [6, 7] we calculate the directional derivative of the cost functional and we formulate necessary optimality condition for this problem. The present paper extends the authors' results contained in [12].

We shall use the following notation : $\Omega \in R^2$ will denote the bounded domain with Lipschitz continuous boundary Γ. The time variable will be denoted by t and the time interval $I = (0, \mathcal{T}), \mathcal{T} > 0$. By $H^k(\Omega)$, $k \in (0, \infty)$ we will denote the Sobolev space of functions having derivatives in all directions of the order k belonging to $L^2(\Omega)$ [1]. For an interval I and a Banach space B $L^p(I; B)$, $p \in (1, \infty)$ denotes the usual Bochner space [2]. $u_t = du/dt$ and $u_{tt} = d^2u/dt^2$ will denote first and second order derivatives, respectively, with respect to t of function u. u_{tN} and u_{tT} will denote normal and tangential components, respectively, of function u_t. $Q = I \times \Omega$, $\gamma_i = I \times \Gamma_i$, $i = 1, 2, 3$ where Γ_i are pieces of the boundary Γ.

2. Contact problem formulation

Consider deformations of an elastic body occupying domain $\Omega \in R^2$. The boundary Γ of domain Ω is Lipschitz continuous. The body is subjected to body forces $f = (f_1, f_2)$. Moreover surface tractions $p = (p_1, p_2)$ are applied to a portion Γ_1 of the boundary Γ. We assume that the body is clamped along the portion Γ_0 of the boundary Γ and that the contact conditions are prescribed on the portion Γ_2 of the boundary Γ. Moreover $\Gamma_i \cap \Gamma_j = \emptyset$, $i \neq j, i, j = 0, 1, 2$, $\Gamma = \Gamma_0 \cup \Gamma_1 \cup \Gamma_2$.

We denote by $u = (u_1, u_2)$, $u = u(t, x)$, $x \in \Omega$, $t \in [0, \mathcal{T}]$, $\mathcal{T} > 0$ the displacement of the body and by $\sigma = \{\sigma_{ij}(u(t, x))\}, i, j = 1, 2$, the stress field in the body. We shall consider elastic bodies obeying Hooke's law

[2, 3, 5, 17] :

$$\sigma_{ij}(u) = c^0_{ijkl}(x)e_{kl}(u) + c^1_{ijkl}(x)e_{kl}(u_t) \ \ x \in \Omega, \quad e_{kl} = \frac{1}{2}(u_{k,l} + u_{l,k}) \quad (1)$$

$i, j, k, l = 1, 2$, $u_{k,l} = \partial u_k / \partial x_l$. We use here the summation convention over repeated indices [2]. $c^0_{ijkl}(x)$ and $c^1_{ijkl}(x)$, $i, j, k, l = 1, 2$ are components of Hooke's tensor. It is assumed that elasticity coefficients c^0_{ijkl} and c^1_{ijkl} satisfy usual symmetry, boundedness and ellipticity conditions [2, 3, 5]. In an equilibrium state a stress field σ satisfies the system [2, 3, 6, 7] :

$$u_{tti} - \sigma_{ij}(x)_{,j} = f_i(x), \ (t, x) \in (0, \mathcal{T}) \times \Omega \ \ i, j = 1, 2 \qquad (2)$$

where $\sigma_{ij}(x)_{,j} = \partial \sigma_{ij}(x)/\partial x_j$, $i, j = 1, 2$. There are given the following boundary conditions :

$$u_i(x) = 0 \ \text{ on } (0, \mathcal{T}) \times \Gamma_0 \quad i = 1, 2,$$

$$\sigma_{ij}(x) n_j = p_i \text{ on } (0, \mathcal{T}) \times \Gamma_1 \quad i, j = 2; \qquad (3)$$

$$u_{tN} \le 0, \ \sigma_N \le 0, \ u_{tN}\sigma_N = 0, \qquad \text{on } (0, \mathcal{T}) \times \Gamma_2; \qquad (4)$$

$$u_{tT} = 0 \quad \Rightarrow \ | \sigma_T | \le \mathcal{F} | \sigma_N |;$$

$$u_{tT} \ne 0 \quad \Rightarrow \quad \sigma_T = -\mathcal{F} | \sigma_N | \frac{u_{tT}}{| u_{tT} |}. \qquad (5)$$

Here we denote : $u_N = u_i n_i$, $\sigma_N = \sigma_{ij} n_i n_j$, $(u_T)_i = u_i - u_N n_i$, $(\sigma_T)_i = \sigma_{ij} n_j - \sigma_N n_i$ $i, j = 1, 2$, $n = (n_1, n_2)$ is the unit outward vector to the boundary Γ. There are given the following initial conditions:

$$u_i(0, x) = u_0 \quad u_{ti}(0, x) = u_1, \quad i = 1, 2, \quad x \in \Omega. \qquad (6)$$

We shall consider problem (2)–(6) in the variational form. Let us assume,

$$f \in H^{1/4}(I; (H^1(\Omega; R^2))^*) \cap L^2(Q; R^2),$$

$$p \in L^2(I; (H^{1/2}(\Gamma_1; R^2))^*),$$

$$u_0 \in H^{3/2}(\Omega; R^2) \quad u_1 \in H^{3/2}(\Omega; R^2), \quad u_{1|\Gamma_2} = 0, \qquad (7)$$

$$\mathcal{F} \in L^\infty(\Gamma_2; R^2) \quad \mathcal{F}(., x) \text{ is continuous for a.e. } x \in \Gamma_2$$

be given. The space $L^2(Q; R^2)$ and the Sobolev spaces $H^{1/4}(I; (H^1(\Omega; R^2))^*)$ as well as $(H^{1/2}(\Gamma_1); R^2)$ are defined in [1, 2]. Let us introduce :

$$F = \{z \in L^2(I; H^1(\Omega; R^2)) \ : \ z_i = 0 \ \text{ on } (0, \mathcal{T}) \times \Gamma_0 \ , i = 1, 2\} \qquad (8)$$

$$K = \{z \in F \; : \; z_{tN} \leq 0 \quad \text{on } (0, \mathcal{T}) \times \Gamma_2 \}. \tag{9}$$

The problem (1) - (6) is equivalent to the following variational problem [6, 7]: find $u \in L^\infty(I; H^1(\Omega; R^2)) \cap H^{1/2}(I; L^2(\Omega; R^2)) \cap K$ such that $u_t \in L^\infty(I; L^2(\Omega; R^2)) \cap H^{1/2}(I; L^2(\Omega; R^2)) \cap K$ and $u_{tt} \in L^\infty(I; H^{-1}(\Omega; R^2)) \cap (H^{1/2}(I; L^2(\Omega; R^2)))^*$ satisfying the following inequality [6, 7],

$$\begin{gathered}\int_Q u_{tti} dx d\tau + \int_Q \sigma_{ij}(u) e_{ij}(v_i - u_{ti}) dx d\tau + \\ \int_{\gamma_2} \mathcal{F} \mid \sigma_N(u) \mid (\mid v_T \mid - \mid u_{tT} \mid) dx d\tau \geq \int_Q f_i(v_i - u_{ti}) dx d\tau + \\ \int_{\gamma_1} p_i(v_i - u_{ti}) dx d\tau \quad \forall v \in H^{1/2}(I; H^1(\Omega; R^2)) \cap K.\end{gathered} \tag{10}$$

Note, that from (2) as well as from Imbedding Theorem of Sobolev spaces [1] it follows that u_0 and u_1 in (6) are continuous on the boundary of cylinder Q. The existence of solutions to system (1) - (6) was shown in [6, 7]:

Theorem 2.1 *Assume : (i) The data are smooth enough, i.e. (2) is satisfied. (ii) Γ_2 is of class $C^{1,1}$. (iii) The friction coefficient is small enough. Then there exists a unique weak solution to the problem (1) - (6).*

Proof. The proof is based on penalization of the inequality (10), friction regularization and employment of localization and shifting technique due to Lions and Magenes. For details of the proof see [7].

□

For the sake of brevity we shall consider the contact problem with prescribed friction, i.e., we shall assume

$$\mathcal{F} \mid \sigma_N \mid = \sigma_T \leq 1. \tag{11}$$

The condition (4) is replaced by the following one,

$$u_{tT}\sigma_T + \mid u_{tT} \mid = 0, \; \mid \sigma_T \mid \leq 1 \quad \text{on } I \times \Gamma_2. \tag{12}$$

Let us introduce the space

$$\Lambda = \{\lambda \in L^2(I; L^\infty(\Gamma_2)) \; : \mid \lambda \mid \leq 1 \quad \text{on } I \times \Gamma_2\}. \tag{13}$$

Taking into account (12) the system (10) takes the form : Find $u \in K$ and $\lambda \in \Lambda$ such that

$$\int_Q u_{tti} dx d\tau + \int_Q \sigma_{ij}(u) e_{ij}(v_i - u_{ti}) dx d\tau - \int_{\gamma_2} \lambda_T (v_T - u_{tT}) dx d\tau$$

$$\geq \int_Q f_i(v_i - u_{ti})dxd\tau + \int_{\gamma_1} p_i(v_i - u_{ti})dxd\tau \tag{14}$$

$$\forall v \in H^{1/2}(I; H^1(\Omega; R^2)) \cap K$$

$$\int_{\gamma_2} \sigma_T u_{tT} dsd\tau \leq \int_{\gamma_2} \lambda_T u_{tT} dsd\tau \quad \forall \lambda_T \in \Lambda. \tag{15}$$

3. Formulation of the shape optimization problem

We consider a family $\{\Omega_s\}$ of the domains Ω_s depending on parameter s. For each Ω_s we formulate a variational problem corresponding to (10). In this way we obtain a family of the variational problems depending on s and for this family we shall study a shape optimization problem , i.e., we minimize with respect to s a cost functional associated with the solutions to (10).

The domain Ω_s we shall consider as an image of a reference domain Ω under a smooth mapping $\mathbf{T}_s$. To describe the transformation $\mathbf{T}_s$ we shall use the speed method [16]. Let us denote by $V(s,x)$ an enough regular vector field depending on parameter $s \in [0,\vartheta), \vartheta > 0$:

$$V(.,.) \; : \; [0,\vartheta) \times R^2 \to R^2$$

$$V(s,.) \in C^2(R^2,R^2) \;\; \forall s \in [0,\vartheta), \;\; V(.,x) \in C([0,\vartheta),R^2) \;\; \forall x \in R^2. \tag{16}$$

Let $\mathbf{T}_s(V)$ denotes the family of mappings : $\mathbf{T}_s(V) \, : \, R^2 \ni X \to x(t,X) \in R^2$ where the vector function $x(.,X) = x(.)$ satisfies the systems of ordinary differential equations :

$$\frac{d}{d\tau}x(\tau,X) = V(\tau,x(\tau,X)), \tau \in [0,\vartheta), \quad x(0,X) = X \in R. \tag{17}$$

We denote by $D\mathbf{T}_s$ the Jacobian of the mapping $\mathbf{T}_s(V)$ at a point $X \in R^2$. We denote by $D\mathbf{T}_s^{-1}$ and ${}^*D\mathbf{T}_s^{-1}$ the inverse and the transposed inverse of the Jacobian $D\mathbf{T}_s$, respectively. $J_s = \det D\mathbf{T}_s$ will denote the determinant of the Jacobian $D\mathbf{T}_s$. The family of domains $\{\Omega_s\}$ depending on parameter $s \in [0,\vartheta), \vartheta > 0$, is defined as follows : $\Omega_0 = \Omega$

$$\Omega_s = \mathbf{T}_s(\Omega)(V) = \{x \in R^2 \; : \; \exists X \in R^2 \text{ s. th. } x = x(s,X), \text{ where the function } x(.,X) \text{ satisfies (17) for } 0 \leq \tau \leq s\}. \tag{18}$$

Let us consider problem (14) - (15) in the domain Ω_s. Let F_s, K_s, Λ_s be defined, respectively, by (8, (9), (13) with Ω_s instead of Ω. We shall write $u_s = u(\Omega_s)$, $\sigma_s = \sigma(\Omega_s)$. The problem (14) - (15) in the domain Ω_s takes the form : find $u_s \in K_s$ and $\lambda_s \in \Lambda_s$ such that,

$$\int_{Q_s} u_{ttsi} v_i dx d\tau + \int_{Q_s} \sigma_{ij}(u_s) e_{ij}(v_i - u_{tsi}) dx d\tau -$$
$$\int_{\gamma_{s2}} \lambda_{sT}(v_T - u_{tsT}) dx d\tau \geq \int_{Q_s} f_i(v_i - u_{tsi}) dx d\tau + \tag{19}$$
$$\int_{\gamma_{s1}} p_i(v_i - u_{tsi}) dx d\tau \quad \forall v \in H^{1/2}(I; H^1(\Omega_s; R^2)) \cap K$$

$$\int_{\gamma_{s2}} \sigma_{sT} u_{tsT} ds d\tau \leq \int_{\gamma_{s2}} \lambda_{sT} u_{tsT} ds d\tau \quad \forall \lambda_{sT} \in \Lambda_s. \tag{20}$$

We are ready to formulate the optimization problem. By $\hat{\Omega} \subset R^2$ we denote a domain such that $\Omega_s \subset \hat{\Omega}$ for all $s \in [0, \vartheta), \vartheta > 0$. Let $\phi \in M$ be a given function. The set M is determined by :

$$M = \{\phi \in L^\infty(I; H_0^2(\hat{\Omega}; R^2)) \ : \phi \leq 0 \text{ on } I \times \hat{\Omega}, \ \| \phi \|_{L^\infty(I; H_0^2(\hat{\Omega}; R^2))} \leq 1\} \tag{21}$$

Let us introduce, for given $\phi \in M$, the following cost functional :

$$J_\phi(u_s) = \int_{\gamma_{s2}} \sigma_{sN} \phi_{tNs} dz d\tau, \tag{22}$$

where ϕ_{tNs} and σ_{sN} are normal components of ϕ_{ts} and σ_s, respectively, depending on parameter s. Note, that the cost functional (22) approximates the normal contact stress [3, 8, 11]. We shall consider such a family of domains $\{\Omega_s\}$ that every Ω_s, $s \in [0, \vartheta)$, $\vartheta > 0$, has constant volume $c > 0$, i.e. : every Ω_s belongs to the constraint set U given by :

$$U = \{\Omega_s \ : \ \int_{\Omega_s} dx = c\}. \tag{23}$$

We shall consider the following **shape optimization problem** :

For given $\phi \in M$, find the boundary Γ_{2s}
of the domain Ω_s occupied by the body, (24)
minimizing the cost functional (22) subject to $\Omega_s \in U$.

The set U given by (23) is assumed to be nonempty. $(u_s, \lambda_s) \in K_s \times \Lambda_s$ satisfy (19) - (20). Note, that the goal of the shape optimization problem (24) is to find such boundary Γ_2 of the domain Ω occupied by the body that the normal contact stress is minimized. Remark, that the cost functional (22) can be written in the following form [3, 17] :

$$\int_{\gamma_{2s}} \sigma_{sN}\phi_{tN} ds d\tau = \int_{Q_s} u_{tts}\phi_{ts} dx d\tau + \int_{Q_s} \sigma_{sij}(u_s) e_{kl}(\phi_{ts}) dx d\tau - \tag{25}$$
$$\int_{Q_s} f\phi_{ts} dx d\tau - \int_{\gamma_{1s}} p_s\phi_{ts} ds d\tau - \int_{\gamma_{2s}} \sigma_{sT}\phi_{tTs} ds d\tau.$$

We shall assume there exists at least one solution to the optimization problem (24). It implies a compactness assumption of the set (23) in suitable topology. For detailed discussion concerning the conditions assuring the existence of optimal solutions see [3, 16].

4. Shape derivatives of contact problem solution

In order to calculate the Euler derivative (44) of the cost functional (22) we have to determine shape derivatives $(u', \lambda') \in F \times \Lambda$ of a solution $(u_s, \lambda_s) \in K_s \times \Lambda_s$ of the system (19)–(20). Let us recall from [16] :

Definition 4.1 *The shape derivative $u' \in F$ of the function $u_s \in F_s$ is determined by :*

$$(\tilde{u_s})_{|\Omega} = u + su' + o(s), \tag{26}$$

where $\| o(s) \|_F /s \to 0$ for $s \to 0$, $u = u_0 \in F$, $\tilde{u}_s \in F(R^2)$ is an extension of the function $u_s \in F_s$ into the space $F(R^2)$. $F(R^2)$ is defined by (8) with R^2 instead of Ω.

In order to calculate shape derivatives $(u', \lambda') \in F \times \Lambda$ of a solution $(u_s, \lambda_s) \in K_s \times \Lambda_s$ of the system (19),(20) first we calculate material derivatives $(\dot{u}, \dot{\lambda}) \in F \times \Lambda$ of the solution $(u_s, \lambda_s) \in K_s \times \Lambda_s$ to the system (19),(20). Let us recall the notion of the material derivative [16]:

Definition 4.2 *The material derivative $\dot{u} \in F$ of the function $u_s \in K_s$ at a point $X \in \Omega$ is determined by :*

$$\lim_{s \to 0} \| [(u_s \circ \mathbf{T}_s) - \sigma]/s - \dot{u} \|_F = 0, \tag{27}$$

where $u \in K$, $u_s \circ \mathbf{T}_s \in K$ is an image of function $u_s \in K_s$ in the space F under the mapping $\mathbf{T}_s$.

Taking into account Definition 4.2 we can calculate material derivatives of a solution to the system (19),(20) :

Lemma 4.1 *The material derivatives $(\dot{u}, \dot{\lambda}) \in K_1 \times \Lambda$ of a solution $(u_s, \lambda_s) \in K_s \times \Lambda_s$ to the system (19)–(20) are determined as a unique* *solution to the following system :*

$$\int_Q \{(\dot{u}_{tt}\eta + u_{tt}\dot{\eta} + u_{tt}\eta div V(0)(DV(0)u)_{tt}\eta + u_{tt}(DV(0)\eta)$$
$$-\dot{f}\eta - f\dot{\eta} + (\sigma_{ij}(u)e_{kl}(\eta) - f\eta)divV(0)\}dxd\tau - \qquad (28)$$
$$\int_{\gamma_1}(\dot{p}\eta + p\dot{\eta} + p\eta D)dxd\tau - \int_{\gamma_2}\{(\dot{\lambda}\eta_{tT} + \lambda\dot{\eta}_{tT} +$$
$$\lambda\nabla\eta_T V(0)n + \lambda\eta_{tT}D\}dxd\tau \geq 0 \quad \forall\eta \in K_1,$$

$$\int_{\gamma_2}(\dot{\lambda} - \mu)u_{tT} + (\lambda - \dot{\mu})u_{tT} + (\lambda - \mu)\dot{u}_{tT} + \lambda u_{tT}D\}dxd\tau \quad \forall\mu \in L_1, \quad (29)$$

where V(0) = V(0,X), DV(0) denotes the Jacobian matrix of the matrix V(0). Moreover :

$$K_1 = \{\xi \in F \ : \ \xi = u - DVu \ \ on\ \gamma_0,\ \xi n \geq nDV(0)u \ \ on\ A_1,$$
$$\xi n = nDV(0)u \ \ on\ A_2\ \}, \quad (30)$$

$$A_0 = \{x \in \gamma_2 \ : \ u_{tN} = 0\},\ \ A_1 = \{x \in B \ : \ \sigma_N = 0\},$$
$$A_2 = \{x \in B \ : \ \sigma_N < 0\}, \qquad (31)$$

$$B_0 = \{x \in \gamma_2 \ : \ \lambda_T = 1,\ \ u_{tT} \neq 0\},$$
$$B_1 = \{x \in \gamma_2 \ : \ \lambda_T = -1,\ \ u_{tT} = 0\}, \qquad (32)$$
$$B_2 = \{x \in \gamma_2 \ : \ \lambda_T = 1,\ :\ u_{tT} = 0\}.$$

$$L_1 = \{\xi \in \Lambda \ : \ \xi \geq 0 \ on\ B_2,\ \xi \leq 0 \ on\ B_1,\ \xi = 0 \ on\ B_0\ \} \qquad (33)$$

and D is given by

$$D = div\, V(0) - (DV(0)n, n). \qquad (34)$$

Proof: It is based on approach proposed in [16]. First we transport the system (19)–(20) to the fixed domain Ω. Let $u^s = u_s \circ \mathbf{T}_s \in F$, $u = u_0 \in F$, $\lambda^s = \lambda_s \circ \mathbf{T}_s \in \Lambda$, $\lambda = \lambda_0 \in \Lambda$. Since in general $u^s \notin K(\Omega)$ we introduce a new variable $z^s = D\mathbf{T}_s^{-1}u^s \in K$. Moreover $\dot{z} = \dot{u} - DV(0)u$ [7, 15]. Using this new variable z^s as well as the formulae for transformation of the function and its gradient into reference domain Ω [15, 16] we write the system (19)–(20) in the reference domain Ω. Using the estimates on time derivative of function u [7] the Lipschitz continuity of u and λ satisfying (19) - (20) with respect to s can be proved. Applying to this system the result concerning the differentiability of solutions to variational inequality [15, 16] we obtain that the material derivative $(\dot{u}, \dot{\lambda}) \in K_1 \times \Lambda$ satisfies the system (28)-(29). Moreover from the ellipticity condition of the elasticity coefficients by a standard argument [15] it follows that $(\dot{u}, \dot{\lambda}) \in K_1 \times \Lambda$ is a unique solution to the system (28)-(29).

□

Recall [16], that if the shape derivative $u' \in F$ of the function $u_s \in F_s$ exists, then the following condition holds :

$$u' = \dot{u} - \nabla u V(0), \tag{35}$$

where $\dot{u} \in F$ is material derivative of the function $u_s \in F_s$.

From regularity result in [7] it follows that :

$$\nabla u V(0) \in F, \quad \nabla \lambda_T V(0) \in \Lambda, \tag{36}$$

where the spaces F and Λ are determined by (8) and (13) respectively.

Integrating by parts system (28),(29) and taking into account (35),(36) we obtain the similar system to (28),(29) determining the shape derivative $(u', \lambda'_T) \in F \times L$ of the solution $(u_s, \lambda_{sT}) \in K_s \times L_s$ of the system (19) - (20) :

$$\int_Q u'_{tt}\eta + u_t t\eta' + (DV(0) +^* DV(0))u_{tt}\eta dx d\tau + \int_\gamma u_{tt}\eta V(0)n$$
$$\int_Q \sigma_{ij}(u')e_k l\eta - \int_{\gamma_2} \lambda'\eta_{tT} + \lambda\eta'_{tT}\} dx d\tau +$$
$$I_1(u_t, \eta) + I_2(\lambda, u, \eta) \geq 0 \quad \forall \eta \in N_1, \tag{37}$$

$$\int_{\gamma_2} [u'_{tT}(\mu - \lambda) - u_{tT}\lambda'] dx d\tau + I_3(u, \mu - \lambda) \geq 0 \quad \forall \mu \in L_1, \tag{38}$$

$$N_1 = \{\eta \in F \,:\, \eta = \lambda - DuV(0), \;\; \lambda \in K_1\}, \tag{39}$$

$$I_1(\varphi, \phi) = \int_\gamma \{\sigma_{ij}(\varphi)e_k l\phi - f\phi -$$
$$((\nabla pn)\phi + (p\nabla\phi)n + p\phi H)V(0)n\} dx d\tau, \tag{40}$$

$$I_2(\mu, \varphi, \phi) = \int_{\gamma_2} \{(\nabla\mu)n\nabla\phi + \mu(\nabla(\nabla\varphi n))\varphi +$$
$$\mu\nabla\varphi_{tT}H + \mu\nabla\varphi n\} V(0) n dx d\tau, \tag{41}$$

$$I_3(\varphi, \mu - \lambda) = \int_{\gamma_2} (\varphi n)(\mu - \lambda) + \varphi(\nabla\mu n) - \varphi(\nabla\lambda n) +$$
$$\varphi(\mu - \lambda)H] V(0) n dx d\tau, \tag{42}$$

where H denotes a mean curvature of the boundary Γ [16].

5. Necessary optimality condition

Our goal is to calculate the directional derivative of the cost functional (22) with respect to the parameter s. We will use this derivative to formulate necessary optimality condition for the optimization problem (24). First, let us recall from [16] the notion of Euler derivative of the cost functional depending on domain Ω :

Definition 5.1 . *Euler derivative $dJ(\Omega; V)$ of the cost functional J at a point Ω in the direction of the vector field V is given by :*

$$dJ(\Omega; V) = \lim_{s \to 0} \sup[J(\Omega_s) - J(\Omega)]/s. \tag{43}$$

The form of the directional derivative $dJ_\phi(u; V)$ of the cost functional (22) is given in :

Lemma 5.1 *The directional derivative $dJ_\phi(u; V)$ of the cost functional (22), for $\phi \in M$ given, at a point $u \in K$ in the direction of vector field V is determined by :*

$$dJ_\phi(u; V) = \int_Q [u'_{tt}\eta + u_{tt}\eta' + (DV(0) +^* DV(0))u_{tt}\eta] dx d\tau +$$

$$\int_\gamma u_{tt}\eta V(0)n + \int_Q (\sigma'_{ij} e_{kl}(\phi) dx +$$

$$\int_\Gamma (\sigma_{ij} e_{kl}(\phi) - f\phi) V(0) n ds - \int_{\Gamma_1} (\nabla p \phi V(0) +$$

$$p \nabla \phi V(0) + p\phi D) ds - \int_{\Gamma_2} \sigma'_T \phi_T ds + I_1(u, \phi) - I_2(\lambda, u, \phi), \tag{44}$$

where σ' is a shape derivative of the function σ_s with respect to s. This derivative is defined by (26). ∇p is a gradient of function p with respect to x. Moreover V(0) = V(0,X), ϕ_T and σ_T are tangent components of functions ϕ and σ, respectively, as well as D is given by (34). DV(0) denotes the Jacobian matrix of the matrix V(0) and div denotes divergence operator.

Proof : Taking into account (22),(25) as well as formulae for transformation of the gradient of the function defined on domain Ω_s into the reference domain Ω [16] and using the mapping (16)– (17) we can express the cost functional (22) defined on domain Ω_s in the form of the functional $J_\phi(u^s)$ defined on domain Ω, determined by :

$$J_\phi(u^s) = \int_Q (D\mathbf{T}_s u^s)_{tt} D\mathbf{T}_s \phi_t^s det D\mathbf{T}_s dx d\tau +$$

$$\int_Q [\sigma_{ij} D\mathbf{T}_s e_{kl}(D\mathbf{T}_s \phi_t^s) - f^s D\mathbf{T}_s \phi) \det D\mathbf{T}_s dx -$$
$$\int_{\gamma_1} p^s D\mathbf{T}_s \phi \parallel \det D\mathbf{T}_s{}^\star D\mathbf{T}_s^{-1} n \parallel ds - \tag{45}$$
$$\int_{\gamma_2} \lambda_{sT} D\mathbf{T}_s \phi_T \parallel \det D\mathbf{T}_s{}^\star D\mathbf{T}_s^{-1} n \parallel ds,$$

where $u^s = u_s \circ \mathbf{T}_s \in F$, $u = u_0 \in F$ and $\lambda = \lambda_0 \in \Lambda$. By (43) we have :

$$dJ_\phi(u; V) = \limsup_{t \to 0} [J_\phi(u^s) - J_\phi(u)]/s. \tag{46}$$

Remark, it follows by standard arguments [3] that the pair $(\sigma_s, u_s) \in Q_s \times K_s$, $s \in [0, \vartheta)$, $\vartheta > 0$, satisfying the system (19)–(20) is Lipschitz continuous with respect to the parameter s. Passing to the limit with $s \to 0$ in (46) as well as taking into account the formulae for derivatives of $D\mathbf{T}_s^{-1}$ and $\det D\mathbf{T}_s$ with respect to the parameter s [16] and (26) we obtain (44).

□

In order to eliminate the shape derivative (u', λ') from (44) we introduce an adjoint state $(r, q) \in K_2 \times L_2$ defined as follows :

$$\int_Q r_{tt} \zeta dx d\tau + \int_Q \sigma_{ij}(\zeta) e_{kl}(\phi + r) dx d\tau +$$
$$\int_{\gamma_2} \zeta_{tT}(q - \lambda) \zeta dx d\tau = 0 \quad \forall \zeta \in K_2, \tag{47}$$

with

$$r(\mathcal{T}, x) = 0, \quad r_t(\mathcal{T}, x) = 0,$$

$$\int_{\gamma_2} (r_{tT} + \phi_{tT} - u_{tT}) \delta dx d\tau = 0 \;\; \forall \delta \in L_2, \tag{48}$$

$$K_2 = \{\zeta \in K_1 \; : \; \zeta n = 0 \text{ on } A_0\}, \tag{49}$$

$$L_2 = \{\delta \in \Lambda \; : \; \delta = 0 \;\text{ on } A_0 \cap B_0 \;\}. \tag{50}$$

Since $\phi \in M$ is a given element, then by the same arguments as used to show the existence of solution $(u, \lambda) \in K \times L$ to the system (19)–(20) we can show the existence of the solution $(r, q) \in K_2 \times L_2$ to the system (47),(48).

From (44),(37),(38), (47),(48) we obtain :

$$dJ_\phi(u; V) = I_1(u, \phi + r) + I_2(\lambda, u, \phi + r) + I_3(u, q - \lambda). \tag{51}$$

The necessary optimality condition has a standard form :

Theorem 5.1 *There exists a Lagrange multiplier $\mu \in R$ such that for all vector fields V determined by (16),(17) the following condition holds :*

$$dJ_{\phi}(u;V) + \mu \int_{\Gamma} V(0)nds \geq 0, \tag{52}$$

where $dJ_{\phi}(u;V)$ is given by (51).

Proof : It is given in [3, 4, 5, 16, 17].

6. Conclusions

In the paper the necessary optimality condition for the shape optimization problem for the dynamical contact problem was formulated. Preliminary numerical results can be found in [13] where the continuous optimization problem was discretized by piecewise linear and piecewise constant functions on each finite element. The discretized problem was numerically solved by an Augmented Lagrangian Algorithm combined with active set strategy and updating of the dual variables.

References

[1] R.A. Adams, *Sobolev Spaces*, Academic Press, New York, 1975.

[2] G.Duvaut and J.L. Lions, *Les inequations en mecanique et en physique*, Dunod, Paris, 1972.

[3] J. Haslinger, Neittaanmaki P., *Finite Element Approximation for Optimal Shape Design. Theory and Application.*, John Wiley& Sons, 1988.

[4] E.J. Haug, K.K Choi, V. Komkov, *Design Sensitivity Analysis of Structural Systems*, Academic Press, 1986.

[5] I. Hlavacek, J. Haslinger, J. Necas, J.Lovisek, *Solving of Variational Inequalities in Mechanics*(in Russian), Mir, Moscow, 1986.

[6] J. Jarusek and C. Eck, Dynamic Contact Problems with Small Coulomb Friction for Viscoelastic Bodies. Existence of Solutions, *Mathematical Models and Methods in Applied Sciences*, **9**, pp. 11 - 34, 1999.

[7] J. Jarusek, Dynamical Contact Problem with Given Friction for Viscoelastic Bodies, *Czech. Math. Journal* , **46**, pp. 475 - 487, 1996.

[8] A. Klabring, J. Haslinger, On almost Constant Contact Stress Distributions by Shape Optimization, *Structural Optimization*, **5**, pp. 213-216, 1993.

[9] A. Myśliński, Mixed Variational Approach for Shape Optimization of Contact Problem with Prescribed Friction, in : Numerical Methods for Free Boundary Problems, P.Neittaanmaki ed., *International Series of Numerical Mathematics*, Birkhäuser, Basel,**99**, pp. 286-296, 1991.

[10] A. Myśliński, Shape Optimization of Contact Problems Using Mixed Variational Formulation, *Lecture Notes in Control and Information Sciences*, Springer,Berlin, **160**, pp. 414 - 423, 1992.

[11] A. Myśliński, Mixed Finite Element Approximation of a Shape Optimization Problem for Systems Described by Elliptic Variational Inequalities, *Archives of Control Sciences*, **3**, No 3-4, pp. 243 - 257, 1994.

[12] A. Myśliński, Shape Optimization for Dynamic Contact Problems, *Discussiones Mathematicae, Differential Inclusions, Control and Optimization*, **20**, pp. 79 - 91, 2000.

[13] A. Myśliński, Augmented Lagrangian Techniques for Shape Optimal Design of Dynamic Contact Problems, *Preprint*, System Research Institute, Warsaw, 2001 - to be published in Proceedings of WCSMO4 Conference, 2001.

[14] J. Necas, *Les Methodes Directes en Theorie des Equations Elliptiques*, Masson, Paris, 1967.

[15] J. Sokolowski and J.P. Zolesio, Shape sensitivity analysis of contact problem with prescribed friction, *Nonlinear Analysis, Theory, Methods and Applications* , **12**, pp. 1399 - 1411, 1988.

[16] J. Sokolowski and J.P. Zolesio, *Introduction to Shape Optimization. Shape Sensitivity Analysis.* Springer, Berlin, 1992.

[17] J. Telega, Variational Methods in Contact Problems of Mechanics(in Russian), *Advances in Mechanics*, **10**, pp. 3-95, 1987.

OPTIMAL SHAPE DESIGN USING DOMAIN TRANSFORMATIONS AND CONTINUOUS SENSITIVITY EQUATION METHODS

Lisa Stanley
Department of Mathematical Sciences
Montana State University
Bozeman, Montana *
stanley@math.montana.edu

Abstract In this paper, we consider two approaches to solving an optimization based design problem where "shape" is the design parameter. Both methods use domain transformations to compute gradients. However, they differ in that the second method is based on solving a transformed optimization problem completely in the computational domain. We illustrate the methods using a simple 1D problem and discuss the benefits and drawbacks of each approach.

Keywords: Continuous Sensitivity Equation Methods, Optimal Design

1. Introduction

The focus of the paper is an optimal design problem where the design parameter determines the shape of the domain of the constraint equation. The cost function is given in terms of an integral expression describing the L_2 difference between some target function and the state variable. The constraint equation, or state equation, takes the form of an elliptic partial differential equation defined on a parameter dependent domain. Under the assumption that each point in the design space determines a unique state variable through the solution of the state equation, we pose the unconstrained optimal design problem.

*Work supported by the National Science Foundation under grant DMS-0072438 and the Air Force Office of Scientific Research under AASERT grant F49620-97-1-0329.

Since the domain of the constraint equation changes with perturbations in the design, numerical solution of the optimal design problem is often hampered by burdensome grid generation requirements at each iteration of an optimization algorithm. One technique that can be used to avoid this problem is to transform the domain of the constraint equation to one that is fixed and no longer depends on the shape parameter. An equivalent *transformed constraint equation* is posed on this fixed, computational domain, see [4, 8], for example. In this paper, we present two approaches to the optimal design problem. Each approach uses the transformation technique mentioned above along with CSEMs (Continuous Sensitivity Equation Methods) in order to solve the optimal design problem. The main difference between the two methods is that one solves the optimal design problem using the parameter dependent domain of the constraint equation while the second approach applies a mapping technique in order to transform both the cost function and the constraint equation to a fixed computational space. This results in a transformed optimization problem. In each case, gradient based optimization is applied, and CSEMs are used to supply gradient information.

One of the major topics of concern for using CSEMs with optimal design is the issue of *consistent derivatives.* Within the optimization literature, the assumption is usually made that the gradient information is the derivative (with respect to the design parameter) of the numerical approximation of the cost function. There is a great deal of concern that convergence and robustness are compromised if the derivative approximations are computed using techniques which do not account for truncation and roundoff errors implicitly contained in the cost function. In [1, 2], the notion of *asymptotically consistent derivatives* is introduced, and CSEMs, when coupled with a trust region method, are shown to be applicable within optimal design algorithms. More precise definitions are introduced in Section 5.1. We first pose an example optimal design problem, and the computational approaches mentioned above are sketched out in the context of this example. Numerical results are shown, and we conclude with some general remarks concerning these approaches in Section 6.

2. A 1D Optimal Design Problem

Let $\mathcal{Q} = [1, +\infty)$ denote the design space, and for $q \in \mathcal{Q}$, let $\Omega_q = (0, q)$. Consider the boundary value problem

$$-\frac{d^2}{dx^2}\mathbf{w}(x\,; q) = \mathbf{f}(x), \quad x \in \Omega_q \tag{1}$$

with homogeneous Dirichlet boundary conditions

$$\mathbf{w}(0) = 0, \quad \mathbf{w}(q) = 0. \tag{2}$$

The forcing function, $\mathbf{f} : (0, +\infty) \to \mathbb{R}$, is the piecewise continuous function defined by

$$\mathbf{f}(x) = \begin{cases} 0, & 0 < x < 1 \\ -1, & 1 \leq x < +\infty. \end{cases} \tag{3}$$

For each $q \in \mathcal{Q}$, (1)-(2) has a unique solution $\mathbf{w}(\cdot\ ; q) \in H^2(0, q) \cap H_0^1(0, q)$. Thus, we define a cost function $F \colon \mathcal{Q} \to \mathbb{R}$ by

$$F(q) = \frac{1}{2} \int_0^1 [\mathbf{w}(x; q) + \sin(\pi x)]^2 \, dx, \tag{4}$$

and we focus on the optimal design problem:

$$\min_{q \in \mathcal{Q}} \ F(q). \tag{5}$$

Observe that the state equation, (1) - (2), is defined on the "physical" space Ω_q, and the cost function, $F(\cdot)$, is defined over a fixed subset of this space. For this simple example, q can be interpreted as a "shape" parameter in the sense that it determines the length of the interval over which the state $\mathbf{w}(\cdot; q)$ is defined.

2.1 Domain Transformations

For large scale problems where the shape of the domain of the state equation is parameter dependent, grid generation often poses a major difficulty in the optimal design process. As mentioned earlier, one way to overcome this obstacle is to apply a domain transformation from the physical space to the fixed, computational space. For the model problem discussed in this paper, transforming is clearly very simple. We note that determining the domain transformation for any given two-dimensional or three-dimensional set can be much more complicated. Moreover, this calculation often requires the application of a numerical method. In order to focus on the issues related to sensitivity computation and the resulting gradient approximations, the application of an algebraic domain mapping to the model problem is justified.

Here we describe the transformation of the parameter dependent domain $[0, q]$ to the fixed computational domain, $[0, 1]$. Once this mapping is constructed, the *transformed state equation* is defined accordingly. For $\alpha > 0$, let $\Omega_\alpha = (0, \alpha)$, and for each fixed $q > 1$, define the transformation $M(\cdot\ ; q) \colon \overline{\Omega}_1 \to \overline{\Omega}_\alpha$ by

$$M(\xi, q) = \xi q = x. \tag{6}$$

Note that the spatial variable on the fixed domain is ξ, and we use x to denote the spatial variable on Ω_q. The transformations given above are used to define the "transformed" functions. Let $\xi \in \Omega_1$ and $q > 1$, and for any function $\mathbf{u} \in H_0^1(\Omega_q)$, define the transformed function $\hat{\mathbf{u}} \in H_0^1(\Omega_1)$ as follows

$$\hat{\mathbf{u}}(\xi; q) = \mathbf{u}(M(\xi, q); q) = \mathbf{u}(x; q). \tag{7}$$

It can be shown that for a given value of q, if $\mathbf{w}(\cdot\ ; q)$ is a solution to the boundary value problem given in (1)-(2), then the corresponding function $\hat{\mathbf{w}}(\cdot\ ; q)$ is a solution to the boundary value problem

$$-\frac{d^2}{d\xi^2}\hat{\mathbf{w}}(\xi) = q^2\hat{\mathbf{f}}(\xi, q), \quad \xi \in (0, 1) \tag{8}$$

with boundary conditions

$$\hat{\mathbf{w}}(0) = 0, \quad \hat{\mathbf{w}}(1) = 0. \tag{9}$$

The forcing function $\hat{\mathbf{f}}(\xi; q)$ is obtained by using the mapping M and the relation $\hat{\mathbf{f}}(\xi; q) = \mathbf{f}(M(\xi, q), q) = \mathbf{f}(x)$ and has the form

$$\hat{\mathbf{f}}(\xi, q) = \begin{cases} 0, & 0 < \xi < \frac{1}{q} \\ -1, & \frac{1}{q} \le \xi < 1. \end{cases} \tag{10}$$

Henceforth, the boundary value problem (8)-(9) is referred to as the *transformed state equation*, and it is used in each of the computational approaches described in the following sections.

3. Computational Approach 1

In this section, we describe one approach for solving the optimal design problem in (5). This approach can be described as a "differentiate-then-map" scheme. Observe that the gradient of the cost function has the form

$$\nabla F(q) = \frac{d}{dq}F(q) = \int_0^1 [\mathbf{w}(x; q) + \sin(\pi x)]\, \mathbf{s}(x; q) dx, \tag{11}$$

where the sensitivity is defined as follows

$$\mathbf{s}(\cdot\ ; q) \triangleq \frac{\partial}{\partial q}\mathbf{w}(\cdot\ ; q). \tag{12}$$

In order to compute the sensitivity, we use the CSEM approach. We derive a *sensitivity equation*, an equation for which the sensitivity in

(12) is a solution. Formally speaking, this equation is derived by implicit "differentiation" of the state equation and boundary conditions in (1)-(2). For the model problem considered here, it can be shown that the sensitivity equation and associated boundary conditions are given by

$$-\frac{d^2}{dx^2}\mathbf{s}(x) = 0, \quad x \in \Omega_q \tag{13}$$

with boundary conditions

$$\mathbf{s}(0) = 0, \quad \mathbf{s}(q) = -\frac{d}{dx}\mathbf{w}(x)\bigg|_{x=q} = -\frac{d}{dx}\mathbf{w}(q). \tag{14}$$

Observe that the normal derivative of $\mathbf{w}$ appears in the right boundary condition in (14). This is typical for shape sensitivity problems, and these boundary conditions are tricky to derive correctly for more complicated problems.

Gradient based optimization requires that we numerically approximate both the cost function and its gradient for a given value of the parameter q. Aside from the implementation of a quadrature rule, each iteration of the optimization algorithm involves a numerical calculation of both the state and the sensitivity for a given design parameter value. The following section describes the numerical scheme employed for these computations.

3.1 State and Sensitivity Calculations

Here we illustrate the use of the mapping technique discussed in Section 2.1. Both a transformed state equation and a transformed sensitivity equation are constructed on the computational domain Ω_1. The derivation of the transformed state equation is presented in detail in Section 2.1 and is given explicitly in equations (8)-(9). In a similar fashion, we define the transformed sensitivity

$$\hat{\mathbf{s}}(\xi; q) = \mathbf{s}(M(\xi, q); q) = \mathbf{s}(x; q), \tag{15}$$

and the *transformed sensitivity equation* is constructed. This boundary value problem has the form

$$-\hat{\mathbf{s}}''(\xi) = 0, \quad \xi \in (0, 1), \tag{16}$$

with boundary conditions

$$\hat{\mathbf{s}}(0) = 0, \quad \hat{\mathbf{s}}(1) = -\left(\frac{1}{q}\right) \cdot \frac{d}{d\xi}\hat{\mathbf{w}}(\xi)\bigg|_{\xi=1} = -\left(\frac{1}{q}\right) \cdot \frac{d}{d\xi}\hat{\mathbf{w}}(1). \tag{17}$$

Once the transformed equations are constructed, a discretization is applied. For the numerical approximations presented here, we apply a piecewise linear finite element method to (8)-(9) and to (16)-(17). For the sake of brevity, the details of the implementation are omitted; however, the interested reader can refer to [3] for a more complete exposition. Once the numerical calculations are performed, the recovery of the state and sensitivity approximations (defined on the physical space Ω_q) is achieved through the relations in (7) and (15).

4. Computational Approach 2

In this section, we present an approach to the optimal design problem which is similar to an idea considered in [5]. Like the previous scheme, both domain transformations and CSEMs are used in this strategy. The fundamental difference between the following approach and the one presented in Section 3 is the order in which these techniques are applied. In this section, the domain transformation is applied to the cost function as well as the state equation. First, we construct a *transformed optimal design problem* which is equivalent to the original in (4) and which uses information from the transformed state equation. A CSEM is then used to supply gradient information for the *transformed cost function.*

Before presenting the transformed optimal design problem, we remark that under the mapping M defined in (6), the following equality holds

$$\int_0^1 g(x)dx = \int_{M(0;q)}^{M(1;q)} g(M(\xi,q)) \left|\frac{dM}{d\xi}\right| d\xi = q\int_0^{\frac{1}{q}} g(M(\xi,q))d\xi,$$

where g is any C^1-function defined on Ω_1. Along with the previous equality, the definitions in (6) and (7) give rise to the transformed cost function

$$\hat{F}(q) = \frac{1}{2}q\int_0^{\frac{1}{q}} \left[\hat{\mathbf{w}}(\xi,q) + \sin(q\pi\xi)\right]^2 d\xi. \tag{18}$$

Here $\hat{\mathbf{w}}(\xi,q)$ is the solution to the transformed state equation given by the boundary value problem (8)-(9) for each $q \in \mathcal{Q}$. Hence, the transformed optimal design problem is given by

$$\min_{q\in\mathcal{Q}} \hat{F}(q), \tag{19}$$

where the design space, $\mathcal{Q}$, remains the same as in Section 2. Observe that a factor of q appears in the expression (18). Recall that the mapping, M, depends explicitly on the parameter q. Hence, the absolute value of the derivative of the mapping (and more generally, the determinant of the Jacobian matrix) is also parameter dependent, and this term appears explicitly in (18). For this particular example, the derivative is very simple, but we remark that the issue of parameter dependent derivatives is more complicated for two-dimensional and three-dimensional domains. A two-dimensional illustration can be found on pages 365-366 of [4].

Once the transformed optimal design problem is constructed, we proceed in much the same manner as previously discussed. Using Leibnitz' formula, the gradient of the transformed cost function has the following form

$$\nabla \hat{F}(q) = \frac{1}{2}\int_0^{\frac{1}{q}} [\hat{\mathbf{w}}(\xi, q) + \sin(q\pi\xi)]^2 \, d\xi + \left(\frac{q}{2}\right)[\hat{\mathbf{w}}(\frac{1}{q}; q)]^2 \left(\frac{-1}{q^2}\right)$$
$$+ \left(\frac{q}{2}\right)\int_0^{\frac{1}{q}} 2[\hat{\mathbf{w}}(\xi; q) + \sin(q\pi\xi)]\left(\frac{\partial}{\partial q}\hat{\mathbf{w}}(\xi; q) + \pi\xi\cos(q\pi\xi)\right) d\xi,$$

which can be simplified to the expression

$$\begin{aligned} \nabla \hat{F}(q) &= q\int_0^{\frac{1}{q}} [\hat{\mathbf{w}}(\xi, q) + \sin(q\pi\xi)]\,[\mathbf{p}(\xi, q) + \pi\xi\cos(q\pi\xi)]d\xi \\ &\quad + \frac{1}{2q}\left(2\hat{F}(q) - [\hat{\mathbf{w}}(\frac{1}{q}; q)]^2\right). \end{aligned} \tag{20}$$

In the equation above, the notation $\mathbf{p}(\xi; q)$ is used to denote the *sensitivity of the transformed state*; that is, we define

$$\mathbf{p}(\xi; q) \triangleq \frac{\partial}{\partial q}\hat{\mathbf{w}}(\xi; q). \tag{21}$$

It is important to note that the sensitivity of the transformed state, $\mathbf{p}(\xi; q)$, is related to, but not the same function as, the *transformed sensitivity*, $\hat{\mathbf{s}}(\xi; q)$. The notation used above reflects this important distinction. The following section describes the techniques used for obtaining numerical approximations for the transformed state and the sensitivity, $\mathbf{p}(\xi; q)$.

4.1 State and Sensitivity Calculations

For this approach, the optimization algorithm requires that we compute a numerical approximation to the transformed state, $\hat{\mathbf{w}}(\cdot; q)$, and the sensitivity of the transformed state, $\mathbf{p}(\cdot; q)$. As in the previous section, a piecewise linear finite element method is used to approximate the transformed state $\hat{\mathbf{w}}(\xi, q)$.

In order to calculate an approximation for $\mathbf{p}(\cdot; q)$, we derive a sensitivity equation for which $\mathbf{p}(\cdot; q)$ is a solution. In particular, the transformed state equation in (8)-(9) is "differentiated". Although the parameter appears explicitly in the right hand side of equation (8) and determines a point of discontinuity of the forcing function, one can still derive the sensitivity equation in a mathematically precise fashion. A rigorous mathematical construction is presented in [6] and references therein. Here we simply state that the sensitivity, $\mathbf{p}(\cdot;q)$, satisfies the second order, linear elliptic boundary value problem given by

$$-\frac{d^2}{d\xi^2}\mathbf{p}(\xi,q) = 2q\hat{\mathbf{f}}(\xi,q) - \delta_{\frac{1}{q}}(\xi), \tag{22}$$

$$\mathbf{p}(0) = 0, \qquad \mathbf{p}(1) = 0. \tag{23}$$

Here $\delta_{\frac{1}{q}}(\xi)$ is the Dirac delta function with mass at $\xi = \frac{1}{q}$. Since the domain Ω_1 does not depend on q, the boundary conditions are clear. Observe that the sensitivity equation is decoupled from the transformed state equation, but we caution the reader that this decoupling is merely a phenomena of the linearity of the transformed state equation. We also note that the linear elliptic problem (22)-(23) does not have a solution in $H_0^2(\Omega_1)$, and the system must be interpreted in the weak sense, that is, in integral form. For the results presented in this paper, a piecewise linear finite element method is used to approximate both $\hat{\mathbf{w}}(\cdot\,;q)$ and $\mathbf{p}(\cdot\,;q)$. For the sake of brevity, the details of the finite element implementations are omitted, and we proceed directly to the computational results.

5. Computational Results

In this section, numerical results are presented for two cases. The first is a comparison using a four-point Gauss quadrature rule for both the cost function approximations and the gradient approximations. From the second we make an interesting anecdotal comment concerning the importance of choosing a quadrature rule with the appropriate degree of accuracy.

Recall that each computational approach involves discretizing and numerically computing an approximation to the transformed state equation (8)-(9). The distinction between the calculations is the fact that Computational Approach 1 recovers an approximation to the original state through the mapping, M, and implements the quadrature rule in the physical space while Computational Approach 2 applies the quadrature rule in the computational space. Since M is a straightforward algebraic manipulation which can be "hard-wired", there is no loss in accuracy for the state approximation during the recovery process of Computational Algorithm 1. We briefly note that a four-point quadrature rule is sufficient to obtain an extremely accurate approximation to the true cost function in each case. Figures 1 and 2 show the respective cost function approximations plotted against the graph of the true cost function. The step in the parameter is $\Delta q = 0.1$ over the parameter range given, and the transformed state approximations are obtained using $N = 3$ grid points for these graphs. We also note that the error (measured in the vector norm, $\|\cdot\|_\infty$) in the cost function approximations is on the order of 10^{-4} for each of the computational algorithms. Now we move to the more interesting issue of gradient approximations.

Figure 1. True Cost Function and Approximations for Computational Approach 1

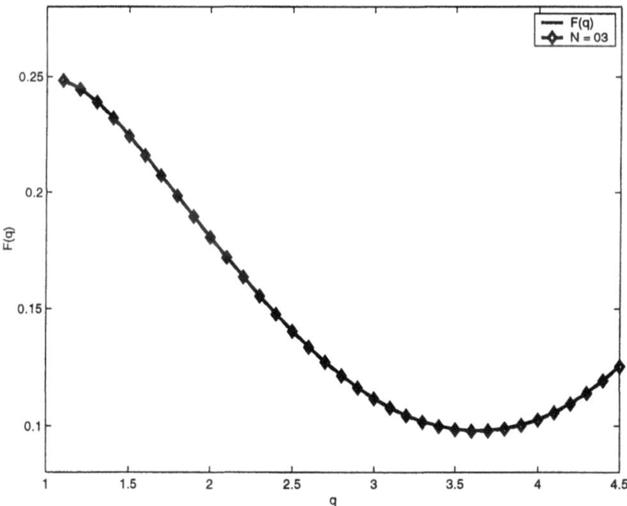

5.1 Gradient Approximations

This section briefly addresses the issue of gradient approximations for each computational approach. We preface the numerical results with two definitions regarding gradient approximations. The following discussion and definitions are taken from [2, 1]. We remark that the notation in [1] is slightly different because they explore the issue of applying different approximation schemes to obtain the state and sensitivity approximations. For our results, the discretization applied to compute the state approximations, and subsequently the cost function approximations, is exactly the same as that applied to compute the sensitivity approximations and the subsequent gradient approximations.

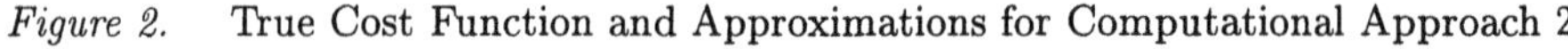

Figure 2. True Cost Function and Approximations for Computational Approach 2

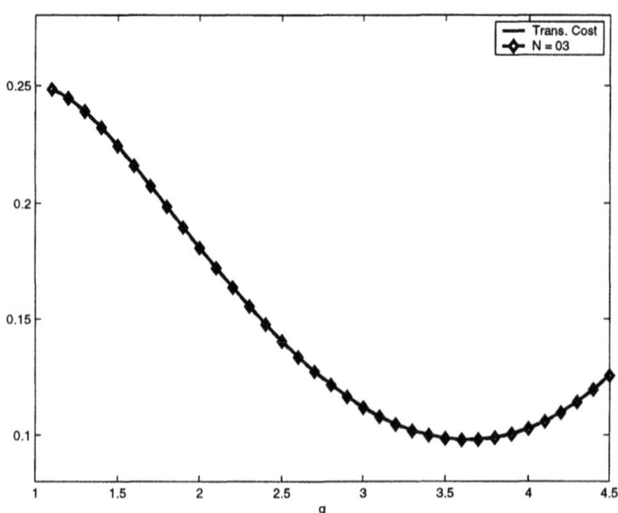

In the following discussion, we also refer to the discretization as an N-discretization in the sense that N refers to the number of grid points in the finite element mesh. To be more precise, we should include notation identifying the quadrature rule here as well. However, since we are comparing approximations using a four point quadrature rule in both cases, we choose to simplify the notation as much as possible. Furthermore, J denotes an arbitrary cost function which depends on the design parameter q. A sensitivity approach is said to produce *consistent derivatives* with respect to the state and sensitivity approximations using the N-discretization if

$$\nabla J^N(q) = [\nabla J(q)]^N \quad \forall\, q \in Q. \tag{24}$$

This definition states that the gradient of the approximate cost function is the same as the approximation of the true gradient. A more relaxed definition stipulates that the difference between the two gradient approximations should approach 0 with grid refinement. In particular, a sensitivity approach is said to produce *asymptotically consistent derivatives* with respect to the state and sensitivity approximations using the N-discretization if

$$\left|\nabla J^N(q) - [\nabla J(q)]^N\right| \to 0 \quad \forall\, q \in Q, \tag{25}$$

Figure 3. True Gradient and Approximations for Computational Approach 1

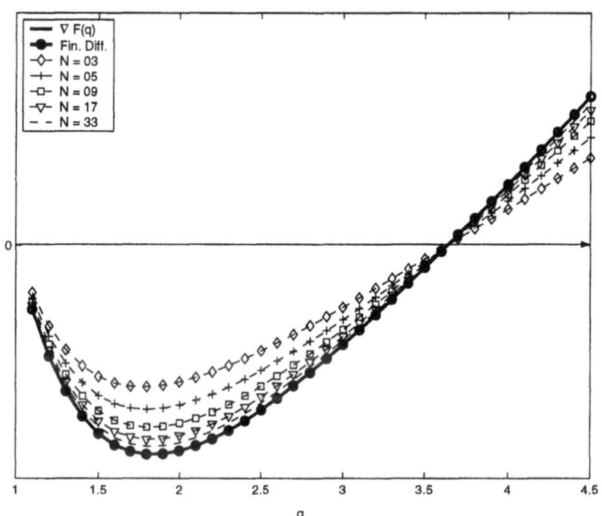

as $N \to \infty$, that is, as the grid is refined. The computational approaches presented in Sections 3 and 4 fall into the category of an approximation of the true gradient. Hence, Computational Approach 1 produces $[\nabla F(q)]^N$, and Computational Approach 2 yields $\left[\nabla \hat{F}(q)\right]^N$.

In the following figures, we present a sample of the gradient approximations obtained using each computational approach. The gradient approximations are compared with both a centered difference gradient approximation (solid curve with o's, representing ∇F^N and $\nabla \hat{F}^N$, respectively) and the true gradient (solid curve). In Figure 3, the gradient approximations generated using Computational Approach 1 converge to the finite difference gradient (and to the true gradient) with mesh refinement. Hence, Computational Approach 1 yields asymptotically consistent derivatives. Figure 4 indicates that Computational Approach 2 produces consistent derivatives.

5.2 Anecdotal Observation

In the case where a three-point Gauss quadrature rule is used, the quadrature rule is insufficient for convergence of the cost function approximations as the mesh is refined. That is, if we use three quadrature points for the integral approximations, then the cost function approxima-

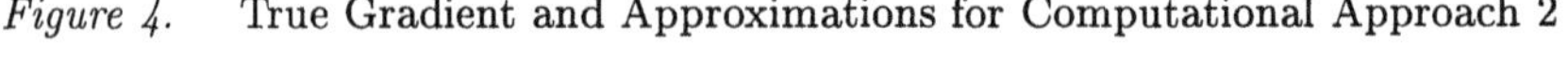

Figure 4. True Gradient and Approximations for Computational Approach 2

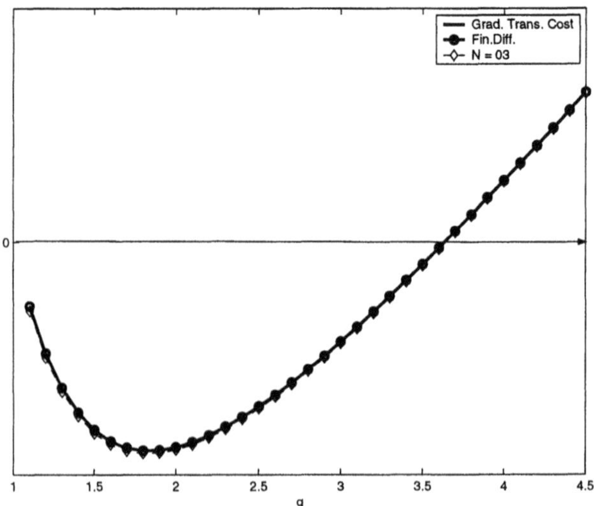

tions for each approach are given in Figures 5 and 6. We have included the graphs for $N = 3$ grid points and $N = 33$ grid points to show that the accuracy of the approximations does not improve with mesh refinement, and the error (using $||\cdot||_\infty$) in these approximations is on the order of 10^{-3}. All of the approximations generated using values of N between 3 and 33 exhibit exactly the same behavior. The gradient approximations for this case are somewhat interesting. In particular, Figures 7 and 8 suggest that Computational Algorithm 1 produces asymptotically consistent gradients while Computational Approach 2 produces inconsistent or "non-consistent" gradient approximations. This behavior may be a result of the fact that the we use the transformed cost function approximation during the gradient calculation on the computational space, recall the expression in (20). The gradient expression for Computational Algorithm 1, in (11), does not explicitly involve the cost function, $F(q)$.

6. Computational Issues

We conclude with some observations gathered during the course of the research. Since the domain transformations depend explicitly on the parameter, spatial derivatives are also parameter dependent and appear ex-

Figure 5. True Cost Function and Approximations for Computational Approach 1 using three-point quadrature rule

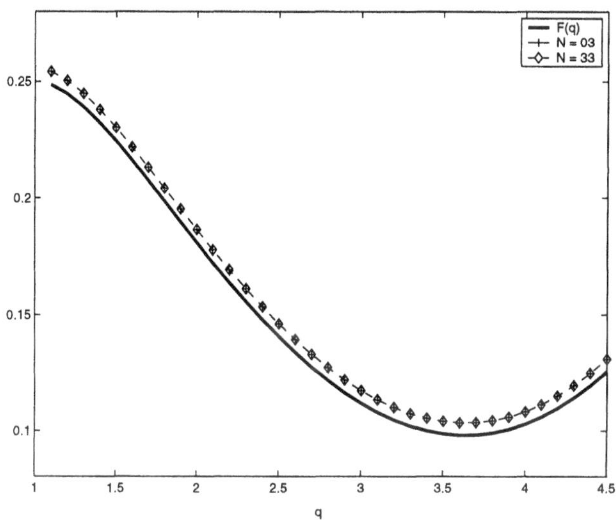

Figure 6. True Cost Function and Approximations for Computational Approach 2 using three-point quadrature rule

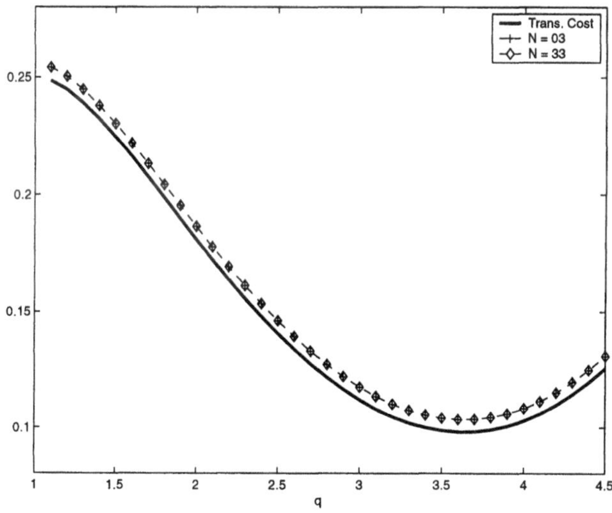

Figure 7. True gradient, finite difference gradient and approximations for Computational Approach 1 using three-point quadrature rule

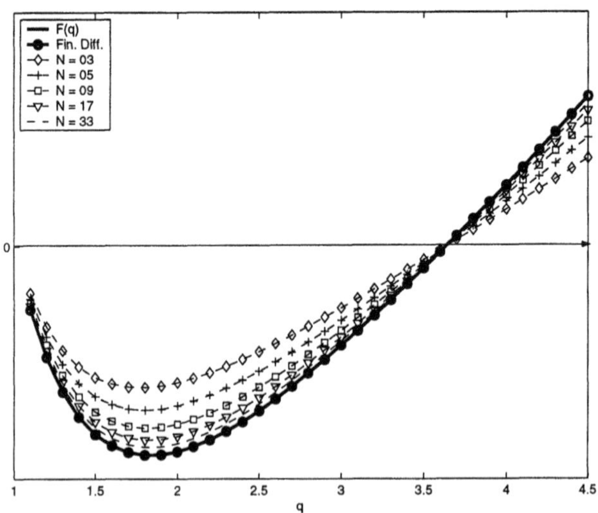

Figure 8. True gradient, finite difference gradient and approximations for Computational Approach 2 using three-point quadrature rule

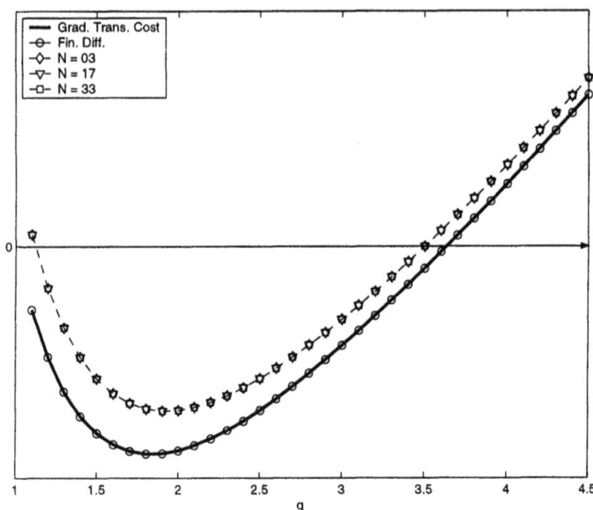

plicitly in both the transformed cost function and the transformed state equation. As a result, the derivation of a gradient expression is tedious and involves several terms including the transformed cost function, $\hat{F}(q)$. We approximate $\hat{F}(q)$ at each iteration of the optimization algorithm, and the approximation is reused in the gradient approximation routine. However, this may require good judgement for the quadrature rules as shown in Section 5.2. The results given here indicate that CSEMs can yield accurate, consistent gradients provided that the numerical schemes are chosen with care. Using the domain transformations is advantageous for the rigorous mathematical derivation of sensitivity equations. However, the issue of differentiability of the mappings becomes an important question for both gradient derivation and sensitivity analysis in Computational Approach 2 for problems with complicated geometries.

In Computational Approach 1, the derivation of the sensitivity equation is somewhat ad hoc; however, differentiation of the domain mapping is not required. One must also be willing to accept the asymptotically consistent derivatives that this method produces. For many problems, we observe that the gradient approximations for this approach tend to accurately pinpoint the location of the root of the gradient even on coarse meshes. Finally, the results given in Section 5.2 indicate that for certain problems, CSEMs can produce asymptotically consistent gradients even if the cost function approximations are inaccurate. Each computational algorithm exhibits specific characteristics that can be view as advantageous. Further research to determine which computational approach best fits a given problem is needed.

Acknowledgments

The author wishes to acknowledge Dr. John Burns and Dr. Eugene Cliff for raising questions which led to this research and Dr. Jeff Borggaard for many helpful conversations concerning optimization.

References

[1] J.T. Borggaard and J.A. Burns. (1997). Asymptotically Consistent Gradients in Optimal Design. *Multidisciplinary design optimization, State of the Art*, 303–314, 1997.

[2] J.T. Borggaard (1994). The Sensitivity Equation Method for Optimal Design. PhD thesis, Virginia Polytechnic Institute and State University, Blacksburg, Virginia. Mathematics PhD.

[3] J.A. Burns and L. G. Stanley. (2001). A Note on the Use of Transformations in Sensitivity Computations for Elliptic Systems. *Mathematical and Computer Modelling*, 33, 101–114.

[4] K.A. Hoffman and S.T. Chiang. (1993). Computational Fluid Dynamics for Engineers. Engineering Education System.

[5] M. Laumen. (2000). Newton's Method for a Class of Optimal Shape Design Problems. *SIAM Journal on Optimization*, 10(2), 503–533.

[6] L.G. Stanley. (2001) Sensitivity Equation Methods for Parameter Dependent Elliptic Equations. *Numerical Functional Analysis and Optimization*, 10(5&6), 721–748.

[7] L.G. Stanley. (2001). Shape Sensitivities for Optimal Design: A Case Study on the Use of Continuous Sensitivity Equation Methods. David Gao, Ray Ogden and Georgios Stavroulakis, editors, *Nonsmooth / Nonconvex Mechanics: Modeling, Analysis and Numerical Methods*, pages 369-384. Kluwer Academic Publishers, Nonconvex Optimization and Its Applications Series (50), The Netherlands, 2001.

[8] J.F. Thompson and Z.U. Warsi and C.W. Mastin. (1985). Numerical Grid Generation Foundations and Applications. Elsevier Publishing Company, 1985.

ADJOINT CALCULATION USING TIME-MINIMAL PROGRAM REVERSALS FOR MULTI-PROCESSOR MACHINES

Andrea Walther
Institute of Scientific Computing
Technical University Dresden
awalther@math.tu-dresden.de

Uwe Lehmann
Center for High Performance Computing
Technical University Dresden
lehmann@zhr.tu-dresden.de

Abstract For computational purposes such as debugging, derivative computations using the reverse mode of automatic differentiation, or optimal control by Newton's method, one may need to reverse the execution of a program. The simplest option is to record a complete execution log and then to read it backwards. As a result, massive amounts of storage are normally required. This paper proposes a new approach to reversing program executions. The presented technique runs the forward simulation and the reversal process at the same speed. For that purpose, one only employs a fixed and usually small amount of memory pads called checkpoints to store intermediate states and a certain number of processors. The execution log is generated piecewise by restarting the evaluation repeatedly and concurrently from suitably placed checkpoints. The paper illustrates the principle structure of time-minimal parallel reversal schedules and quotes the required resources. Furthermore, some specific aspects of adjoint calculations are discussed. Initial results for the steering of a Formula 1 car are shown.

Keywords: Adjoint calculation, Checkpointing, Parallel computing

1. Introduction and Notation

For many industrial applications, rather complex interactions between various components have been successfully simulated with computer models. This is true for several production processes, e.g. steel manufacturing with regards to various product properties, for example stress distribution. However, the simulation stage can frequently not be followed by an optimization stage, which would be very desirable. This situation is very often caused by the lack or inaccuracy of derivatives, which are needed in optimization algorithms. Hence, enabling the transition from simulation to optimization represents a challenging research task.

The technique of algorithmic or automatic differentiation (AD), which is not yet well enough known, offers an opportunity to provide the required derivative information [5]. Therefore, AD can contribute to overcoming the step from pure simulation and hence "trial and error"-improvements to an exact analysis and systematic derivative-based optimization.

The key idea of algorithmic differentiation is the systematic application of the chain rule. The mathematical specification of many applications involves nonlinear vector functions

$$F : \mathbb{R}^n \to \mathbb{R}^m, \qquad x \mapsto F(x),$$

that are typically defined and evaluated by computer programs. This computation can be decomposed into a (normally large) number of very simple operations, e.g. additions, multiplications, and trigonometric or exponential function evaluations. The derivatives of these elementary operations can be easily calculated with respect to their arguments. A systematic application of the chain rule yields the derivatives of a hierarchy of intermediate values. Depending on the starting point of this methodology, either at the beginning or at the end of the sequence of operations considered, one distinguishes between the forward mode and the reverse mode of AD. The reverse mode of algorithmic differentiation is a discrete analog of the adjoint method known from the calculus of variations.

The gradient of a scalar-valued function is yielded by the reverse mode in its basic form for no more than five times the operations count of evaluating the function itself. This bound is completely independent of the number of independent variables. More generally, this mode allows the computation of Jacobians for at most five times the number of dependents times the effort of evaluating the underlying vector function. However, the spatial complexity of the basic reverse mode, i.e. its memory requirement, is proportional to the temporal complexity of the evaluation of the function itself. This behaviour is caused by the fact that one has to record a complete *execution log* onto a data structure called *tape* and subsequently read this tape backward. For each arithmetic operation, the execution log contains a code and the addresses of the arguments as well as the computed value. It follows that the practical exploitation of the advantageous temporal complexity bound for the reverse mode is severely limited by the amount of memory required.

The reversal of a given function F is already being extensively used to calculate hand-coded adjoints. In particular, there are several contributions on weather data assimilation (e.g. [11]). Here, the desired gradients can be obtained with a low temporal complexity by integrating the linear co-state equation backwards along the trajectory of the original simulation. This well-known technique is closely related to the reverse mode of AD [3]. Moreover, debugging and interactive control may require the reconstruction of previous states by some form of running the program backwards that evaluates F. The need for some kind of logging arises whenever the process described by F is not invertible or ill conditioned. In these cases one cannot simply apply an inverse process to evaluate the inverse mapping F^{-1}. Consequently, the reversal of a program execution within a reasonable memory requirement has received some (but only perfunctory) attention in the computer science literature (see e.g. [12]).

This paper presents a new approach to reversing the calculation of F. For that reason, in the remainder of this section, the structure of the function F is described in detail. The reversal technique proposed in this article only employs a fixed and usually small amount of memory pads to store intermediate states and a certain number of processors for reversing F in minimal time. The corresponding time-minimal parallel reversal schedules are introduced in Section 2. The simulation of a Formula 1 car is considered in Section 3. The underlying ODE system is introduced. Then two different ways to calculate adjoints are discussed. Subsequently, the initial numerical results are presented. Finally, some conclusions are drawn in Section 4.

Throughout it is assumed that the evaluation of F comprises the evaluation of subfunctions F_i, $1 \leq i \leq l$, called physical steps that act on state x^{i-1} to calculate the subsequent intermediate state x^i for $1 \leq i \leq l$ depending on a control u^{i-1}. Hence, one has

$$x^i = F_i(x^{i-1}, u^{i-1}) .$$

Therefore, F can be thought of as a discrete *evolution*. The intermediate states of the evolution F represented by the counter i should be thought of as vectors of large dimensions. The physical steps F_i describe mathematical mappings that in general cannot be reversed at a reasonable cost even for given u^{i-1}. Hence, it is impossible to simply apply the inverses F_i^{-1} in order to run the program backwards from state l to state 0. It will also be assumed that due to their size, only a limited number of intermediate states can be kept in memory.

Furthermore, it is supposed that for each $i \in \{1,\ldots,l\}$, there exist functions $\hat{F}_i$ that cause the recording of intermediate values generated during the evaluation of F_i onto the tape and corresponding functions $\bar{F}_i$ that perform the reversal of the ith physical step using this tape. More precisely, one has the reverse steps

$$(\bar{x}^{i-1}, \bar{u}^{i-1}) \quad = \quad \bar{x}^i\, F_i'(x^{i-1}, u^{i-1}) \;\equiv\; \bar{F}_i(x^{i-1}, u^{i-1}, \bar{x}^i)\,,$$

where F_i' denotes the Jacobian of F_i with respect to x^{i-1} and u^{i-1}. The calculation of adjoints using the basic approach is depicted in Figure 1. Applying a checkpointing technique, the execution log is gen-

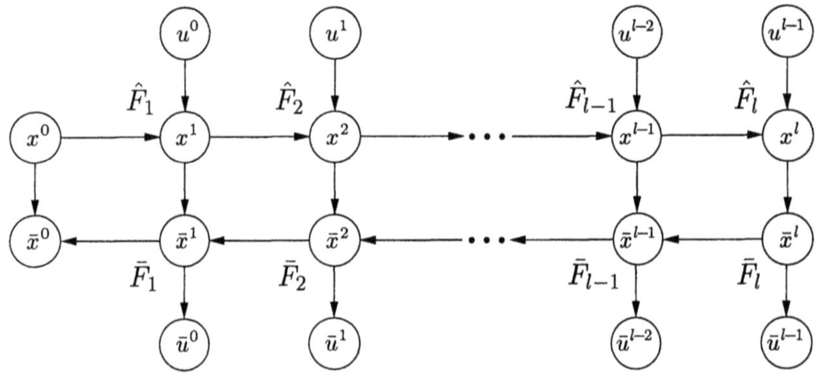

Figure 1. Naïve approach to calculate Adjoints

erated piecewise by restarting the evaluation repeatedly from suitably placed checkpoints, according to requests by the reversal process. Here, the checkpoints can be thought of as pointers to nodes representing intermediate states i. Using a checkpointing strategy on a uni-processor machine, the calculation of F can be reversed, even in such cases where the basic reverse mode fails due to excessive memory requirement (see e.g. [7, 6]). However, the runtime for the reversal process increases compared to the naïve approach. For multi-processor machines, this paper presents a checkpointing technique with concurrent recalculations that reverses the program execution in minimal wall-clock time.

2. Time-minimal Parallel Reversal Schedules

To derive an optimal reversal of the evaluation procedure F, one has to take into account four kinds of parameters, namely:

1.) the number l of physical steps to be reversed;

2.) the number p of processors that are available;

3.) the number c of checkpoints that can be accommodated; and

4.) the step costs: $\tau = TIME(F_i)$, $\hat{\tau} = TIME(\hat{F}_i)$, $\bar{\tau} = TIME(\bar{F}_i)$.

Well known reversal schedules for serial machines, i.e. $p = 1$, and constant step costs τ allow an enormous reduction of the memory required to reverse a given evolution F in comparison with the basic approach (see e.g. [7, 6]). Even if the step costs $\tau_i = TIME(F_i)$ are not constant it is possible to compute optimal serial reversal schedules [13]. However, one has to pay for the improvements in the form of a greater temporal complexity because of repeated forward integrations.

If no increase in the time needed to reverse F is acceptable, the use of a sufficiently large number of additional processors provides the possibility to reverse the evolutionary system F with drastically reduced spatial complexity and still minimal temporal complexity. Corresponding parallel reversal schedules that are optimal for given numbers l of physical steps, $p > 1$ processors, c checkpoints, and constant step costs were presented for the first time in [13]. For that purpose, it is supposed that $\tau = 1$, $\hat{\tau} \geq 1$, and $\bar{\tau} > 1$, with $\hat{\tau}, \bar{\tau} \in \mathbb{N}$. Furthermore, it is always assumed that the memory requirement for storing the intermediate states is the same for all i. Otherwise, it is not clear whether and how parallel reversal schedules can be constructed and optimized. The techniques developed in [13] can certainly not be applied. In practical applications, nonuniform state sizes might arise, for example as result of adaptive grid refinements, or function evaluations that do not conform naturally to our notion of an evolutionary system on a state space of fixed dimension.

Finding a time-minimal parallel reversal schedule can be interpreted as a very special kind of scheduling problem. The general problem class is known to be NP-hard (e.g. [4]). Nevertheless, it is possible to specify suitable time-minimal parallel reversal schedules for a arbitrary number l of physical steps because the reversal of a program execution has a very special structure. For the development of these time-minimal and resource-optimal parallel reversal schedules, first an exhaustive search algorithm was written. The input parameters were the number p of available processors and the number c of available checkpoints with both $\hat{\tau}$ and $\bar{\tau}$ set to 1. The program then computed a schedule that reverses the maximal number of physical steps $l(p, c)$ in minimal time using no more than the available resources p and c for $p + c \leq 10$. Here, minimal time means the wall clock equivalent to the basic approach of recording all needed intermediate results. Examining the corresponding parallel reversal schedules, one obtained that for $p > c$, only the resource number $\varrho = p + c$ has an influence on $l(p,c) \equiv l_\varrho$. Therefore, the development of time-minimal parallel reversal schedules, that are also resource-optimal, is focused on a given resource number ϱ under the tacit assumption $p > c$. The results obtained for $\varrho \leq 10$ provided sufficient insight to deduce the general structure of time-minimal parallel reversal schedules for arbitrary combinations of $\hat{\tau} \geq 1$, $\bar{\tau} \geq 1$, and $\varrho > 10$. Neglecting communication cost, the following recurrence is established in [13]:

Theorem: *Given the number of available resources $\varrho = p + c$ with $p > c$ and the temporal complexities $\hat{\tau} \in \mathbb{N}$ and $\bar{\tau} \in \mathbb{N}$ of the recording steps $\hat{F}_i$ and the reverse steps $\bar{F}_i$, then the maximal length of an evolution that can be reverted in parallel without interruption is given by*

$$l_\varrho = \begin{cases} \varrho & \text{if } \varrho < 2 + \hat{\tau}/\bar{\tau} \\ l_{\varrho-1} + \bar{\tau}\, l_{\varrho-2} - \hat{\tau} + 1 & \text{else.} \end{cases} \tag{1}$$

In order to prove this result, first an upper bound on the number of physical steps that can be reversed with a given number ϱ of processors and checkpoints was established. Subsequently, corresponding reversal schedules that attain this upper bound were constructed recursively. For this purpose, the resource profiles of the constructed parallel reversal schedules were analyzed in detail. In addition to the recursive construction of the desired time-minimal reversal schedules, the resource profiles yield an upper bound for the number p of processors needed during the reversal process. To be more precise, for reversing l_ϱ physical steps, one needs no more than

$$p_u = \begin{cases} \left\lceil \frac{\varrho+1}{2} \right\rceil & \text{if } \bar{\tau} \geq \hat{\tau} \\ \left\lceil \frac{\varrho+1}{2} \right\rceil + \left\lceil \frac{1}{2} \left\lfloor \frac{\hat{\tau}-1}{\bar{\tau}} \right\rfloor \right\rceil & \text{else} \end{cases}$$

processors [13]. Hence, roughly half of the resources have to be processors. This fact offers the opportunity to assign one checkpoint to each processor.

A time-minimal reversal schedule for $l = 55$ is depicted in Figure 2. Here, vertical bars represent checkpoints and slanted bars represent running processes. The shading indicates the physical steps F_i, the recording steps $\hat{F}_i$ and the reverse steps $\bar{F}_i$ to be performed.

Based on the recurrence (1), it is possible to describe the behaviour of l_ϱ more precisely. For $\hat{\tau} = \bar{\tau} = 1$, one finds that the formula for l_ϱ is equal to the Fibonacci-number $f_{\varrho-1}$. Moreover, for other combinations of $\hat{\tau}, \bar{\tau} \in \mathbb{N}$, the recurrence (1) produces generalized Fibonacci-numbers

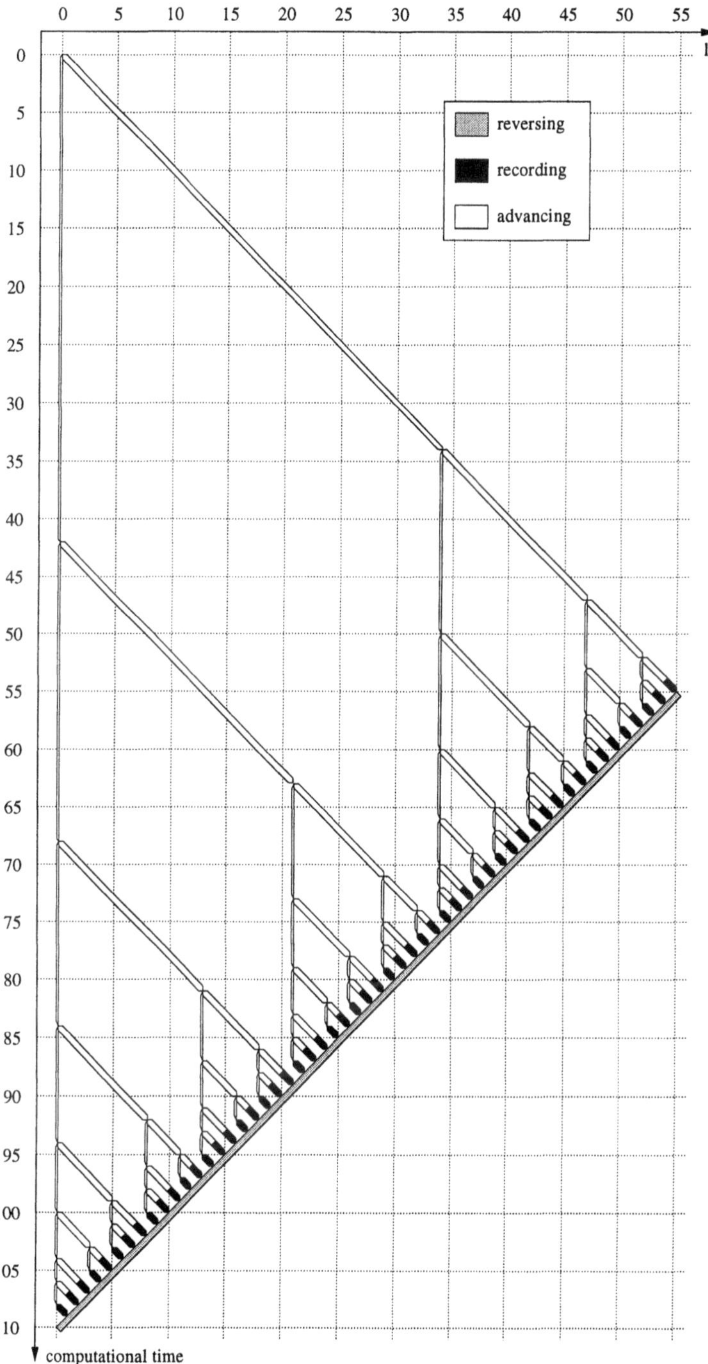

Figure 2. Time-minimal Parallel Reversal Schedule for $l = 21$ and $\hat{\tau} = \bar{\tau} = 1$.

(see e.g. [9]). More specifically, one finds that

$$l_\varrho \;\sim\; \frac{1}{2}\left(1+\frac{3}{\sqrt{1+4\bar{\tau}}}\right)\left[\frac{1}{2}(1+\sqrt{1+4\bar{\tau}}\,)\right]^{\varrho-1},$$

in the sense that the ratio between the two sides tends to 1 as ϱ tends to infinity. In the important case $\bar{\tau}=1$ even their absolute difference tends to zero. Thus, $l=l_\varrho$ grows exponentially as a function of $\varrho \approx 2p$ and conversely $p \approx c$ grows logarithmically as a function of l. In order to illustrate the growth of l_ϱ, assume 16 processors and 16 checkpoints are available. These resources suffice to reverse an evolution of $l = 2\,178\,309$ physical steps when $\hat{\tau}=\bar{\tau}=1$ and even more steps if $\hat{\tau}=1$ and $\bar{\tau}>1$.

For $\bar{\tau}=1$, i.e., if the forward simulation and the reversal of the time steps can be performed at the same speed, the implementation of this theory was done using the distributed memory programming model [10]. It is therefore possible to run the parallel reversal schedules framework on most parallel computers independent of their actual memory structure. To achieve a flexible implementation, the MPI routines for the communication are used. The parallel reversal schedules are worked off in a process-oriented manner instead of a checkpoint-oriented manner (see [10] for details). This yield the optimal resource requirements of Theorem 1.

In order to apply the parallel reversal schedules framework, one has to provide interfaces and define the main data structures for computing the adjoint. The data structures required are the checkpoints, the traces or tapes, as a result of the recording step $\hat{F}_i$, and the adjoint values. The structure and complexity of this data is independent of the framework since the framework only calls routines such as

- forward(..) for the evaluation of one physical step F_i,
- recording(..) for the evaluation of one recording step $\hat{F}_i$,
- reverse(..) for the evaluation of one reverse step $\bar{F}_i$,

provided by the user. These functions are equivalent to the functions used for a sequential calculation of the adjoint. The index i is an argument of each of the modules. The function recording(..) generates the trace or tape. The function reverse(..) obtains the trace of the last recording step and the adjoint computed so far as arguments. Furthermore, if $i=l$, the function reverse(..) may initialize the adjoints.

Additionally, the user must code communication modules, for example sendCheckpoint(..) and receiveCheckpoint(..). All user-defined routines have to be implemented applying MPI routines. The required process identifications and message tags are arguments of routines provided by the parallel reversal schedules framework.

3. Model Problem: Steering a Formula 1 Car

In order to test the implementation of parallel reversal schedules, the simulation of an automobile is considered. The aim is to minimize the time needed to travel along a specific road. A simplified model of a Formula 1 racing car [1] is employed. It is given by the ODE system:

$$\begin{aligned}
\dot{x}_1 &= x_2 \\
\dot{x}_2 &= \frac{(F_{\eta_1}(x,u_2) + F_{\eta_2}(x,u_2))l_f - (F_{\eta_3}(x,u_2) + F_{\eta_4}(x,u_2))l_r}{I} \\
\dot{x}_3 &= \frac{F_{\eta_1}(x,u_2) + F_{\eta_2}(x,u_2) + F_{\eta_3}(x,u_2) + F_{\eta_4}(x,u_2))}{M} - x_2 x_4 \\
\dot{x}_4 &= \frac{F_\xi(x,u_2) - F_a(x)}{M} + x_2 x_3 \\
\dot{x}_5 &= x_4 \sin(x_1) + x_3 \cos(x_1) \\
\dot{x}_6 &= x_4 \cos(x_1) - x_3 \sin(x_1) \\
\dot{x}_7 &= u_1 .
\end{aligned}$$

Hence, a go-kart model with rigid suspension and a body rolling about a fixed axis is considered. There are seven state variables representing the yaw angle and rate (x_1, x_2), the lateral and longitudinal velocity (x_3, x_4), global position (x_5, x_6), and the vehicle steer angle (x_7) as shown in Figure 3. The control variables are u_1 denoting the front steer rate and u_2 denoting the longitudinal force as input. The lateral and longitudinal vehicle forces F_η and F_ξ are computed using the state and the control variables as well as the tire forces given by a tire model described in [2]. The force F_a represents the aerodynamic drag depending on the longitudinal velocity. All other values are fixed car parameters such as mass M and length of the car given by l_f and l_r.

In order to judge the quality of the driven line, the cost functional

$$J(s_l) = \int_0^{s_l} S_{cf}(x,s)\big(1 + g(x,s)\big)ds \tag{2}$$

is used. The scaling factor $S_{cf}(x,s)$ changes the original time integration within the cost function to distance integration. Therefore, an integration over the arc length is performed. This variable change has to be done because the end time t_l of the time integration is the value one actually wants to minimise. Hence, t_l is unknown. The computation of the scaling factor $S_{cf}(x,s)$ is described in [1]. The function $g(x,s)$ measures whether or not the car is still on the road. The road is defined by

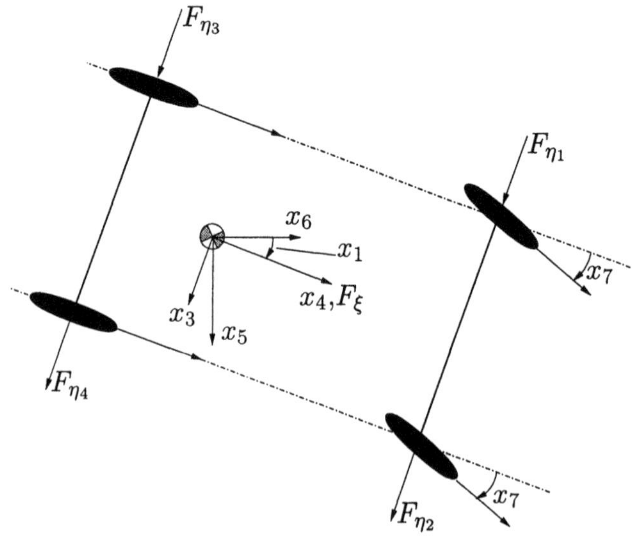

Figure 3. Model of Formula 1 Car.

the road centre line and road width. In the example presented here, the road width is constant a 2.5 m along the whole integration path. The function $g(x, s)$ returns zero as long as the car drives within the road boundaries. If the car leaves the road then $g(x, s)$ returns the distance from the car to the road boundary squared.

3.1. The Forward Integration

For the numerical results presented here, a discretization has to be applied. Therefore, an appropriate initial vector x^0 and the starting position $s^0 = 0$ were chosen. The route is divided equidistantly with a step size of $h = 10\ cm$. The well known four-stage Runge-Kutta scheme

$$\begin{aligned} k_1 &= f(x^{i-1}, u(s^{i-1})) \\ k_2 &= f(x^{i-1} + hk_1/2, u(s^{i-1} + h/2)) \\ k_3 &= f(x^{i-1} + hk_2/2, u(s^{i-1} + h/2)) \\ k_4 &= f(x^{i-1} + hk_3, u(s^{i-1} + h)) \\ x^i &= x^{i-1} + h(k_1 + 2k_2 + 2k_3 + k_4)/6 \end{aligned} \tag{3}$$

serves as physical step F_i for $i = 1, \ldots, 1000$.

The calculations of a physical step F_i form the forward(..)-routine needed by the time-minimal parallel reversal schedules. As mentioned above, in addition to this, one has to provide two further routines, namely recording(..) and reverse(..). The content of these two modules is described in the next subsection.

3.2. Calculating Adjoints

There are two basic alternatives for calculating the adjoints of a given model. Firstly, one may form the adjoint of the continuous model equation and discretize the continuous adjoint equation. Secondly, one may use automatic differentiation (AD), or hand-coding, to adjoin the discrete evaluation procedure of the model. Both ways do not commute in general (see e.g. [8]). Therefore, one has to be careful when deciding how to calculate the desired adjoints. For the computations shown below, the second option was applied, namely the adjoining of the discretized equation (3). Application of AD in reverse mode amounts to the following adjoint calculation $\bar{F}_i$ (see e.g. [5]):

$$\begin{array}{rclrcll}
\tilde{k}_j &=& \dfrac{\partial}{\partial x} k_j & \hat{k}_j &=& \dfrac{\partial}{\partial u} k_j & 1 \le j \le 4 \\
a_4 &=& h\bar{x}^i/6 & b_4 &=& a_4 \tilde{k}_4 & \\
a_3 &=& h\bar{x}^i/3 + hb_4 & b_3 &=& a_3 \tilde{k}_3 & \\
a_2 &=& h\bar{x}^i/3 + hb_3/2 & b_2 &=& a_2 \tilde{k}_2 & \qquad (4) \\
a_1 &=& h\bar{x}^i/3 + hb_2/2 & b_1 &=& a_1 \tilde{k}_1 & \\
\bar{u}^i &=& \bar{u}^i + a_4 \hat{k}_4 & \bar{u}^{i-1} &=& a_1 \hat{k}_1 & \\
\bar{x}^{i-1} &=& \dfrac{\partial J}{\partial x^{i-1}} + \bar{x}^i + b_1 + b_2 + b_3 + b_4, & & & &
\end{array}$$

for $i = l, \ldots, 1$, where the functions k_j, $1 \le j \le 4$, are defined as in (3). Here, $\bar{u}^i$ denotes the adjoint of the control u at s^i. Note that the integration of the adjoint scheme (4) has to be performed in reverse order starting at $i = l$. One uses $\bar{x}_i^l = \partial J/\partial x^l$, $1 \le i \le 7$ and $\bar{u}_i^l = 0$, $i = 1, 2$ as initial values because of the influence on the cost functional (2). After the complete adjoint calculation, each value $\bar{u}^i$ denotes the sensitivity of the cost functional J with respect to the value u_i.

Now the return value of the routine reverse(..) is clear. It has to contain the computations needed to perform an adjoint step $\bar{F}_i$ according to (4). However, there are two ways to implement the interface between the modules recording(..) and reverse(..). One can either store the stages k_j, $1 \le j \le 4$, during the evaluation of the recording step $\hat{F}_i$. Then the corresponding reverse step $\bar{F}_i$ comprises all calculations shown in (4), i.e. also the computation of the Jacobians $\tilde{k}_j$, $1 \le j \le 4$. As an alternative, one can compute the Jacobians $\tilde{k}_j$, $1 \le j \le 4$ in the recording step $\hat{F}_i$ and store this information on the tape. Then the appropriate reverse step $\bar{F}_i$ only has to evaluate the last three statements of Equation (4). The runtimes represented here are based on the second approach in order to achieve $\bar{\tau} = 1$. As a result, $\hat{\tau}$ equals 5. This implementation has the advantage that the value of $\bar{\tau}$ and hence the wall clock time are reduced at the expense of $\hat{\tau}$. This can be seen for example in Figure 2, where an increase of $\bar{\tau}$ would result in an bigger slope of the bar describing the adjoint or reverse computations.

As mentioned above, one has to be careful about the adjoint calculation because of the lack of commutativity between adjoining and discretizing in general. Therefore, it is important to note that the Runge-Kutta scheme (3) belongs to a class of discretizations, for which both possibilities of adjoint calculation coincide, giving the same result [8].

3.3. Numerical Results

To test the parallel reversal schedule framework, one forward integration of the car model shown in Figure 4 and one adjoint calculation were performed. As previously mentioned, the integration distance was 100 m and the step size 10 cm. Hence, there are 1000 forward steps F_i. The Figure 5(a) shows the growth of the cost functional for which

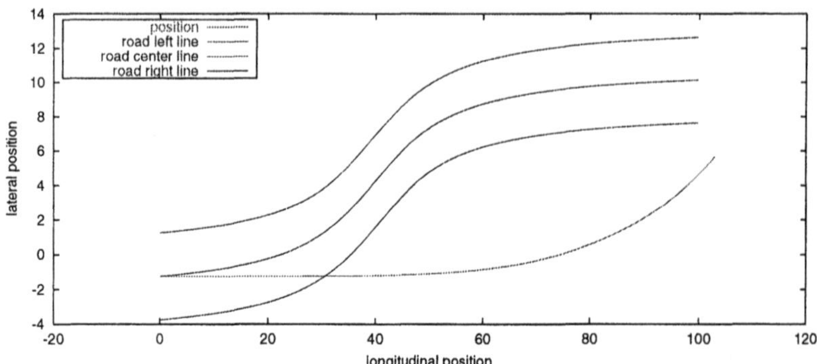

Figure 4. Position of Formula 1 Car.

we computed the sensitivities of the control variables u_1 (Figure 5(b)) and u_1 (Figure 5(c)). However, the resource requirements are of primary interest. One integration step in the example is relatively small in terms of computing time. In order to achieve reasonable timings 18 integration steps form one physical step of the parallel reversal schedule. The remaining 10 integration steps were spread uniformly. Hence, one obtains 55 physical steps. Therefore, five processors were needed for the corresponding time-minimal parallel reversal schedule for $\hat{\tau} = \bar{\tau} = 1$. This reversal schedule is with small modifications also nearly optimal for the

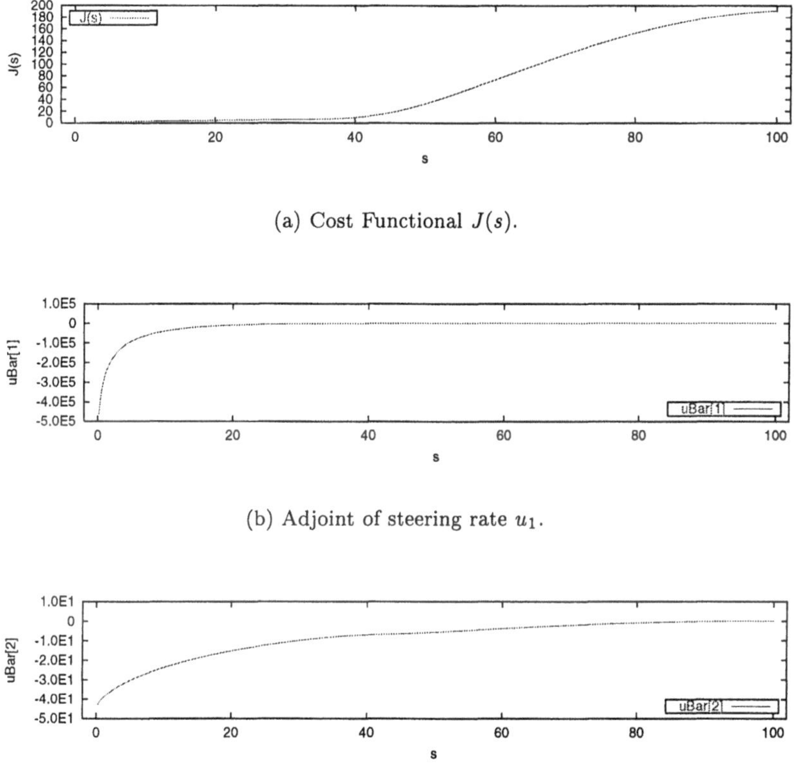

(a) Cost Functional $J(s)$.

(b) Adjoint of steering rate u_1.

(c) Adjoint of longitudinal force u_2.

Figure 5. Cost Functional and Adjoint of Control Variables.

considered combination $\hat{\tau} = 5$ and $\bar{\tau} = 1$. A sixth processor (master) was used to organise the program run.

		naïve approach	parallel checkpointing
double variables needed		266010	5092
memory	in kByte	2128.1	40.7
required	in %	100.0	1.9

Table 1. Memory Requirement

The main advantage of the parallel reversal schedules is the enormous reduction in memory requirement as illustrate in Table 1. It shows that for this example, less than a fiftieth of the original memory requirement is needed, i.e., less than 2%. On the other hand, only six times the original computing power, i.e., processors, is used.

The theoretical runtime is also confirmed by the example as can be seen in Table 2. Due to the slower memory interface on a Cray T3E, the usage of less memory in parallel causes an enormous decrease in runtime. On the other hand the problem is too small and the SGI Origin 3800 too fast to show this effect. Nevertheless, one obtains that the assumption of negligible communication cost is reasonable. This is caused by the fact that the processors have the duration of one full physical step to send and receive a checkpoint because the checkpoint is not needed earlier. Only if the send and receive of one checkpoint needs more time than one physical step the communication cost becomes critical.

		naïve approach	parallel checkpointing
T3E	in sec.	20.27	18.91
	in %	100.0	93.3
Origin 3800	in sec.	6.71	6.04
	in %	100.0	90.0

Table 2. Runtime results

4. Conclusions

The potentially enormous memory requirement of program reversal by complete logging often causes problems despite the ever increasing size of memory systems. This paper proposes an alternative method, where the memory requirement can be drastically reduced by keeping at most c intermediate states as checkpoints. In order to avoid an increase in runtime, p processors are used to reverse evolutions with minimal wall clock time. For the presented time-minimal parallel reversal schedules, the number l of physical steps that can be reversed grows exponentially as a function of the resource number $\varrho = c + p$. A corresponding software tool has been coded using MPI. Initial numerical tests are reported. They confirm the enormous reduction in memory requirement. Furthermore, the runtime behaviour is studied. It is verified that the wall clock time of the computation can be reduced compared to the logging-all approach if the memory access is comparatively costly. This fact is caused by the reduced storage in use. If the memory access is comparatively cheap, the theoretical runtime of time-minimal parallel reversal schedules is also confirmed.

The following overall conclusion can be drawn. For adjoining simulations $\log_{a(\bar{\tau})}$(# physical steps) processors and checkpoints are wall clock equivalent to 1 processor and (# physical steps) checkpoints with $a(\bar{\tau}) = \frac{1}{2}(1 + \sqrt{1 + 4\bar{\tau}})$ and $\bar{\tau}$ the temporal complexity of a reverse step.

Acknowledgments

The authors are indebted to Daniele Casanova for his support during the numerical experiments and to Andreas Griewank for many fruitful discussions.

References

[1] J. Allen. Computer optimisation of cornering line. Master's thesis, School of Mechanical Engineering, Cranfield University, 1997.

[2] E. Bakker, H. Pacejka, and L. Lidner. A new tire model with an application in vehicle dynamics studies. *SAE-Paper*, 890087, 1989.

[3] Y. Evtushenko. Automatic differentiation viewed from optimal control. In G. F. Corliss and A. Griewank, editors, *Computational Differentiation: Techniques, Implementations, and Application*, Philadelphia, 1991. SIAM.

[4] M. Garey and D. Johnson. *Computers and intractability: Aguide to the theory of NP-completeness.* Freeman and Company, New York, 1980.

[5] A. Griewank. *Evaluating Derivatives: Principles and Techniques of Algorithmic Differentiation.* Frontiers in Applied Mathematics. SIAM, Philadelphia, 1999.

[6] A. Griewank and A. Walther. Revolve: An implementation of checkpointing for the reverse or adjoint mode of computational differentiation. *ACM Trans. Math. Software*, 26, 2000.

[7] J. Grimm, L. Pottier, and N. Rostaing-Schmidt. Optimal time and minimum space-time product for reversing a certain class of programs. In M. Berz, C. Bischof, G. Corliss, and A. Griewank, editors, *Computational Differentiation: Techniques, Applications, and Tools*, Philadelphia, 1996. SIAM.

[8] W. Hager. Runge-kutta methods in optimal control and the transformed adjoint system. *Numer. Math.*, 87:247–282, 2000.

[9] P. Hilton and J. Petersen. A fresh look at old favourites: The fibonacci and lucas sequences revisited. *Australian Mathematical Society Gazette*, 25:146–160, 1998.

[10] U. Lehmann and A. Walther. The implementation and testing of time-minimal and resource-optimal parallel reversal schedules. Technical Report ZHR-IR-0109, Tech. Univ. Dresden, Center for High Perf. Comp., 2001.

[11] O. Talagrand. The use of adjoint equations in numerical modeling of the atmospheric circulation. In G. F. Corliss and A. Griewank, editors, *Computational Differentiation: Techniques, Implementations, and Application*, Philadelphia, 1991. SIAM.

[12] J. van de Snepscheut. *What computing is all about.* Texts and Monographs in Computer Science. Springer, Berlin, 1993.

[13] A. Walther. *Program Reversal Schedules for Single- and Multi-processor Machines.* PhD thesis, Tech. Univ. Dresden, Inst. for Sci. Comp., 1999.

www.ingramcontent.com/pod-product-compliance
Ingram Content Group UK Ltd.
Pitfield, Milton Keynes, MK11 3LW, UK
UKHW012155240726
13966UKWH00002B/358
* 9 7 8 1 4 7 5 7 6 6 6 8 4 *